IEE CIRCUITS, DEVICES AND SYSTEMS SERIES 10
Series Editors: Dr D. G. Haigh
　　　　　　　Dr R. S. Soin
　　　　　　　Dr J. Wood

SPIE Press Monograph Volume PM52

Distributed feedback semiconductor lasers

Other volumes in the IEE Circuits, Devices and Systems series:

Volume 1 **GaAs technology and its impact on circuits and systems**
D. G. Haigh and J. Everard (Editors)
Volume 2 **Analogue IC design: the current-mode approach**
C. Toumazou, F. J. Lidgey and D. G. Haigh (Editors)
Volume 3 **Analogue-digital ASICs** R. S. Soin, F. Maloberti and
J. Franca (Editors)
Volume 4 **Algorithmic and knowledge-based CAD for VLSI**
G. E. Taylor and G. Russell (Editors)
Volume 5 **Switched-currents: an analogue technique for digital technology**
C. Toumazou, J. B. Hughes and N. C. Battersby (Editors)
Volume 6 **High frequency circuits** F. Nibler and co-authors
Volume 7 **MMIC design** I. D. Robertson (Editor)
Volume 8 **Low-power HF microelectronics** G. A. S. Machado (Editor)
Volume 9 **VLSI testing: digital and mixed analogue/digital techniques**
S. L. Hurst

Distributed feedback semiconductor lasers

John Carroll,
James Whiteaway
& Dick Plumb

The Institution of Electrical Engineers
SPIE Optical Engineering Press

Published by: The Institution of Electrical Engineers, London,
United Kingdom

© 1998: The Institution of Electrical Engineers

This publication is copyright under the Berne Convention and the Universal Copyright Convention. All rights reserved. Apart from any fair dealing for the purposes of research or private study, or criticism or review, as permitted under the Copyright, Designs and Patents Act, 1988, this publication may be reproduced, stored or transmitted, in any forms or by any means, only with the prior permission in writing of the publishers, or in the case of reprographic reproduction in accordance with the terms of licences issued by the Copyright Licensing Agency. Inquiries concerning reproduction outside those terms should be sent to the publishers at the undermentioned address:

The Institution of Electrical Engineers,
Michael Faraday House,
Six Hills Way, Stevenage,
Herts. SG1 2AY, United Kingdom
www: http://www.iee.org

Copublished by:
SPIE—The International Society for Optical Engineering
PO Box 10, Bellingham, Washington 98227-0010
Phone: 360/676-3290
Fax: 360/647-1445
E-mail: spie@spie.org
www: http://www.spie.org

While the authors and the publishers believe that the information and guidance given in this work is correct, all parties must rely upon their own skill and judgment when making use of it. Neither the author nor the publishers assume any liability to anyone for any loss or damage caused by any error or omission in the work, whether such error or omission is the result of negligence or any other cause. Any and all such liability is disclaimed.

The moral right of the authors to be identified as authors of this work has been asserted by him/her in accordance with the Copyright, Designs and Patents Act 1988.

British Library Cataloguing in Publication Data

A CIP catalogue record for this book
is available from the British Library

IEE ISBN 0 85296 917 1

SPIE ISBN 0 8194 2660 1

Printed in England by Redwood Books, Trowbridge

Contents

Preface	xiii
Acknowledgments	xv
Principal abbreviations	xvii
Principal notation	xix

1	**The semiconductor-diode laser**	**1**
1.1	Background	1
1.2	Early developments	1
	1.2.1 The first semiconductor lasers	1
	1.2.2 Fabry–Perot gain and phase requirements	2
	1.2.3 Some characteristics of diode lasers	4
1.3	Improvements to reduce operating currents	6
	1.3.1 Heterojunctions: carrier confinement	6
	1.3.2 Heterojunctions: photon confinement	7
	1.3.3 Structures for 'horizontal' confinement	9
	1.3.4 Degree of confinement—the confinement factor	11
1.4	Variations on conventional Fabry–Perot laser design	12
	1.4.1 High-low reflective facets	12
	1.4.2 External cavities	13
	1.4.3 External grating	14
1.5	System requirements for single-frequency lasers	15
1.6	Introduction to lasers based on Bragg gratings	18
	1.6.1 Introduction to Bragg gratings	18
	1.6.2 Fabrication of gratings inside lasers	22
1.7	Some principal forms of grating laser	25
	1.7.1 The distributed Bragg reflector laser	25
	1.7.2 The distributed feedback (DFB) laser	26
	1.7.3 More complex grating-based lasers	28
1.8	Summary	30
1.9	Bibliography	31
	1.9.1 Semiconductor lasers	31
	1.9.2 Optical communication systems	32
1.10	References	32

vi *Contents*

2 Gain, loss and spontaneous emission — 37
  2.1 Introduction — 37
  2.2 Electronic processes in semiconductors — 37
    2.2.1 Energy states — 37
    2.2.2 Occupation probabilities — 40
    2.2.3 Radiative recombination and absorption — 41
    2.2.4 Transitions and transition rates — 43
    2.2.5 Auger recombination — 44
  2.3 Absorption, emission rates and spectra — 46
    2.3.1 Absorption, stimulated and spontaneous emission in a semiconductor — 46
    2.3.2 Stimulated-gain spectra in semiconductors — 50
    2.3.3 Homogenous and inhomogeneous broadening — 53
    2.3.4 Spontaneous-emission spectra from semiconductors — 54
  2.4 Semiconductor interactions with the lasing mode — 55
    2.4.1 Spontaneous-coupling factor — 55
    2.4.2 Petermann's 'K factor' — 58
    2.4.3 Gain saturation in semiconductors — 58
    2.4.4 Spectral hole burning and carrier heating — 59
    2.4.5 Scattering losses — 60
    2.4.6 Free-carrier absorption — 60
  2.5 Henry's α factor (or linewidth enhancement factor) — 61
  2.6 Temperature-induced variations in semiconductor lasers — 64
  2.7 Properties of quantum-well-laser active regions — 66
    2.7.1 Introduction to quantum wells — 66
    2.7.2 Gain saturation and the need for multiple quantum wells — 68
    2.7.3 Strained quantum wells — 69
    2.7.4 Carrier transport — 70
  2.8 Summary — 72
  2.9 Bibliography — 72
  2.10 References — 73

3 Principles of modelling guided waves — 76
  3.1 Introduction — 76
    3.1.1 Vertical and horizontal guiding — 77
    3.1.2 Index and gain guiding — 77
    3.1.3 Effective area and confinement factor — 79
  3.2 The slab guide — 81
    3.2.1 TE and TM guided waves — 81
    3.2.2 Multilayer slab guides — 82
  3.3 Wave equations for the TE and TM guided waves — 83
  3.4 Solving multislab guides — 84
    3.4.1 Effective refractive index — 84
    3.4.2 Reflection coefficient calculation — 86
    3.4.3 Gain-guiding example — 89

3.5	Scaling	90
3.6	Horizontal guiding: effective-index method	90
3.7	Orthogonality of fields	92
3.8	Far fields	92
3.9	Waveguiding with quantum-well materials	93
3.10	Summary and conclusions	95
3.11	References	96

4 Optical energy exchange in guides — 97
- 4.1 The classic rate equations — 97
  - 4.1.1 Introduction — 97
  - 4.1.2 Rate of change of photon density — 98
  - 4.1.3 Rate of change of electron density — 100
- 4.2 Some basic results from rate-equation analysis — 103
  - 4.2.1 Simplifying the rate equations — 103
  - 4.2.2 Steady-state results — 104
  - 4.2.3 Dynamic analysis — 106
  - 4.2.4 Problems of particle balance — 110
- 4.3 Field equations and rate equations — 111
  - 4.3.1 Introduction — 111
  - 4.3.2 Wave propagation — 111
  - 4.3.3 Decoupling of frequency and propagation coefficient — 114
- 4.4 Field equations with a grating — 116
  - 4.4.1 The periodic permittivity — 116
  - 4.4.2 Phase matching — 117
  - 4.4.3 Second-order gratings — 120
  - 4.4.4 Shape of grating — 124
- 4.5 Summary — 125
- 4.6 References — 126

5 Basic principles of lasers with distributed feedback — 128
- 5.1 Introduction — 128
- 5.2 Coupled-mode equations for distributed feedback — 129
  - 5.2.1 Physical derivation of the coupling process — 129
  - 5.2.2 Complex gratings — 131
- 5.3 Coupled-mode solutions and stopbands — 132
  - 5.3.1 Eigenmodes — 132
  - 5.3.2 The dispersion relationship and stopbands — 133
- 5.4 Matrix solution of coupled-mode equations for uniform grating laser — 135
  - 5.4.1 The field input–output relationships — 135
  - 5.4.2 Reflections and the observed stopband — 137
- 5.5 DFB lasers with phase shifts — 139
  - 5.5.1 Phase shifts — 139
  - 5.5.2 Insertion of phase shifts: the transfer-matrix method — 142

viii *Contents*

5.6	Longitudinal-mode spatial-hole burning	145
	5.6.1 The phenomena	145
	5.6.2 The stopband diagram	147
	5.6.3 Influence of κL product on spatial-hole burning	148
	5.6.4 Influence of phase shifts on spatial-hole burning	148
	5.6.5 Spectrum and spatial-hole burning	150
5.7	Influence of series resistance	152
5.8	Simulating the static performance of DFB lasers	155
	5.8.1 Light/current characteristics	155
	5.8.2 Simulation of emission spectrum	159
5.9	Summary	161
5.10	References	162

6 More advanced distributed feedback laser design — **165**

6.1	Introduction	165
6.2	Linewidth	166
	6.2.1 General	166
	6.2.2 Calculation of linewidth under static conditions	168
	6.2.3 Linewidth enhancement	171
	6.2.4 Effective linewidth enhancement	172
	6.2.5 Effective dynamic linewidth enhancement	176
	6.2.6 Linewidth rebroadening	176
6.3	Influence of reflections from facets and external sources	177
	6.3.1 Reflections and stability	177
	6.3.2 Facet reflectivity and spectral measurements	179
	6.3.3 Influence of facet reflectivity on SMSR for DFB lasers	180
6.4	Complex grating-coupling coefficients	183
	6.4.1 General	183
	6.4.2 Techniques for introducing complex grating-coupling coefficients	183
	6.4.3 Influence of complex grating-coupling coefficient on static performance	184
	6.4.4 Influence of complex grating-coupling coefficient on dynamic performance	187
	6.4.5 Influence of facet reflectivity	189
6.5	High-power lasers with distributed feedback	189
	6.5.1 General	189
	6.5.2 Techniques for obtaining high front-to-back emission ratios	190
	6.5.3 Laser-amplifier structures with distributed feedback	191
6.6	Dynamic modelling of DFB lasers	194
	6.6.1 Uniform-grating DFB laser with reflective rear facet	194
	6.6.2 Large signal performance of $2 \times \lambda_m/8$ DFB lasers with strong and weak carrier-transport effects	197
6.7	Summary	202

		Contents	ix

	6.8	References	202
7	**Numerical modelling for DFB lasers**		**209**
	7.1	Introduction	209
	7.2	Ordinary differential equations	211
		7.2.1 A first-order equation	211
		7.2.2 Accuracy	213
	7.3	First-order wave equations	214
		7.3.1 Introduction	214
		7.3.2 Step lengths in space and time—central-difference method	215
		7.3.3 Numerical stability	216
		7.3.4 Gain and phase	218
	7.4	Coupled reflections	219
		7.4.1 Kappa coupling but no gain or phase changes	219
		7.4.2 Matrix formulation	219
		7.4.3 Phase jumps replacing scattering	221
		7.4.4 Fourier checks	221
	7.5	A uniform Bragg laser: finite difference in time and space	222
		7.5.1 Full coupled-wave equations	222
		7.5.2 MATLAB code	223
		7.5.3 Analytic against numeric solutions	224
	7.6	Spontaneous emission and random fields	226
		7.6.1 Spontaneous noise and travelling fields	226
		7.6.2 Null correlation for different times, positions and directions	228
		7.6.3 Spontaneous magnitude	229
		7.6.4 Tutorial programs	229
	7.7	Physical effects of discretisation in the frequency domain	230
		7.7.1 Discretisation process—integrals to sums	230
		7.7.2 Fast Fourier transform (FFT)	232
	7.8	Finite-element strategies for a spectral filter	233
		7.8.1 Lorentzian filter	233
		7.8.2 Numerical implementation	235
	7.9	Application of the filter theory to gain filtering	237
		7.9.1 General	237
		7.9.2 Filtering the gain in the travelling-wave equations	238
		7.9.3 Numerical implementation	240
	7.10	Basic DFB laser excited by spontaneous emission	241
		7.10.1 Introduction and normalisation	241
		7.10.2 Field equations	243
		7.10.3 Charge-carrier rate equation	243
		7.10.4 Numerical programs	246
	7.11	Summary	248
	7.12	References	249

x   *Contents*

8 Future devices, modelling and systems analysis   **252**
   8.1   Introduction   252
   8.2   Systems analysis   252
       8.2.1   Introduction   252
       8.2.2   Component modelling   253
       8.2.3   System modelling   255
       8.2.4   10 Gbit/s power amplification   256
       8.2.5   Direct modulation: recapitulation   258
       8.2.6   Simulation of integrated DFB laser and electroabsorption modulator   259
       8.2.7   Cross-gain and four-wave-mixing wavelength conversion in an SOA   260
       8.2.8   Simulation of cross-phase wavelength conversion in a Mach–Zehnder interferometer incorporating two SOAs   263
   8.3   The push–pull laser   265
       8.3.1   Introduction: push-pull electronics   265
       8.3.2   Symmetrical push–pull DFB laser   266
       8.3.3   Asymmetry and the push–pull DFB laser   269
       8.3.4   Speed of response for a push–pull DFB laser   272
   8.4   Tunable lasers with distributed feedback   274
       8.4.1   Introduction   274
       8.4.2   Simple multicontact tunable lasers   277
       8.4.3   Wide-tuning-range lasers with nonuniform gratings   279
       8.4.4   Other tunable-laser structures   282
       8.4.5   Tunable-laser linewidth and modulation   284
       8.4.6   Modelling tunable semiconductor lasers   284
       8.4.7   Multiple DFB lasers with optical couplers for WDM   285
   8.5   Surface-emitting lasers   286
       8.5.1   Introduction to surface-emitting lasers   286
       8.5.2   Operating parameters of VCSELs compared with edge emitters   287
       8.5.3   Construction of VCSELs   291
       8.5.4   Additional features of VCSELs   294
   8.6   Summary   294
   8.7   References   295

Appendix 1 Maxwell, plane waves and reflections   **304**
A1.1   The wave equation   304
A1.2   Linearly polarised plane waves (in a uniform 'infinite' material)   304
A1.3   Snell's law and total internal reflection   305
A1.4   Reflection amplitudes at surfaces: TE fields   308
A1.5   TE reflection amplitudes: three special cases   309
A1.6   Reflection amplitudes at surfaces: TM fields   309

A1.7	TM reflection amplitudes at surfaces: four special cases	310
A1.8	Reflection for waveguide modes at facets	311
A1.9	References	312

Appendix 2 Algorithms for the multilayer slab guide — 313
- A2.1 TE slab modes — 313
- A2.2 TM slab modes — 317
- A2.3 Far fields — 319
- A2.4 Slab waveguide program — 322
- A2.5 References — 323

Appendix 3 Group refractive index of laser waveguides — 324
- A3.1 References — 328

Appendix 4 Small-signal analysis of single-mode laser — 329
- A4.1 Rate equations: steady-state and small-signal — 329
- A4.2 Carrier-transport effects — 334
- A4.3 Small-signal FM response of single-mode laser — 336
- A4.4 Small-signal FM response and carrier transport — 337
- A4.5 Photonic and electronic equations for large-signal analysis — 339
- A4.6 Reference — 340

Appendix 5 Electromagnetic energy exchange — 341
- A5.1 Dielectric polarisation and energy exchange — 341
- A5.2 Electromagnetic-energy exchange and rate equations reconciled — 344
- A5.3 Electromagnetic-energy exchange and guided waves: field equations — 348
- A5.4 References — 351

Appendix 6 Pauli equations — 352
- A6.1 Reference — 356

Appendix 7 Kramers–Krönig relationships — 357
- A7.1 Causality — 357
- A7.2 Cauchy contours and stability — 359
- A7.3 A proper physical basis builds in causality — 360
- A7.4 Refractive index of transparent quaternary alloys — 362
- A7.5 References — 364

Appendix 8 Relative-intensity noise (RIN) — 366
- A8.1 References — 371

Appendix 9 Thermal, quantum and numerical noise — 372
- A9.1 Introduction — 372
- A9.2 Thermal and quantum noise — 373

xii *Contents*

A9.3 Ideal amplification	374
A9.4 The attenuator	377
A9.5 Einstein treatment: mode counting	378
A9.6 Aperture theory	379
A9.7 Numerical modelling of spontaneous noise	380
A9.8 Higher-order noise statistics	384
A9.9 References	385

Appendix 10 Laser packaging — **386**

A10.1 Introduction	386
A10.2 Electrical interfaces and circuits	386
A10.3 Thermal considerations	388
A10.4 Laser monitoring	388
A10.5 Package-related backreflections and fibre coupling	389
A10.6 References	391

Appendix 11 Tables of device parameters and simulated performance for DFB laser structures — **392**

Appendix 12 About MATLAB programs — **396**

A12.1 Instructions for access	396
A12.2 Introduction to the programs	398

Index — **405**

Preface

The authors have had the pleasure of working for many years within one of the exciting areas of research and development: sources for optical communications, an area which has seen phenomenal growth over three decades. The semiconductor laser diode has proved to be an essential device and one form or another of a distributed feedback semiconductor laser (DFB laser) is a key component for modern optical communication systems. Technically, the term DFB implies a particular form of laser but to the authors the term has come to mean *any* laser in which there is distributed feedback or where Bragg gratings are integrated within the device.

This is a research monograph which combines a high level of tutorial material with a research review. It will be useful for postgraduate courses in optoelectronics studying either systems or devices, research workers in optical systems, lasers and optoelectronic devices, and also be of use for final-year-project students in BEng and MEng degree courses. Lecturers and research workers in the field of laser diodes will find this a useful source of ideas, references and tutorials which reach the explanations other books do not reach. Combined with this tutorial material are accounts of the practical operation and design of DFB lasers as used by the communications industry along with discussions of the operational reasons for requiring key modifications within the laser in order to meet specific performance targets.

The tutorial material is distributed throughout the text and the appendices. The different orders of operation of a Bragg grating are outlined, novel concepts of the stopband diagram are explained, and the modelling of laser interaction is based on sound fundamentals of energy balance so that both particle models of lasers and electromagnetic-field models can be tied together. There are many different ways of estimating the spontaneous emission, and again the consistency between the different methods is reconciled. The relationship between gain and phase demanded by causality is also examined, tying together different approaches; this is vital to understand the role of the line width enhancement factor. A consistent approach to the numerical modelling of the wave processes within a DFB laser is extensively discussed in Chapter 7. Practical programs are supplied through the World Wide Web (WWW). The use of MATLAB as a user-friendly tool is emphasised with a number of sample programs.

The emphasis in the book is on the physical and numerical modelling of laser diodes, and DFB lasers in particular. The treatment of semiconductor materials and fabrication technology is necessarily limited by the length of the book and the authors' expertise and strengths, but within a whole chapter on this important topic enough should have been said to ensure that the reader appreciates the synergy that is required between material science, technology and device physics in order to make successful DFB lasers. Material parameters which are important for getting good physical representations of DFB operation are listed.

For similar reasons of length, the discussion about optical communication systems is not extensive but should be sufficient to set the context of the DFB laser within such systems. System modelling where the interactions between a number of sources, amplifiers, modulators and so on is numerically modelled is a growing area and this is touched on in a significant section in Chapter 8.

There are always idiosyncratic choices made by authors in their selection of material and selection of references. There will be many friends and colleagues who will be 'mortally offended' by the omission of their best work from the reference list or from the material that is discussed. All that one can say is that no offence was intended and it is inevitable that with so much material published within this area a selection has to be made. If the reader gains some of the terrific enthusiasm and understanding that has gone into the work on DFB lasers by research workers from all over the world, then it will all have been worthwhile.

J.E. Carroll
J.E.A. Whiteaway
R.G.S. Plumb

Acknowledgments

The authors would like to begin by acknowledging the help and support of their respective establishments:

The Department of Engineering, University of Cambridge and Nortel (Northern Telecom), Harlow Laboratories; along with British Telecom, BT&D Technologies Ltd., the Department of Trade and Industry, the Science and Engineering Research Council (now the Engineering and Physical Sciences Research Council), the Ministry of Defence (now the Defence Evaluation Research Agency) and various branches of the European Commission, all of whom have supported our work on semiconductor lasers over many years.

There are numerous colleagues, students and research associates who have either directly or indirectly contributed to our joint understanding of the physics and modelling of laser diodes. It is impossible to assess all their relative contributions, but each in his or her own way will known that they have had an influence on our writing or ability to do research for which we are indebted. Avoiding invidious comparisons, we mention many by name below:

Mike Adams, Laurent Bacik, Peter Batchelor, Mark Bray, Andy Carter, Andrew Collar, Phil Couch, John Devlin, Catharine Ewart, Martyn Fice, Barry Flanigan, Kent Gardiner, Brian Garrett, Bob Goodfellow, David Greene, Clive Irving, Jessica James, David Jones, Ping Chek Koh, David Kozlowski, Ian Lealman, Dominique Marcenac, Paul Morton, Mark Nowell, David Nugent, Jeff Offside, Greg Pakulski, David Parker, Eddie Pratt, Penny Probert, Dave Robbins, Mike Robertson, Charlanne Scahill, Alan Shore, Tim Simmons, Andy Skeats, Alan Steventon, Robin Thompson, Chi Tsang, Les Westbrook, Ian White, Yuc Lun Wong, John Young, Si Fung Yu, Li Ming Zhang.

The names of all our friends and colleagues who have participated in Laser Workshops, COST240, and key laser conferences over relevant years are too numerous to be mentioned individually but they all have had an influence on our thoughts about lasers and all are thanked. We ask forgiveness of those who feel they should have been listed.

Thanks are also given to the IEE and in particular Jonathan Simpson for support in publishing this book.

The support from our families, while apparently taken for granted, has never been forgotten by us.

Principal abbreviations

AR	antireflection
CCCH	Auger process involving three conduction-band electrons and one heavy hole
CNR	carrier-to-noise ratio
DBR	distributed Bragg reflector
DFB	distributed feedback
EDFA	erbium-doped fibre amplifier
FP	Fabry–Perot
GaAs	gallium arsenide
$GaAs_xP_{1-x}$	gallium arsenide phosphide
$Ga_xAl_{1-x}As$	gallium aluminium arsenide
$Ga_xIn_{1-x}As_yP_{1-y}$	gallium indium arsenide phosphide
GAVCF	grating-assisted vertical-coupler filter
HOFC	higher-order Fourier coefficients
InP	indium phosphide
IVBA	intervalence-band absorption
L–I	light–current
MATLAB	registered trade mark of a MathWorks computing environment
MBE	molecular-beam epitaxy
MOCVD	metal-organic chemical-vapour deposition
MQW	multiquantum well
SCH	separate-confinement heterostructure
RIN	relative intensity noise
SiO_2	silicon dioxide
Si_3N_4	silicon nitride
SMSR	side-mode-suppression ratio
SNR	signal-to-noise ratio
SOA	semiconductor optical amplifier
TDM	time-division multiplexing

xviii *Principal abbreviations*

TE	transverse electric (often used as subscripts on fields) (Chapter 4)
TIR	total internal reflection
TLLM	transmission-line laser modelling
TM	transverse magnetic (often used as subscripts on fields) (Chapter 4)
WDM	wavelength-division multiplexing

Principal notation

a	arbitrary distance, atomic spacing
a, b, c, d	parameters in a transfer matrix
a_m	field losses per unit length
A	linear-recombination coefficient $\rightarrow AN$ (usually nonradiative)
A_m	optical power absorption per unit time
$\mathcal{A}$	effective area or aperture of guide
α_H	linewidth broadening (enhancement) factor $= (\mathrm{d}n_{real}/\mathrm{d}N)/(\mathrm{d}n_{imag}/\mathrm{d}N)$
b	coefficient for intervalence-band absorption
B	bipolar recombination constant $\rightarrow$ $B(NP) \sim BN^2$ spontaneous bipolar emission
$\beta = 2\pi/\lambda_m = n_r \omega/c$	propagation coefficient in material with wavelength λ_m
$\beta_o, \beta_e, \beta_b, \beta_y, \beta_z$	propagation coefficient associated with the subscript (e.g. b for Bragg)
β_{sp}	spontaneous-coupling coefficient
c	velocity of light in free space
CN^3	approximation for $C'(N^2P) + C''(NP^2)$; Auger recombination
d, d_1 etc.	layer thicknesses
δ	$\beta_o - \beta_b$, a measure of the frequency offset from the Bragg frequency

xx   *Principal notation*

δf	signal bandwidth
Δf	system bandwidth, gain bandwidth
$\Delta_{LW} f$	linewidth
$\Delta_{sp} f$	spontaneous bandwidth for laser
$\Delta_g f$	change in frequency with change in gain
$\delta(z_1 - z_2)$	Dirac delta function of position
$\Delta \rho$	elemental reflection per Bragg period
ΔT	elemental transmission coefficient per Bragg period
δt	computational step length in time
Δx	uncertainty in position
Δp	uncertainty in momentum
δz	space step related to $\delta t = \delta z / v_g$
e, exp	exponential
E	optical electric field also as a function of $z \omega$ time, etc.: $E(z), E(\omega), E(t), E(y, z)$
$E_o, E_{out}, E_x, E_{rad}, E_{sp f(r)}$	subscripted versions of electric field
E_f, E_r	forward, reverse electric field
$\mathscr{E}$	energy of particle
$\mathscr{E}_c, \mathscr{E}_v, \mathscr{E}_{Q1}, \mathscr{E}_{Q2}$	energy levels (e.g.: c, v $\to$ conduction, valence band)
$\mathscr{E}_g, \mathscr{E}_{act}, \mathscr{E}_{clad}$	energy gaps appropriate to different materials
$\mathscr{E}_{fN}, \mathscr{E}_{fP}$	quasi-Fermi levels for electrons (holes)
ε_o	permittivity of free space (Chapter 4)
$\varepsilon_r, \varepsilon_{r1}, \varepsilon_{r2}$	relative permittivity, same in layers 1 and 2
$\varepsilon_{r\,eff}$	effective relative permittivity
$\varepsilon_{rr}, \varepsilon_{ri}, \varepsilon_{rr\,eff}, \varepsilon_{ri\,eff}$	relative and effective, real and imaginary permittivities
ϵ	gain-saturation parameter where $G(N) = G'(N - N_{tr})/(1 + \epsilon S)$
f	optical frequency of operation $= \omega / 2\pi$
f_{bragg}, f_{max}	specific frequencies, e.g. central Bragg, maximum power etc.
f_M	modulation frequency, $f_{M\,max}$ = maximum modulation frequency
F	forward field: different functionalities, e.g. $F(t, z), F(z), F(T, Z)$
F_{in}, F_L	forward-field values at input or $z = L$

Principal notation xxi

$\mathbf{F} = \begin{bmatrix} F \\ R \end{bmatrix}$	ordered forward- and reverse-field components
$F_N(\mathscr{E})$, $(F_P(\mathscr{E}))$	Fermi–Dirac function for electrons (holes)
ϕ	angle, phase shift, radians
Φ_0	$= \mathscr{V} S_0 / \tau_p$, steady-state photon output
g	net field gain/unit distance $= n_i \omega / c$
g'	differential gain $= dg/dN$
$g(N)$, $g(\omega)$	net field gain/unit length: function of N, ω etc.
$g(n)$	net field gain/unit length in section n (Chapter 7)
g_m, $g_m(\lambda)$, $g_m(N)$	material optical field gain per unit length as function of λ, N etc.
$G(N)$	net power gain per unit time (as a function of N)
$G_m(N)$	material power gain/unit time
G'_m	differential optical-power gain $= dG/dN$
G_{m1}	small-signal dynamic changes in G_m
G_{thr}	round-trip, threshold gain
G_{aerial}	gain of an aerial or radiating aperture $= 4\pi\lambda^2/\mathscr{A}$
G_{round}	round trip gain
γ_y	vertical imaginary (evanescent) propagation coefficient
Γ	confinement factor
h	Planck's constant
η_{ext}	external quantum efficiency
i_n^2	mean-square shot noise
i_{sp}, i_{spf}, i_{spr}	spontaneous excitation, forward, reverse
I, I_o, I_1, I_{th}	current: total, steady state, small-signal, threshold
J	current density into laser's active region
k	Boltzmann's constant

xxii *Principal notation*

$\boldsymbol{k}$	electron's wave vector
k_o	$=\omega_o/c$ propagation coefficient in free space
K_p, K_{tr}	Petermann transverse spontaneous emission factor, subscripts $_p$ or $_{tr}$
K_z	Petermann longitudinal spontaneous emission factor
K_{tot}	$K_z\,K_{tr}$ product = total spontaneous emission factor
K_{sp}	spontaneous emission coefficient in two-level model
K_{stim}	stimulated emission coefficient in two-level model
K	normalised coefficient for spectral filter
κ, κ_{fr}, κ_{rf}	coupling coefficients for grating
$\kappa_{index}+j\kappa_{gain}$	coupling coefficient for complex guide
L	laser length
λ, λ_o	free-space wavelength
λ_b	free-space Bragg wavelength
$\lambda_{electron}$	electron wavelength
λ_m, λ_1, λ_2	wavelength in material/guide, material 1, 2 etc.
λ_{peak}	wavelength of peak gain
Λ, Λ'	Bragg-period first-order grating, period of second-order grating $\Lambda'=2\Lambda$
m_c, m_v, m_r	effective mass for: conduction, valence electrons, reduced mass
M	integer (e.g. diffraction order)
M_{12}, M_{22} (etc.)	matrix elements: row 1 columns 2 and row 2 columns 2 (etc.) of matrix $\boldsymbol{M}$
$\boldsymbol{M}$, $\boldsymbol{M}_1$, $\boldsymbol{M}_\phi$	transfer matrix for section 1, phase shift section etc.
μ_o	permeability of free space
μ_r	relative permeability ($=1$ in this text)
$n=n_r+jn_i$	complex-refractive-index real + imaginary parts
n_1, n_2, n_{eff}	$=\sqrt{\varepsilon_{r1}}$, $\sqrt{\varepsilon_{r2}}$, $\sqrt{\varepsilon_{r\,eff}}$, refractive index in layer 1 and 2, effective index
n_g	group index

Principal notation xxiii

n_{inv}	inversion factor (Appendix 9, Chapter 6)
n	integer indicating segment number
N, $N(z)$	electron density (as function of z)
N_0, N_1	steady-state, small-signal electron density
N_{ref}, N_{th}, N_{tr}	reference, threshold, transparency values of electron density
N	integers
$N_d(\mathcal{E})\,d\mathcal{E}$	density of electron states over an energy range $d\mathcal{E}$
N_1, N_2	electron states at energy $\mathcal{E}_1$, $\mathcal{E}_2$
N_a	acceptor density in p-doped material
N_c, N_v	overall density of states in conduction and valence bands
Ne	$=N/N_{tr}$, a normalised electron density
p	number of photons emitted per second; also probability; also integer defining grating order
p	electron's momentum
P	density of holes
$P_{out}(\omega)\,d\omega$	optical power in spectral range ω to $\omega+d\omega$
P_{fin}, P_{fout}	forward power at input, output (subscript r for reverse)
P_{net}, $P_{net\,sp}$, $P_{net\,out}$	net output power, net spontaneous power
P_{SM}	side-mode optical power
q	electronic charge
Q	cavity Q ratio of stored energy to ($2\pi \cdot$ energy lost per cycle)
θ, θ'	angles giving measure of index, gain, coupling coefficients through $\sin\theta$ and $\sinh\theta'$
θ_{air}, θ_{int}	half-cone angle of emission outside/inside laser
θ_i, θ_d, θ_c	incident, diffracted, critical, angle
$\Theta = (\beta_y\,d/2)$	optical half thickness of layer
$\Theta = \exp(j\,\overline{\omega}_{offset})$	phase factor in filter—for producing offset frequency
r_{sp}^2	mean-square spontaneous excitation
R_t, R_{source}, R_{laser}	resistances

Principal notation

R, $R(T, Z)$	reverse field amplitude as a function of time and space steps
R_L	reverse field at position $z = L$
$\mathcal{R} = \mathcal{X} + j\mathcal{Y}$	complex random (normally distributed) number such that $\overline{\mathcal{R}^2} = 1$
ρ	amplitude reflectivity or reflectivity measured for electric fields
ρ_{left}, ρ_{right}	amplitude of reflection at left and right facets
ρ_c	contact-layer resistivity
$S_{sp\,f(r)}$	spontaneous power coupling to forward (reverse) fields
s_t	normalised 'temporal' step $= \nu_{mg}\,\delta t$
s	step length in space for computation
S, $S(x, y, z)$	optical photon density, as function of position
$S_{max}(z)$	maximum photon density at z, over an equivalent area $\mathcal{A}$
S_{spont}, S_{stim}	density of spontaneous/stimulated photons
$S'(f)\,df$	photon density emitted over a range of frequencies df
S_0, S_1	photon density in the steady state
S_f, S_r	photon density associated with forward, reverse flux
t	time
t_c	thickness of the contact layer
T	transmission coefficient at facets
T, T_n	absolute temperature, noise temperature
T	integer giving time steps
T_1, T_2	transistor labels
τ	time constant of spectral filter
τ_p, τ'_p	effective photon lifetime including the distributed losses
τ_{sp}	lifetime for spontaneous radiative emission
τ_r	spontaneous recombination time constant including all mechanisms
τ_{ro}	recombination time constant when $N = N_{tr}$
$t_{intraband}$	intraband relaxation time $\sim 10^{-12}$ s

Principal notation xxv

$u(x, y)$, $u_n(x, y)$	transverse-field distribution and for nth mode
U	stored electromagnetic energy
U_e, U_m	electric and magnetic energy
$\mathcal{V} = L\mathcal{A}$	effective volume for photons
V	parameter for guide of width d where $V = (\omega d/c)\sqrt{(\delta\varepsilon_r)}$
$v(x, y)$	transverse-field pattern for radiation fields, $v(y)$ in one dimension
V_{con}	contact voltage
V_J, ΔV_J	junction voltage, change in junction voltage
v_g	group velocity $= d\omega/d\beta$
v_p	phase velocity $= \omega/\beta$
$w_{a\,vert}$	active-layer effective vertical width (i.e. thickness)
$w_{eff\,vert}(w_{eff\,hor})$	effective vertical (horizontal) width of optical intensity
W_v	vertical-layer thickness
$\overline{\omega}_M$, $\overline{\omega}_{offset}$	normalised modulation and offset angular frequency
ω, ω_0, ω_{offset}	$= 2\pi f$ angular (optical) frequency, central frequency, offset frequency
ω_M, ω_r	modulation and photon–electron-resonance angular frequencies $\ll \omega_o$
Ω, Ω'	solid angle for either 'forward' or reverse spontaneous emission inside and outside guide
y	co-ordinate distance
z	axial distance
Z	integer giving space steps

Chapter 1
The semiconductor-diode laser

1.1 Background

The invention of semiconductor diode lasers [1–4] along with their ability to be directly modulated with information [5,6] and the invention of optical-fibre communication [7–9] in the 1960s–1970s have together had far-reaching effects on telecommunications leading to extensive optical-fibre communication networks and systems [10–12]. This book explores one aspect of this success story. High-performance optical-fibre systems need sources with a number of features:

(i)   stable single-frequency output;
(ii) modulation capability of gigabit/s;
(iii) stable operating lifetimes measured in $10^6 \rightarrow 10^8$ hours at room temperature; and
(iv) manufacturability.

The distributed-feedback (DFB) laser has these qualities and is now a standard device. The design and development of these lasers has proved to be fascinating work for those involved, and the physical and numerical modelling which has gone into developing these lasers is the topic of this book. Although the text concentrates on lasers with distributed feedback, the modelling work presented is more widely applicable and it is hoped that the reader will be as excited as the authors by the unfolding story.

1.2 Early developments

1.2.1 The first semiconductor lasers

The earliest semiconductor diode lasers [1–4] were straightforward *p–n homojunctions* formed in crystalline blocks of gallium arsenide

(GaAs), or GaP_xAs_{1-x} [4], with typical dimensions of the order of a few hundred microns. In such lasers, two opposing parallel facets provide reflective optical feedback, hence forming a Fabry–Perot resonator (see, e.g. [13]) but of sufficiently small cross-section that the radial fringes, typically associated with this optical interferometer, are not important. At injected current densities of the order of 100 000 A/cm², the stimulated emission from the electron–hole recombination in the *p–n* junction plane is sufficient to create optical gain at frequencies whose photon energies correspond to the band-gap energy of the material. The optical feedback increases the light levels in the resonator and lasing threshold can be reached, though this is only possible in such simple *p–n* junctions when cooled below room temperature and/or with short pulses to keep the thermal dissipation low. *Single-heterojunction* lasers [14–16] were next developed. These GaAs lasers had $Ga_xAl_{1-x}As$ ($x \sim 0.3$) on one side of the junction and heterojunctions (Section 1.3.1) have the remarkable feature of confining photons and charge carriers to the same region of the laser, thereby reducing threshold-current densities to around 10 000 A/cm² at room temperature. Powers of 10 W or more for 100 ns pulses were now possible, and made the first useful lasers, though their spectral and spatial output characteristics were relatively unstable. The *double-heterojunction* (or $Ga_xAl_{1-x}As$–GaAs–$Ga_xAl_{1-x}As$ sandwich) [17–19] laser further reduced the threshold currents and allowed continuous operation at room temperature with current densities down to around 1000 A/cm². Reducing the broad metal contacts to narrow stripes [20] a few microns wide yet further improved the lasers, so that by the 1980s, Fabry–Perot diode lasers could be used confidently for data transmission at over 100 Mbit/s through optical fibres [21,22] in real systems. With careful design, such lasers emitted from a single small spot, i.e. had a *single 'transverse' spatial mode*, but the spectrum was generally multimoded corresponding to many *'longitudinal'* modes. The remarkable advances in lasers achieving stable *single-frequency* operation with a capability for multigigabit/s direct modulation is a key subject of this book.

1.2.2 Fabry–Perot gain and phase requirements

The earliest diode lasers were all Fabry–Perot (FP) lasers which remain an important class of general-purpose laser. FP lasers, of small enough cross-section, may be initially modelled by considering a resonator which contains plane optical waves travelling back and forth along the length of the laser. These waves will have optical (angular) frequencies

$\omega = 2\pi f$ with an associated propagation coefficient $\beta = 2\pi/\lambda_m$ where λ_m is the wavelength in the material/waveguide. Such a wave starting from the left-hand plane reflector and travelling to the right is referred to as a *forward* wave and has its phase and amplitude written in complex form:

$$E_f(z) = E_0\, e^{(g_m - a_m)z}\, e^{-j\beta z}(e^{j\omega t}) \tag{1.1}$$

The amplitude decays or grows with distance because the wave suffers scattering and other fixed losses a_m per unit length, but also experiences a material optical gain g_m per unit length caused by the holes and electrons being stimulated into recombining at a rate which increases with the injected carrier density. Note that both a_m and g_m vary with wavelength, but in particular that the gain $g_m(\lambda)$ has a limited width of several tens of nm. The detailed mechanisms of optical gain are discussed in Chapter 2 but at present gain is modelled by considering a constant value of $(g_m - a_m) > 0$ along the laser so the electric field at the optical frequencies ω and position z is described by eqn. 1.1.

The laser is of length L and, for simplicity, it is assumed that there is zero phase change on reflection from the partially reflective facets at either end. At the right facet $(z = L)$, the forward optical wave has a fraction ρ_{right} reflected and this fraction now travels back (from right to left), and following the notation of eqn. 1.1, these *reverse* fields are described by

$$E_r(z) = \{E_0\, \rho_{right}\, e^{(g_m - a_m)L}\, e^{-j\beta L}\}\, e^{(g_m - a_m)(L - z)}\, e^{-j\beta(L - z)} \tag{1.2}$$

The time variation $e^{j\omega t}$ occurs in all terms, and will be included implicitly only from now on.

The reverse wave now travels back to the left facet $(z = 0)$ and a fraction ρ_{left} is reflected to form the forward wave. For stable resonance, the amplitude and phase after this single whole round trip must be identical to the phase and amplitude for the wave when it started, or in symbols:

$$E_0 = E_0\, \rho_{left}\, \rho_{right}\, e^{(g_m - a_m)2L}\, e^{-j2\beta L} \tag{1.3}$$

This results in the 'amplitude' condition for stable oscillation (lasing) given by

$$\rho_{left}\, \rho_{right}\, e^{2(g_m - a_m)L} = 1 \tag{1.4}$$

or

4 *Distributed feedback semiconductor lasers*

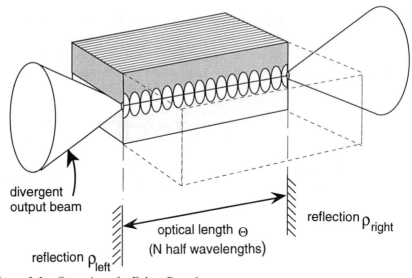

Figure 1.1 Operation of a Fabry–Perot laser

Showing schematic of standing electric-field waves inside the resonator

$$g_m = a_m + \frac{1}{2L} \ln\left(\frac{1}{\rho_{\text{left}}\,\rho_{\text{right}}}\right) \qquad (1.5)$$

where the logarithmic term can be considered as a distributed reflector loss. Simultaneously, the 'phase' condition must be met:

$$e^{-j2\beta L} = 1 \quad \text{or} \quad 2\beta L = 2\Theta = 2N\pi \qquad (1.6)$$

where N is any integer.

This concept of a net round-trip complex gain, G_{round} which has a value of unity for lasing, holds for all lasers and not just for FP lasers where $G_{round} = \rho_{left}\,\rho_{right}\,\exp[(g_m - a_m - j\beta_m)2L]$. It is equivalent to the classic condition for oscillation in electronic feedback amplifiers [23] where the product of gain and positive feedback is unity. Put another way, $1/(1 - G_{round}) \rightarrow$ infinity; giving output without input! Noise in oscillators and spontaneous emission in lasers give a non-zero input keeping $1/(1 - G_{round})$ finite.

1.2.3 Some characteristics of diode lasers

For a FP laser, lasing might be expected only at λ_{peak} where the peak value of gain $g_m(\lambda_{peak})$ satisfies eqn. 1.4. However, as in any electronic oscillator where random electrical noise gives the initial excitation, the

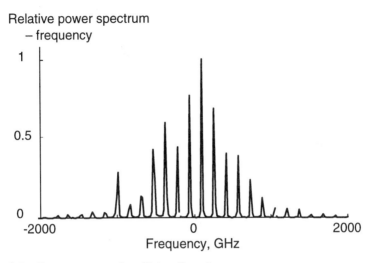

Figure 1.2 Power spectrum for a Fabry–Perot laser

250 μm Fabry–Perot laser operating at a central wavelength of around 1.55 μm

magnitude of the *round-trip* gain G_{round} is ever so slightly less than unity and it is the random *spontaneous emission* (discussed in Chapter 2) which is amplified through gain and feedback, by the factor of $1/(1-G_{round}) \sim 10^3$ to 10^6. Because the gain and spontaneous emission spectra cover several nanometres change of wavelength, the phase condition $\beta L = N\pi$ (N integer) is satisfied at a number of frequencies, or modes, within that band but the gain condition ensures that the spontaneous emission is only amplified significantly at modes where $|1/(1-G_{round})|$ is very large. As $|G_{round}|$ becomes smaller away from λ_{peak}, the amplification of the modes by the random excitation grows weaker. The spectrum of the FP laser in general shows multiple modes with frequencies determined from the phase condition giving a mode spacing $\Delta_{FP}f = (v_g/2L)$ where v_g is the group velocity for the optical fields in the laser's waveguide. Figure 1.2 gives a typical spectrum for a 250 μm Fabry–Perot laser operating at a central wavelength of around 1.55 μm showing these features.

With an ideal design, the laser's output power is proportional to the drive current above threshold, as in Figure 1.3. This is useful for direct amplitude modulation of the optical output by modulating the current drive. This ideal linear behaviour is modified by heating which reduces the optical gain and so increases the threshold. Heating and gain variations can also alter the spatial variations in optical and electronic

6 Distributed feedback semiconductor lasers

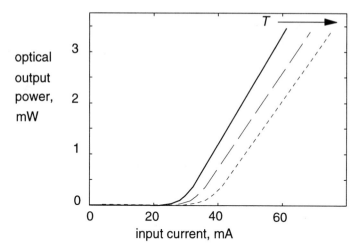

Figure 1.3 Typical optical output-power/current curve for FP laser

— — — and - - - - curves show the effect of increasing temperature T (see Section 2.6, eqn. 2.44)

densities within the laser, thereby also changing the ideal linearity. Modulation at gigabit rates creates additional phenomena such as a change of frequency (called 'chirp'; discussed in Chapter 4).

It follows that lasing action can occur only in semiconductors which exhibit optical gain; this normally implies a *direct-band-gap* semiconductor (see Chapter 2), and as a consequence silicon, so widely used for electronic devices, is unsuited to make lasers. Furthermore, the need for low threshold currents and stable optical modes requires the complex technology of different materials for heterojunction diodes as discussed briefly next.

1.3 Improvements to reduce operating currents

1.3.1 Heterojunctions: carrier confinement

It is found that, for lasing, densities of both electrons and holes, in the conduction and valence bands respectively, are required to be significantly greater than 10^{18} cm^{-3}. A major obstacle to achieving these densities in conventional *p–n* junctions formed within a single material is that carriers rapidly diffuse away from the junction and, to compensate for this diffusion, injected currents of tens of thousands of amperes per square centimetre are required. The previously mentioned double heterojunction controls this difficulty of diffusion.

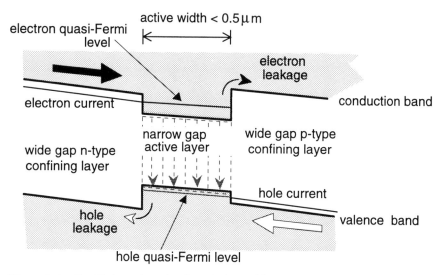

Figure 1.4 Double-heterojunction laser (schematic)
The electron and hole leakage is minimised by having energy barriers which are several kT in height so that the carriers have inadequate energy to surmount them

In Figure 1.4, the electronic-band diagram for a double heterojunction laser in operation is drawn schematically with holes flowing from the right, and electrons from the left, each with the Fermi energy defined by the injecting contact. The small slope on the bands indicates the presence of electric fields. These are required to overcome the ohmic resistance to the current flow. The Fermi levels in the materials adjacent to the contacts are typical of any forward-biased semiconductor diode but, in the active layer where the electrons recombine with the holes, the boundary conditions with the contacts require continuity of the quasi-Fermi levels and so ensure strong nonequilibrium conditions. Electron and hole leakage currents over the heterobarriers are usually negligible at room temperature, whereas the equivalent leakage away from a p–n homojunction would be several orders of magnitude greater. Current densities of the order of 1000 A/cm^2 can give electron densities well in excess of 10^{18} cm^{-3} and so give positive optical gain in the active region (Chapter 2).

1.3.2 Heterojunctions: photon confinement

It is a remarkably convenient feature of semiconductor physics that increasing the bandgap of a semiconductor by changing its composi-

8 Distributed feedback semiconductor lasers

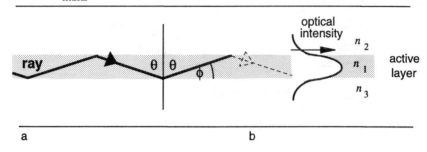

Figure 1.5 *Rays and intensity for confinement of optical fields to an active layer*

    a Ray picture (total internal reflection)
    b Field or modal picture

tion also normally decreases its refractive index. In Figure 1.5, the outer *cladding* or *confining* layers have a slightly lower refractive index than the active or gain layer. The layers act as a slab waveguide confining the photons closely around the active layer through the action of total internal reflection. The simple optical-ray picture is shown in Figure 1.5a. Total internal reflection (Appendix 1) requires that the refractive indexes of each of the outer materials, n_2 and n_3, are smaller than the refractive index n_1 in the active region. A guided ray has to propagate at a sufficiently shallow angle to the active region, and from Snell's law, the angle of incidence $\theta > \theta_{critical}$. One may therefore write:

$$\sin \theta > \sin \theta_{critical} = n_2/n_1 = 1 - \Delta n/n_1 \text{ here } \Delta n = n_1 - n_2 \text{ and } n_2 = n_3 \quad (1.7)$$

The angle ϕ shown in Figure 1.5 then has to be 'glancing' such that

$$\phi = (\pi/2 - \theta) < \cos^{-1}(1 - \Delta n/n_1) \simeq \sqrt{(2\Delta n/n)} \quad (1.8)$$

where n is the mean index. GaAs has a refractive index of 3.6, corresponding to n_1, and $Ga_{1-x}Al_xAs$ is lower by $\Delta n \simeq -0.66x$; typically $x = 0.4$, so Δn is about 0.26.

The conditions in eqns. 1.7 and 1.8 ensure that optical rays which propagate down the active region are reflected from side to side. This ray picture, however, fails to show the evanescent fields in the cladding layers which have to accompany total internal reflection. The more accurate wave-guiding picture, shown schematically in Figure 1.5b, indicates that, although optical power does not propagate outwards from the active layer, an *evanescent* wave penetrates to the order of a wavelength into the cladding layers. The details of waveguiding in multislab guides are discussed in Chapter 3 and Appendices 1 and 2.

The semiconductor-diode laser 9

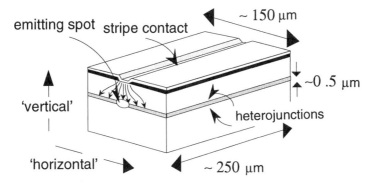

Figure 1.6 Simple oxide-stripe laser (schematic)

        Stripe metal contact ~5–10 μm wide helps to localise current and hence gain and so provide 'horizontal' confinement while the heterojunction provides 'vertical' confinement for both charge carriers and optical fields

1.3.3 Structures for 'horizontal' confinement

The carrier and optical confinement, mentioned so far, have operated perpendicular to the junction plane and this will be referred to as *vertical* confinement. Most applications involving coupling of the laser to an optical fibre require the laser to emit in a single mode with a well defined spot, typically of the order of one wavelength wide which is much less than the chip width. The light must then also be guided and confined across the junction plane and this will be called *horizontal* confinement. A high operating efficiency demands that the light and the injected charge carriers are confined as closely as possible to the same volume, so that the carriers must similarly be confined horizontally as well as vertically.

    Figure 1.6 illustrates one simple way of providing this horizontal optical and electron confinement in a laser [21,22]. Contact is made to the top of the laser through a stripe, etched through an insulating oxide, and although the current spreads out sideways under this contact, peak carrier densities are achieved under the middle of the contact, and so this is where the lasing filament has its maximum strength. The optical guiding in the junction plane is weak but adequate and is determined by a combination of *gain guiding* and weak *index guiding* (Chapter 3). Both of these guiding mechanisms are altered by temperature gradients, carrier-density changes and stress fields. Good waveguide design is essential to avoid the problems of current spreading and weak guiding which lead to high threshold currents, unstable modes and nonlinear light-current characteristics.

10   *Distributed feedback semiconductor lasers*

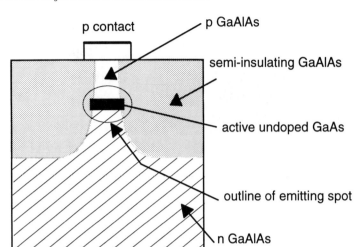

Figure 1.7   Simplified end view of buried-heterostructure (BH) laser

Buried guiding layer (black region) may typically be 2 μm wide and 0.2 μm thick

A theoretically good structure to solve this set of problems is the *buried heterostructure* (BH) [24] where the active filament is surrounded on all four sides by material of lower refractive index (and larger bandgap). Figure 1.7 shows the essential features; however, the technology associated with manufacturing such devices is difficult. In the BH GaAs laser, semi-insulating GaAlAs regions force all the injected current to flow straight down through the active region, which typically may be 2 μm wide and 0.2 μm thick. In this case it can be seen that the active region is surrounded on all four sides by a material with a wider bandgap which also has a lower refractive index, so forming an optical waveguide (analogous to a rectangular optical fibre). Technology and other constraints cause structures to be more complicated in practice, especially in the longer-wavelength InP/GaInAsP-materials system where *p–n–p–n* 'thyristor' current-blocking structures usually take the place of semi-insulating material. These layers contribute substantial stray capacitance which reduces the ability for high speed modulation.

Another solution to the horizontal confinement problem is the *ridge laser* [25] shown in cross-section in Figure 1.8. This needs only one epitaxial semiconductor growth process, and the processing steps are also relatively noncritical, resulting in a robust manufacturable device.

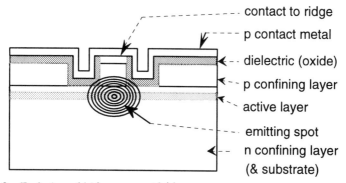

Figure 1.8 End view of 'ridge-waveguide' laser

Ridge typically may be 5–10 μm wide with active layer beneath, ~0.2 μm thick

There are no semiconductor current-blocking structures which add junction capacitances, so the stray capacitance is inherently low, and the device is easy to modulate at high speed. However, it can be seen that the current confinement stops just above the active layer, so sideways loss of drive current occurs and carriers diffuse out from under the ridge. Hence these devices typically require higher drive currents than comparable buried heterostructures. With the ridge guide, the optical fields extend out towards regions where the channels, with their lower refractive index than solid semiconductor, are etched. The effective refractive index (see Section 3.6) is then lower at the edges of the optical field than in the centre and provides the classic method for index waveguiding to occur. However the refractive-index profile which creates the guide depends on the optical-field profile which may change with power level. Careful design is needed to provide sufficient optical guiding over the whole range of required power levels and so avoid optical instability.

1.3.4 Degree of confinement—the confinement factor

Heterojunctions are effective in confining charge carriers to active narrow-bandgap material, as in the structures outlined previously. However, the nature of waveguiding by dielectric steps results in evanescent optical fields extending significantly into the wide-bandgap semiconductor surrounding the active region. This effect is most easily pictured in, say, the vertical dimension (Figure 1.9). If the active layer with an effective thickness $w_{a\,vert}$ is narrowed, a decreasing proportion, known as the *confinement factor*, of the optical power travels within this

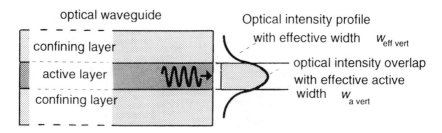

Figure 1.9   Overlap of carriers and optical field

active region. This vertical confinement factor is given by $\Gamma = w_{a\,vert}/w_{e\!f\!f\,vert}$ in Figure 1.9 where $w_{e\!f\!f\,vert}$ is the effective vertical width of the optical intensity. The concept is extended in Chapter 3 to both horizontal and vertical dimensions.

In modelling of diode lasers, the confinement factor Γ is an essential parameter which 'sweeps up' a large number of details of the structure into a single parameter that can be estimated from calculations of the two-dimensional guiding or sometimes from curve fitting between the theoretical and experimental current/light characteristics. Its importance stems from the fact that, with a well designed index guide, the value of Γ remains substantially the same over a wide range of optical-power levels, despite the influence of injected carriers on the local active-region refractive index. In addition, Γ is almost constant over a useful range of frequencies around that of the main mode for which Γ has been calculated.

1.4 Variations on conventional Fabry–Perot laser design

1.4.1 High-low reflective facets

Lasers considered so far have been 'standard' Fabry–Perot devices, which are probably the most straightforward to fabricate. Specific applications of lasers may require features such as:

(a) high output power (tens of milliwatts),
(b) small emitting area ($\sim 1\ \mu m^2$) matching into optical fibres,
(c) narrow ($\Delta\lambda < 1$ nm) spectrum, and
(d) high-speed (Gbit/s) modulation capability through directly modulating the drive current.

More complex structures exist which enhance these useful features, either singly or in combination. There are usually many ways of

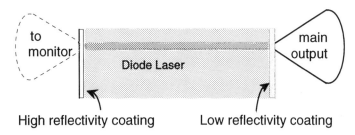

Figure 1.10 High-low reflective facets

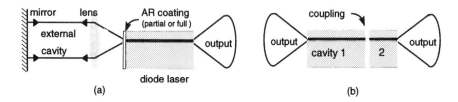

Figure 1.11 External-cavity and cleaved-coupled-cavity lasers

    a External-cavity
    b Cleaved-coupled-cavity

achieving any particular characteristic from a laser, but a few representative variations are briefly reviewed below.

In a conventional Fabry–Perot laser, the light comes out equally from both facets. Higher efficiency is possible by forcing most of the lasing light to come out of one facet, with only a small amount of light, used for monitoring, from the other (Figure 1.10). This can be achieved through the use of a high-reflectivity coating on the rear facet and a reduced-reflectivity coating on the front (high-low facet coatings) [26]. The product of the two field reflectivities ($\rho_{left}\,\rho_{right}$) is approximately the same as in an uncoated laser, so that the net feedback and gain required for oscillation (eqn. 1.5) are retained at about the same level. Extending the round-trip gain calculation in Section 1.2.2, leads to the expression for the ratio of facet power emissions: $P_{left}/P_{right} = (\rho_{right}/\rho_{left})(1 - \rho_{left}^2)/(1 - \rho_{right}^2)$.

1.4.2 External cavities

The external-cavity laser (Figure 1.11*a*) is an interesting device [27, 31] where, by varying the external cavity's dimensions and the reflectivity of the laser facet emitting into the cavity, conditions can be optimised for very narrow-line (<1 MHz) continuous output. The essential features of the oscillation condition are similar to those discussed for the Fabry–Perot laser but now the phase condition has to be satisfied simultaneously within the laser itself and within the external resonator. The length of the external cavity may be adjusted so that this phase condition is met only once within the gain spectrum of the laser's material. Changing temperature will cause such a laser to become two moded unless it is readjusted.

With appropriate device gain and full antireflection coating of the laser so as to produce one long effective cavity, the many modes of the combined system can 'lock' together in phase to create a 'mode-locked' laser giving optical pulses with peak powers at the watt level for durations of the order of a picosecond. Repetition rates of many tens of gigabits per second are possible, determined by the optical-pulse round-trip time within the extended cavity. However, these pulses have a broad spectrum appropriate to the short pulse width [28, 29]. Monolithic mode-locked lasers can now achieve pulse-repetition rates of hundreds of gigabits per second [30].

The cleaved-coupled-cavity [31,32] (or C^3) laser is a variant (Figure 1.11b) on the external cavity laser, where a FP laser is cleaved into two shorter FP lasers of incommensurate lengths with the cleave forming a third short (typically <1 μm-length) air-filled cavity (gap) between the two laser sections. Such lasers have numerous attractive properties in principle, including single-line operation with tunability of that line. However, their manufacturability and stability in operation have proved to be too variable for use in real systems.

Another way to achieve single-frequency operation under modulation is to operate one 'master' laser continuously, and then inject power from this into a 'slave' laser which may be modulated [33]. When carefully adjusted correctly, such a system works very well, but it is again expensive and relatively bulky, and a cheaper and more robust source of single optical frequencies is still required.

1.4.3 External grating

A tunable device which has become popular for many optical measurements is yet another variant on the external-cavity theme, where the external mirror is replaced by a diffraction grating (Figure 1.12) [34]. Single-frequency operation is ensured because feedback only occurs at the wavelength selected by the grating, and a bonus is that the laser may be tuned over the whole of the material-gain spectrum by rotating the grating mechanically. Precision mechanical

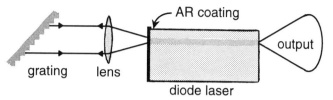

Figure 1.12   Laser with external grating reflector

design allows linewidths of less than 100 kHz and tunability over tens of nanometres to be achieved, but the mechanical assembly is too expensive and delicate for such lasers to be used in the robust environment of commercial communication systems, especially as the maximum frequency for direct modulation is limited.

1.5 System requirements for single-frequency lasers

Before going further into the design and dynamic operation of single-frequency lasers, the commercial push for such requirements should be understood. A schematic long-distance optical-fibre link is sketched in Figure 1.13 for an amplitude-modulated system. To keep costs down, the system should operate with the maximum possible distance between the repeaters which will regenerate the signal and retransmit, because repeaters are expensive.

At one time it was envisaged that adequately long distances might be possible only with 'coherent' detection where, like a television heterodyne receiver, one has a tunable local oscillator (tunable laser) mixing in a photodiode with the incoming optical signal to produce an intermediate-frequency output which is detected and amplified. The advent [35,36] of the low-noise optical erbium-doped fibre amplifier (EDFA) and its subsequent development have demonstrated that the type of system shown in Figure 1.13 is almost as sensitive as the more complex coherent systems; consequently coherent receivers are not discussed here.

The optical fibre [37] is, of course, a vital component, but in the 1960s and 1970s the optical wavelength was determined by the ability to make lasers with layers of $Ga_{1-x}Al_xAs$ giving emission in the 0.7–0.9 μm range. However, it was soon realised that the loss process in silica glass made the fibre's attenuation much lower around two special wavelength 'windows' (1.3 μm with attenuation of ~0.5 dB/km and 1.55 μm with attenuation of ~0.2 dB/km). The challenge of new materials for lasers at these longer wavelengths was met with both

16 *Distributed feedback semiconductor lasers*

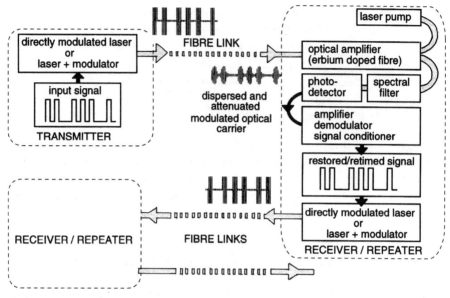

Figure 1.13 *Schematic optical system with repeaters*

Some repeater systems only need the optical amplifier without detection and restoration. The fibre link may have several lasers feeding into it via an optical coupler for WDM systems

standard fibre and laser sources becoming centred around the 1.3 μm and 1.55 μm wavelengths.

Although the attenuation in optical fibres is sufficiently low to allow links of well over 100 km using standard 1 mW FP lasers at 140 Mbit/s, at higher bit rates there is a significant dynamic wavelength shift (or 'chirp') in the laser's emission which causes loss of information, regardless of the laser's output power, because of the *dispersion* in the fibre. Dispersion is the phenomenon whereby, because of wavelength-dependent phenomena within a material, different wavelengths in the fibre have different speeds of propagation. Consequently a pulse which contains a range of wavelengths will spread as it travels. Eventually, given a long enough fibre link, one pulse spreads into an adjacent pulse and causes interference. At 1.55 μm, where the attenuation in standard silica fibre is lowest, dispersion is significant amounting to about 17 ps/km nm. For such standard silica fibre, dispersion is nominally zero at 1.3 μm, but laser wavelengths are very rarely matched exactly to the dispersion minimum of the particular fibre in a link (due to manufacturing and installation tolerances, as well as temperature variations) so low chirp is desirable from lasers, even at 1.3 μm or in special low-dispersion (dispersion-shifted) fibre that can be formed around 1.55 μm.

As an example of transmission in standard fibre at 1.5 μm with dispersion of 17 ps/km nm, a bit rate of 500 Mbit/s over 100 km requires a spectral width ~0.15 nm (~20 GHz) or less if interpulse mixing is to be limited to 10% of the bit period. This rules out the use of the typical multiple-longitudinal-mode Fabry–Perot laser where, for a cavity length of 400 μm, the spacing of the modes is 0.8 nm (~100 GHz). Increasing the bit rate to 5 Gbit/s over the same distance would require a source spectral width of less than 0.015 nm, which is a difficult design problem even for the single-line distributed-feedback (DFB) laser introduced in Section 1.6 and techniques to compensate the dispersion in the fibre caused by chirp may need to be used.

Besides the need for narrow linewidths to reduce the effects of dispersion and so permit longer transmission distances, there is another commercial goal of enhancing the information capacity of optical-communication networks by transmitting different channels of communication at sets of well defined wavelengths. Such a process is called wavelength-division multiplexing (WDM)[38]. As the channel spacing is decreased, one refers to dense WDM but at the time of writing dense WDM is still a gleam in the eye of systems' designers. Another gleam is wavelength routing whereby a particular wavelength is assigned to a particular route or perhaps identified with some particular transmitter.

While wavelength and linewidth control for a laser are of primary importance in communication systems, it must not be thought that these are the only parameters that matter. A key requirement is to achieve an adequate carrier-to-noise ratio (CNR) so that bit-error rates (BER) are less than 1 error in 10^9 bits. While increasing the optical-carrier power is an obvious method of improving the CNR and reducing any BER, the output-power levels for long-distance fibre systems are restricted to the milliwatt range to limit nonlinear mixing and other nonlinear effects in the fibre. The output of a laser will contain amplitude noise (measured by relative-intensity noise or RIN), frequency noise (line width broadening), and unwanted side modes (measured by side-mode suppression ratios or SMSR). The detailed specification of how low these levels must be usually depends on the application, but typically the combined power of all these unwanted signals (the noise) over the system bandwidth needs to give a CNR in excess of 30 dB to maintain adequate BERs. Gowar [39] gives useful power budgets for several optical-communication systems.

18 *Distributed feedback semiconductor lasers*

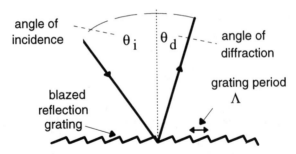

Figure 1.14 Bragg grating in reflection

The Bragg grating buried inside a laser may not be the final answer to the problems of precise frequency determination and high CNR, but it provides an excellent starting point and a fascinating study. This chapter therefore concludes with an introduction to the Bragg grating and how it may be buried inside a practical laser.

1.6 Introduction to lasers based on Bragg gratings

1.6.1 *Introduction to Bragg gratings*

It is likely that most readers have observed the selection of colour (wavelength) on looking at a CD ROM under a white light at certain angles. This is one example of a periodic system selecting the wavelength of light through the phenomena recognised and analysed by Bragg [40] around the 1920s. Figure 1.14 illustrates the conventional reflective diffraction of a plane wave from a periodic grating structure with elements spaced at a period Λ. A plane incoming wave is incident at an angle θ_i, and the 'Mth-order' diffracted plane wave emerges at the angle θ_d. Constructive interference between the diffracted waves emanating from adjacent elements of the grating requires the condition

$$\Lambda(\sin \theta_i - \sin \theta_d) = M\lambda \qquad (1.9)$$

The integer $M = 0, \pm 1, \pm 2, \ldots$ defines the *order* of the diffraction or the phase shift in wavelengths between the diffracted waves from adjacent grating elements, and λ is the wavelength of the radiation.

Because a Bragg grating can select specific frequencies, it is sensible to consider embedding a diffraction grating into a semiconductor waveguide. This is illustrated in Figure 1.15 with concepts similar to those in Figure 1.14, except that there are now different refractive

The semiconductor-diode laser 19

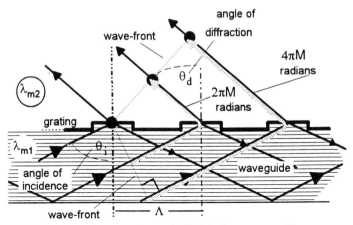

Figure 1.15 Diffraction from grating embedded in laser waveguide

> For a grating embedded inside a laser, $\theta_i \sim \pi/2$ and for feedback the diffracted wave $\theta_d \sim \pi/2$. However, one can also have $\theta_d \sim 0$ when the diffracted wave gives radiation out from the guide. The diffracted wavefront is reinforced because there are $2\pi M$ radians per pitch where M is an integer and is known as the order of diffraction

indices and one is also thinking of obtaining a diffracted wave which feeds backwards. Using the usual Huygen's construction for constructive interference from 'identical' points but at different periods of the grating, one requires the sum (or difference) of the total optical path in radians to be integer multiples of 2π. In terms of distances and the wavelengths appropriate to the different materials as in the diagram of Figure 1.15:

$$\frac{\Lambda \sin \theta_i}{\lambda_{m1}} + \frac{\Lambda \sin \theta_d}{\lambda_{m2}} = M \qquad (1.10)$$

The integer M defines the *diffraction order.*

On applying the result of eqn. 1.10 to a propagating optical mode in a DFB laser waveguide, one notes that the angle of incidence θ_i cannot be less than the critical angle for total internal reflection [41] θ_c so that

$$\sin \theta_i \geq \sin \theta_c = \frac{n_2}{n_1} = \frac{\lambda_{m1}}{\lambda_{m2}} \qquad (1.11)$$

where n_1 and n_2 are the refractive indices in the central and outer regions of the waveguide, respectively. Inserting eqn. 1.11 into eqn. 1.10 gives

$$\sin\theta_d \geq \left(\frac{\lambda_{m2}M}{\Lambda}-1\right) \approx \left(\frac{2M}{p}-1\right) \qquad (1.12)$$

where an integer $p=\Lambda/(\tfrac{1}{2}\lambda_m)$ defines the *grating order*. Because of the small differences in the refractive index, $\lambda_{m2} \sim \lambda_{m1} \sim \lambda_m$, the effective wavelength in the guide. In general, for a laser with feedback along the waveguide, one requires $\theta_d \sim \theta_i \sim \pi/2$ requiring, from eqn. 1.12 using the equality sign and $\sin\theta_d = 1$, that the order p of the grating equals the order M of the diffraction. Given the physical periodicity Λ of the grating, the optimum free-space wavelength λ_b which is most strongly reflected is called the Bragg wavelength and satisfies the Bragg condition:

$$M(\lambda_b/n_{eff}) = 2\Lambda \qquad (1.13)$$

where n_{eff} is a mean value of the refractive index in the guide and relates the free-space Bragg wavelength λ_b with the corresponding guide wavelength $\lambda_m = (\lambda_b/n_{eff})$.

However, it is also possible to find gratings where $\sin\theta_d \sim 0$ (e.g. when p=2M) or $\sin\theta_d$ is significantly less than 1 so that the diffracted wave is then radiated away from the guide. The second-order grating where p=2 is of particular note because this can be used to couple light from a laser out into the direction perpendicular to the junction to form a surface-emitting laser (Section 8.5.1). Table 1.1 summarises the influence of grating order p and diffraction order M on the feedback and radiation loss experienced by a propagating mode.

There is often confusion between what is loosely termed a pth-order grating, and what is correctly a pure pth-order grating. The former expression refers to a grating with a fundamental spatial periodicity of $p\lambda_m/2$, and in general is described by Fourier components at the fundamental and higher-harmonic periodicities, while the pure grating will have only the fundamental component. A *pure* second-order grating will not be able to give feedback.

A first-order grating with a period of one-half a guide wavelength gives optical feedback in the first diffraction order. By comparison, a second-order grating radiates normal to the grating in the first diffraction order but gives feedback in the second diffraction order. Because second-order gratings have larger dimensions than first-order ones, they are often easier to define photolithographically, and so can be preferred by manufacturers. The radiation loss need not be serious unless such a grating is deliberately designed to couple out light as in a surface-emitting laser (Section 8.5.1). A third-order grating gives

Table 1.1 Summary of feedback and radiation loss from DFB laser gratings

Order of diffraction	M=0	M=1	M=2	M=3
First-order grating, p = 1	Feed-forward	Feedback		
Second-order grating, p = 2	Feed-forward	Radiation at $\theta_d \sim 90°$	Feedback	
Third-order grating, p = 3	Feed-forward	Radiation $\theta_d \sim \sin^{-1}(-1/3)$	Radiation $\theta_d \sim \sin^{-1}(1/3)$	Feedback

radiation loss only in the first and second diffraction orders at angles symmetrically disposed about the normal, but gives feedback in the third diffraction order. In general, feedback is obtained from a grating of order N in diffraction order N, but diffraction orders intermediate between 0 and N result in radiation loss for such a grating.

In principle, just as the blaze or shape of the grating in a spectrometer alters the power division between the different orders of diffraction, so also one can change the reflection and radiation [42–44] by tailoring the grating-tooth shape. One may also need to distinguish between the TE and TM polarisation within the guides [45]. In general, the requirement to mass manufacture gratings buried within a laser gives severe constraints on the tooth shape which is usually controlled by the chemical etching and crystal orientation.

Details of wave propagation and gain within such gratings with a distributed reflectivity are covered at the end of Chapter 4 and beginning of Chapter 5. A method of calculating the reflectivity per unit length is discussed in Chapter 4 with the help of a MATLAB program **slabexec**.

1.6.2 Fabrication of gratings inside lasers

The theoretical concepts of lasers with distributed feedback [46] were developed well before the technology could deliver a reliable and useful method of fabrication. The early semiconductor lasers with distributed feedback operated at low temperatures and had short operating lifetimes [47,48] but in the mid 1970s GaAs lasers with distributed feedback could be made to operate at room temperature [49], and then developments with new materials led to room-temperature operation of lasers, with distributed feedback, around 1.3 μm and 1.55 μm [50,51] in the early 1980s. The details of fabrication of semiconductor lasers and the techniques of substrate formation and subsequent epitaxial-layer growth using metal organic chemical vapour-deposition (MOCVD) or molecular-beam epitaxy (MBE) are outside the scope of this book and are far from trivial. The research and development of materials-growth techniques for forming the confinement-layer/active-layer heterojunctions of lasers have been vital. All that can be given here is a mere glimpse of this marvellous interaction of materials and technology that has led to reliable commercial devices. Figure 1.16 illustrates some principal processing steps needed to form a semiconductor laser with a uniform embedded grating which creates the feedback as described in Section 1.6.1.

The first step is to grow, by the methods of epitaxial growth, good laser structures with the substrate followed by the confining layers and

The semiconductor-diode laser 23

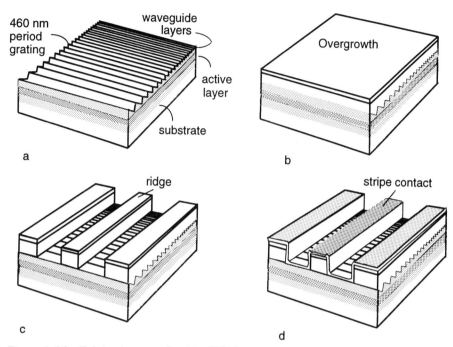

Figure 1.16   *Fabrication steps for ridge-DFB-laser structure*

> *a* Grow MQW SCH structure by MOCVD then define and etch DFB grating; *b* Overgrowth of grating; *c* Etch ridge; *d* Apply stripe contact. Completed device

active layer as discussed in Section 1.3. The active-gain region can be a single layer of material (referred to as a *bulk*-gain region) or it can be formed from one or more 'quantum-well' layers, e.g. a *multiquantum well* (MQW) region which may be ≤100 nm thick, surrounded by separate-confinement-heterostructure (SCH) layers (see Section 2.7.4). While some further discussion of quantum-well material is given in Section 2.7, the details of growth again have to be left for further reading. Many of the steps for constructing a typical DFB laser are similar to those for constructing a good FP laser, and the grating can be etched into epitaxial layers below the gain region or above the gain region (substrate or surface gratings). Here is depicted the scheme where the grating is etched into a confining layer above the active layer. Hence after the confining and active layers have been grown, the grating is lithographically defined in a resist which then acts as a mask for the grating etch. For mass production, techniques are favoured which expose a periodic pattern in a resist using ultraviolet-laser

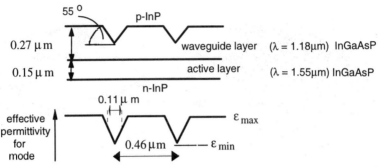

Figure 1.17 Schematic cross-section of DFB laser grating etched in epitaxial-layer structure

radiation. However, electron-beam (E-beam) lithography is an alternative technique which uses special electron-sensitive resists and has the advantage of greater resolution enabling first-order gratings to be written more easily. E-beam lithography also facilitates fabrication of more complex structures such as gratings with a changing periodicity or incorporating 'phase shifts' as discussed shortly. After definition, the grating is chemically wet-etched, or dry-etched using a plasma, so as to transfer the pattern in the resist into the semiconductor. The crystal planes and etching processes usually determine the angles and shapes of the grating teeth.

Following removal of the resist, the grating is then overgrown with material which completes the guiding structure and has a highly doped top layer to make a good ohmic contact. The substrate usually forms the lower contact. The vertical waveguide is then fabricated and is usually either a ridge structure (Figures 1.8 and 1.16) or a buried heterostructure (Figure 1.7). For the ridge guide, the final steps involve depositing an insulator on the *p*-side ridge structure which is often SiO_2 or Si_3N_4. Photolithography is then used to open a window in the insulating layer on the top of the ridge and a metallisation-evaporation process completes the device by applying contact metal (Figure 1.16*d*).

Figure 1.17 illustrates schematically the cross-section of a practical second-order etched grating for operation around 1.55 μm free-space wavelength. The guide wavelength λ_m is 0.46 μm which is the same periodicity as for the grating, so that at first sight it looks as though the feedback would all cancel, but because the teeth have an average length of $\sim \lambda_m/4$ there is net feedback along the laser as in a first-order grating.

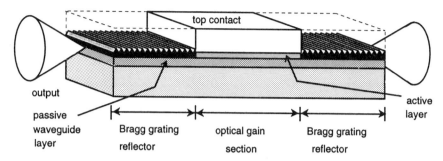

Figure 1.18 Diagram of distributed-Bragg-reflector (DBR) laser

1.7 Some principal forms of grating laser

1.7.1 The distributed Bragg reflector laser

The insertion of a 'Bragg grating' into a laser has proved to be a very good solution to providing low-cost and reliable single-frequency optical sources capable of high-speed modulation. A device with a uniform grating is usually known as a distributed-feedback (DFB) laser or, in a slightly different form, the distributed-Bragg-reflector (DBR) laser. The DBR laser is easier to understand and this is considered first.

Figure 1.18 shows schematically a laser with a centre section of waveguide with gain which is essentially identical to the Fabry–Perot lasers considered earlier. The key change is that, instead of using lumped reflections from the end facets of the laser as in a FP laser, the passive waveguide is extended beyond the gain region into the end sections where the grating is etched and designed to give its multiple tiny reflections adding in-phase at the operating wavelength. The advantage of this structure over the simple Fabry–Perot laser is that the multiple reflections give significant feedback *only* around a frequency—the Bragg frequency—determined by the grating pitch.

The description of a grating as in Section 1.6 gives one straightforward way to consider the effects of a Bragg grating, but another simplified argument, particularly appropriate for a grating in line with a waveguide, is to note that identical reflectors which are spaced $\lambda_m/2$ apart within a guide create reflections which all 'add'. The point to note here is that, between any two reflectors (A and B), the wave travelling forward advances by 180° from A to B but the reflected wave also has to advance by 180° to get from B back to A, giving a 360° round trip determining the reference phase of the reflection. Each reflection therefore reinforces the other reflections.

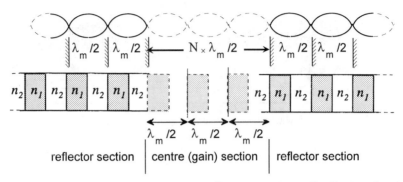

Figure 1.19 Grouping of a grating into $\lambda_m/2$ sections where all reflections $\Delta\rho$ from each section add

Figure 1.19 sketches the concept of reflective blocks formed from alternating sections of refractive indices n_1 and n_2, where each total block is $\lambda_m/2$ thick at the desired 'Bragg' frequency where there is to be maximum reflection or maximum feedback. Each block will give some net reflection $\Delta\rho$ to a wave that is travelling forward. With N pairs there will approximately be a net reflection $N\Delta\rho$ at the Bragg frequency, making no allowance for the loss of forward energy as the wave is reflected back; a more accurate calculation must await until Chapter 4. As the optical frequency moves away from the Bragg frequency, so this reflection rapidly reduces because the reflections no longer reinforce each other in phase. To estimate the maximum magnitude of this effect (see Appendix 1), take each section of high or low refractive index to be $\lambda_m/4$ in length so that $\Delta\rho \sim (n_1 - n_2)/(n_1 + n_2)$ for each reflection, giving a net reflection per unit length of $\kappa \sim 4(n_1 - n_2)/[(n_1 + n_2)\lambda_m]$. The reflection per unit length κ is known as the grating *coupling factor* and, for a grating of length L, κL is a key dimensionless parameter determining the frequency selectivity and performance of the grating with practical values of κL taking typically values around 2.

1.7.2 The distributed feedback (DFB) laser

In principle, to obtain oscillation there is no need to separate the regions of gain and reflection as shown in Figure 1.18. Figure 1.20 indicates schematically the concept of a uniform distributed feedback laser whereby, with distributed reflections and gain within the laser cavity, an optical wave travelling in one direction is continuously

The semiconductor-diode laser 27

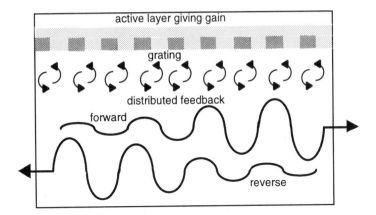

Figure 1.20   Uniform distributed-feedback laser
　　　　The concept of continuous feedback combined with gain

scattered into the optical wave in the reverse direction. Both waves grow as they travel towards their respective facets because of the feedback and the gain. The abrupt changes in refractive-index, with each section of length $\lambda_m/4$, are purely schematic to emphasise the main features. This is the distributed-feedback (DFB) referred to earlier. However, it is found that the DFB laser oscillates at two possible frequencies slightly removed from the Bragg frequency depending on κL (Chapters 4 and 5). The reason for this is that the phase condition for oscillation cannot be met at the Bragg frequency, as discussed below.

Now reconsider the DBR laser. It is found that the presence of gain enhances the reflectivity of the Bragg gratings and this increased reflectivity dramatically reduces the required gain for the FP-gain section. A particular utility of the argument used in Section 1.7.1 based on the $\lambda_m/2$ sections above comes in the ideal design of the central section for a DBR laser required to operate at the Bragg frequency of the grating where there is the maximum reflection. It is possible to ensure that the round-trip phase matching occurs, as in a Fabry–Perot laser, if the two reflectors are identical mirror versions of each other with a real reflection of the same sign, and then these reflectors are separated by a central section of length $N \times \lambda_m/2$ (Figure 1.19). The shortest possible Fabry–Perot section is $\lambda_m/2$ long (N=1) and this then requires the insertion of an additional $\lambda_m/4$, say into the lower-refractive-index region to form a length $\lambda_m/2$ between two mirrored sections of Bragg gratings (Figure 1.21*a*).

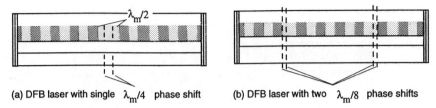

(a) DFB laser with single $\lambda_m/4$ phase shift   (b) DFB laser with two $\lambda_m/8$ phase shifts

Figure 1.21   Schematic of DFB-laser variants

Many commercial lasers simply insert lithographically a physical $\lambda_m/4$ phase shift into the grating mask, and the resultant devices are known as $\lambda_m/4$-*shifted* or *phase-adjusted* DFB lasers and oscillate with a single frequency close to the Bragg frequency of the grating. These $\lambda_m/4$-shifted lasers do not give optimum dynamic wavelength stability during modulation, and so some more complex phase-adjusting regime such as two $\lambda_m/8$ phase shifts may be preferred (Figure 1.21*b*), giving greater stability in the output [52]. More phase shifts have been proposed [53], but these add to the difficulties of manufacture and the simpler construction is usually preferred unless clear advantages are demonstrated. High-low reflectivity facets have also been proposed to give single-frequency operation [54], but the precise position of the high- reflectivity facet with respect to the periodicity in the index grating is important and it is currently impossible to maintain this few-nanometre tolerance in production.

Although lasers with uniform gratings can give good single-mode operation, the selection of the mode, either just above or just below the Bragg wavelength, depends considerably on manufacturing nonuniformities and residual facet reflections so that high yields for one specific frequency may be a problem. These subtleties of grating design are the key to obtaining optimum stable performance under modulation, especially for high-speed-communication applications, and are discussed more fully in Chapter 6.

1.7.3 More complex grating-based lasers

Splitting or branching the contacts allows the laser designer an extra degree of freedom to improve the functionality of the laser. There is an immense variety available and only a brief review is possible. Figure 1.22 shows simplified schematic side views of lasers which can give very low dynamic chirp (i.e. changes of frequency with changes of modulating current). In Figures 1.22*a* and *b* the total drive current to the laser is kept almost constant under modulation, but one end is

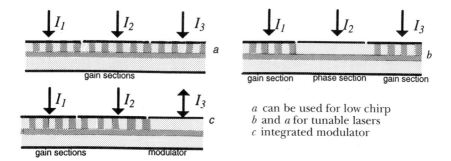

Figure 1.22   Multisection Bragg lasers

driven harder than the other, and under modulation the drives to the ends are reversed, this being known as 'push-pull' operation [55] (see also Chapter 8). The total gain and electron density in the laser remain constant, but optical power is switched from end to end (Section 8.3). Linewidths of less than 0.01 nm spread in wavelength are predicted to be possible at 10 Gbit/s with these lasers. Structures such as that shown in Figure 1.22*b*, with a plain gain region in the middle, can be used to give a wide range of tuning for the laser [56,57].

'DFB' lasers with more complex grating designs have been designed to give wide-range tunability though not allowing high-speed modulation [58,59]. Here the lasers are designed with an additional periodicity of several wavelengths superimposed on the $\lambda/2$ Bragg wavelength. This then allows a different periodic reflection-spectrum from each end of the laser, each of which can be shifted slightly by the small refractive-index reduction caused by carrier injection. 'Vernier' action, discussed in Section 8.4.3, then occurs so that the laser operates only at a wavelength where reflection peaks from *both* of the gratings coincide. This has allowed quasicontinuous tuning over greater than 100 nm centred on 1.5 μm. Vernier action is also possible where the waveguide for the laser is branched into a Y with the favoured mode having a common peak amplification/reflection from each arm of the Y (see Section 8.4.4) [60].

For applications which demand narrow stable linewidths, structures such as that shown in Figure 1.22*c* are sometimes used [61,62]. The main part of the laser is run continuously, and hence has a narrow spectrum, and the modulator section simply switches the output on and off. In practice, the modulator interacts with the master laser, and some excess chirp results, but designs are improving, and many more multisection, multicontact lasers with increased functionality may be expected in the future.

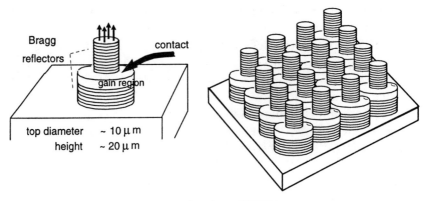

Figure 1.23 Vertical-cavity surface-emitting laser (VCSEL)
Single devices and arrays

Another fascinating laser which uses Bragg gratings is the vertical cavity surface-emitting laser or VCSEL (pronounced vicsel) shown schematically in Figure 1.23 [63,64] and discussed further in Chapter 8. The 'length' of the laser is now only a few microns, and therefore the maximum gain available is typically two orders less than that of a conventional laser and so the mirror reflectivities typically have to be in excess of 99%. Such a high reflectivity requires that the top mirror has to use a multilayer stack, which is a form of Bragg-grating reflector with either deposited multilayer dielectrics or grown semiconductor layers of alternating high and low refractive index. These micron-size lasers may be fabricated side by side on a single chip and, at the time of writing, can form arrays of about 10×10 [65], but significantly larger arrays are expected in future provided that the problems of packaging and connections can be overcome. Such lasers offer many exciting possibilities as sources for optical switching, routing and signal processing, and their technology is discussed in detail in specialist texts such as that by T.E. Sale (see Bibliography).

1.8 Summary

This chapter has set the scene for the operation and fabrication of semiconductor lasers, emphasising why the authors see lasers with grating reflectors as so important. The earliest lasers were examined, showing that a key requirement is to confine the photons and electrons to the same physical region. A geometrical factor Γ, known as the confinement factor, expresses the degree to which this has been achieved. The fact that light is harder to confine closely around a given region than it is to confine the electrons around the same region means that Γ takes values typically ranging from 0.3 to 0.01, depending in detail on the technology that is used. The simplest requirements for lasing were put forward so that the reader understands the importance, just as in electronic feedback oscillators, of the round-trip gain and phase in achieving a stable oscillation. Many factors influence the net efficiency of a laser other than the facet-reflectivities (Section 1.4.1). Nonradiative recombination, waveguide absorption and scattering losses reduce the quantum efficiency for converting electron/hole pairs to photons from the ideal of 100%. Further reductions occur because not all charge carriers, injected at the contacts, reach the active region. Carriers spread away from contacts, leak across blocking layers or over heterobarriers and suffer interfacial recombination. Typical single-facet effieciencies ~10-30% dependant on structure and processing.

The importance of single-mode lasers for modern optical communication along silica fibres over many tens of kilometres was noted. A variety of different types of laser were outlined, and in particular it was observed that to gain a stable single mode the favoured technology is to incorporate some form of Bragg grating. The fundamentals of gratings were introduced and the elements of different laser designs with gratings buried in their structures were presented.

1.9 Bibliography

1.9.1 Semiconductor lasers

KRESSEL, H. and BUTLER, J.K.: 'Semiconductor lasers and heterojunction LEDs' (Academic Press, New York, 1977)

CASEY, H.C. and PANISH, M.B.: 'Heterojunction lasers. Part A—Fundamental properties; Part B—Materials and operating characteristics' (Academic Press, New York, 1978)

THOMPSON, G.H.B.: 'Physics of semiconductor laser devices' (Wiley, Chichester, 1980)

HAUS, H.A.: 'Waves and fields in optoelectronics' (Prentice–Hall, Englewood Cliffs, 1984)

ADAMS, M.J., STEVENTON, A.G., DEVLIN, W.J. and HENNING, I.D.: 'Semiconductor lasers for longwavelength optical fibre communication systems' (Peter Peregrinus, London, 1987)

BUUS, J. 'Single frequency semiconductor lasers' (SPIE Optical Engineering Press, 1991)

PETERMANN, K.: 'Laser diode modulation and noise' (Kluwer Academic Press, Dordrecht, 1991)

YARIV, A.: 'Optical electronics' (Saunders College, Philadelphia, 1991), 4th edn.
AGRAWAL, G.P. and DUTTA, N.K.: 'Semiconductor lasers' (Van Nostrand Rheinhold, New York, 1993), 2nd edn.
OHTSU, M.: 'Highly coherent semiconductor lasers' (Artech House, Boston, 1992)
KAWAGUCHI, H.: 'Bistabilities and nonlinearities in laser diodes' (Artech House, Boston, 1994)
ZORY, P.S. (Ed.): 'Quantum well lasers' (Academic Press, Boston, 1993)
SALE, T.E.: 'Vertical cavity surface emitting lasers' (Research Studies Press, Taunton; Wiley, New York, 1995)
COLDREN, L.A. and CORZINE, S.W.: 'Diode lasers and photonic integrated circuits' (Wiley, 1995)
GHAFOURI SHIRAZ, H. and LO, B.S.K.: 'Distributed feedback laser diodes' (Wiley, 1996)
MORTHIER, G. and WANKWIKEKBERGE, P.: 'Handbook of distributed feedback laser diodes' (Artech House, 1997)

1.9.2 Optical communication systems

GOWAR, J.: 'Optical communication systems' (Prentice Hall International, Englewood Cliffs, 1983) 2nd ed.
POWERS, J.P.: 'An introduction to fiber optic systems' (Irwin, Homewood, 1993)
GREEN, P.E.: 'Fiber optic networks' (Prentice Hall, Englewood Cliffs, 1993)
LAUDE, J.P.: 'Wavelength division multiplexing' (Prentice Hall, New York, 1993)
AGRAWAL, G.P.: 'Fibre optic communication systems' (Academic Press, Boston, 1992)
HOSS, R.J.: 'Fibre optic communications design handbook' (Prentice Hall, Englewood Cliffs, 1990)
SENIOR, J.M.: 'Optical fibre communications' (Prentice Hall International, Englewood Cliffs, 1988)
SANDBANK, C.P. (Ed.): 'Optical fibre communication systems' (Wiley, Chichester, 1980)

1.10 References

1 HALL, R.N., FENNER, G.E., KINGSLEY, J.D., SOLTYS, T.J. and CARLSON, R.O.: 'Coherent light emission form GaAs junctions', *Phys. Rev. Lett.*, 1962, **9**, pp. 366–368
2 NATHAN, M.I., DUMKE, W.P., BURNS, G., DILL, F.H. and LASHER, G.: 'Stimulated emission of radiation from GaAs $p-n$ junctions', *Appl. Phys. Lett.*, 1962, **1**, pp. 62–64
3 QUIST, T.M., REDIKER, R.H., KEYES, R.J., KRAG, W.E., LAX, B., MCWHORTER, A.L. and ZEIGLER, H.J.: 'Semiconductor laser of GaAs', *Appl. Phys. Lett.*, 1962, **1**, pp. 91–92

4 HOLONYAK, N. and BEVACQUA, S.F.: 'Coherent visible light emission from GaAsP junctions', *Appl. Phys. Lett.*, 1962, **1**, pp. 82–84
5 IKEGAMI, T. and SUEMATSU, Y.: 'Direct modulation of semiconductor junction laser', *Electr. Comm. Japan*, 1968, **51**(3), pp. 51–58
6 PAOLI, T. and RIPPER, J.E.: 'Direct modulation of semiconductor lasers', *Proc. IEE*, 1970, **58**, pp 1457–1465
7 KAO, K.C. and HOCKHAM, G.A.: 'Dielectric-fibre surface waveguides for optical frequencies', *Proc. IEE*, 1966, **113**, pp. 1151–1158
8 ADAMS, M.J.: 'An introduction to optical waveguides', (J. Wiley, Chichester, 1981); see historical introduction, pp xiii–xiv
9 CLARRICOATS, P.J.B. (Ed.): 'Progress in optical communications', IEE reprint series 3 (Peter Peregrinus, UK, 1980)
10 SANDBANK, C.P. (Ed.): 'Optical fibre communication systems' (J. Wiley, Chichester, 1980)
11 PAL, B.P. (Ed.): 'Fundamentals of fibre optics in telecommunication and sensor systems' (Wiley Eastern, New Delhi, 1992)
12 GREEN, P.E.: 'Fiber optic networks' (Prentice–Hall, Englewood Cliffs, 1993)
13 BORN, M. and WOLF, E.: 'Principles of optics' (Pergamon Press, Oxford, 1980), 6th ed., Section 7.6.2
14 HAYASHI, I., PANISH, M.B. and FOY, P.W.: 'A low threshold room temperature injection laser', *IEEE J. Quantum Electron*, 1969, **5**, pp. 211–212
15 KRESSEL, H. and NELSON, N.: 'Close confinement gallium arsenide PN junction laser with reduced optical loss at room temperature', *RCA Rev.*, 1969, **30**, pp. 106–113
16 HAYASHI I. and PANISH, M.B.: 'GaAs $Ga_xAl_{1-x}As$ heterojunction injection lasers which exhibit low threshold room temperature operation', *J. Appl. Phys.*, 1970, **41**, pp. 150–163
17 ALFEROV, ZH.I., ANDREEV, V.M., PORTNOI, E.L. and TRETYAKOV, D.N.: 'Injection properties of n-$Al_xGa_{1-x}As$ p-GaAs heterojunctions', *Sov. Phys. Semicond.*, 1969, **2**, pp. 843–845 (translated from *Fiz Tekh Polupovodn*, 1968, **2**, pp. 1016–1018)
18 HAYASHI, I., PANISH, M.B., FOY, P.W. and SUMSKI, S.: 'Junction lasers which operate continuously at room temperature', *Appl. Phys. Lett.*, 1970, **17**, pp. 109–111
19 ALFEROV, ZH. I., ANDREEV, V.M., GARBUZOV, D.Z., ZHILYAEV, Y.V., MOROZOV, E.P., PORTNOI, E.L. and TRIOFIM, V.G.: 'Investigation of the influence of the AlAs-GaAs heterostructure parameters on the laser threshold current and the realization of continuous emission at room temperature', *Sov. Phys. Semicond.*, 1971, **4**, pp. 1573–1576 (translated from *Fiz Tekh Polupovodn*, 1970, **4**, pp. 1826–29)
20 RIPPER, J.E., DYMENT, J.C., D'ASARO, L.A. and POOLE, T.L.: 'Stripe geometry double heterostructure junction lasers: mode structure and CW operation above room temperature', *Appl. Phys. Lett.*, 1971, **18**, pp. 155–167
21 HILL, D.R.: '140 Mbit/s optical fibre field demonstration system', *in* SANDBANK, C.P. (Ed.): 'Optical fibre communication systems' (J. Wiley Chichester, 1980), Chapter 11

22 PAL, B.P. (Ed.): 'Fundamentals of fibre optics in telecommunication and sensor systems' (Wiley Eastern, New Delhi, 1992), Chapters 18–20
23 AHMED, H. and SPREADBURY, P.J.: 'Analogue and digital electronics for engineers' (Cambridge University Press, Cambridge, 1984), 2nd ed., Chapter 6
24 BURNHAM, R.D. and SCIFRES, D.R.: 'Etched buried heterostructure GaAs/$Ga_{1-x}Al_xAs$ injection lasers', *Appl. Phys. Lett.*, 1975, **27**, pp. 510–512
25 LEE, T.P., BURRUS, C.A., MILLER, B.I. and LOGAN, R.A.: '$Al_xGa_{1-x}As$ double heterostructure rib-waveguide injection laser', *IEEE J. Quantum Electron.*, 1975, **11**, pp. 432–435
26 HORIKAWA, H. and ISHII, A.: 'Semiconductor pump laser technology', *J. Lightwave Technol.*, 1993, **11**, pp.167–175
27 LANG, R. and KOBAYASHI, K.: 'External optical feedback effects on semiconductor injection laser properties', *IEEE J. Quantum Electron.* 1980, **16**, pp. 347–355
28 HAUS, H.A.: 'Modelocking of semiconductor laser diodes', *Japan. J. Appl. Phys.*, 1981, **20**, pp. 1007–1020
29 CHEN, Y-K. and WU, M.C.: 'Monolithic colliding pulse mode-locked semiconductor lasers', *IEEE J. Quantum Electron.*, 1992, **28**, pp. 2176–2185
30 MARTINS FILHO, J.F. and IRONSIDE, C.N.: 'Multiple colliding pulse mode locked operation of a semiconductor laser', *Appl. Phys. Lett.*, 1994, **65**, pp. 1894–1896
31 TSANG, W.T.: 'The cleaved coupled-cavity (C^3) laser' *in* 'Semiconductors and semi-metals' (Academic Press, New York 1985), Vol. 22 Pt. B, Chap. 5
32 COLDREN, L.A. and KOCH, T.L.: 'Analysis and design of coupled cavity lasers Parts 1 and 2', *IEEE J. Quantum Electron.*, 1984, **20**, pp. 659–670, 671–682
33 LANG, R.: 'Injection locking properties of a semiconductor laser', *IEEE J. Quantum Electron.*, 1982, **18**, pp. 976–983
34 MATTHEWS, M.R., CAMERON, K.H., WYATT, R. and DEVLIN, W.J.: 'Packaged frequency tuneable 20 kHz linewidth 1.5 μm InGaAsP external cavity laser', *Electron. Lett.*, 1985, **21**, pp. 113–115
35 MEARS, R., JAUNCEY, J. and PAYNE, D.: 'Low-noise erbium doped fibre amplifier operating at 1.54 μm', *Electron. Lett.*, 1987, **23**, pp. 1207–1208
36 POWERS, J.P.: 'An introduction to fibre optic systems' (Irwin, Homewood, 1993), chapter 7
37 KAO, C.K.: 'Optical fibre' (Institution of Electrical Engineers, London, 1988)
38 LAUDE J.P.: 'Wavelength division multiplexing' (Prentice Hall, New York, 1993)
39 GOWAR, J.: 'Optical communication systems' (Prentice Hall International, Englewood Cliffs, 1993), 2nd ed., see chapter 26
40 BRAGG, W.H. and BRAGG, W.L.: 'X-rays and crystal structure' (G.Bell, London, 1915)
41 RAMO, S., WHINNERY, J.R. and VAN DUZER, T.: 'Fields and waves in communication electronics' (John Wiley & Sons, 1994), 3rd ed., p. 310 and pp. 637–676

42 STREIFER, W., SCIFRES, D.R. and BURNHAM, R.D.: 'Coupling coefficients for distributed feedback single- and double-heterostructure diode lasers', *IEEE J. Quantum Electron.*, 1975, **11**, pp. 867–873
43 STREIFER, W., SCIFRES, D.R. and BURNHAM, R.D.: 'Analysis of grating-coupled radiation in GaAs:GaAlAs lasers and waveguides', *IEEE J. Quantum Electron.*, 1976, **12**, pp. 422–428
44 STREIFER, W., BURNHAM, R.D. and SCIFRES, D.R.: 'Analysis of grating-coupled radiation in GaAs:GaAlAs lasers and waveguides–II: Blazing effects', *IEEE J. Quantum Electron.*, 1976, **12**, pp. 494–499
45 STREIFER, W., SCIFRES, D.R. and BURNHAM, R.D.: 'TM-mode coupling coefficients in guided-wave distributed feedback lasers', *IEEE J. Quantum Electron.*, 1976, **12**, pp. 74–78
46 KOGELNIK, H. and SHANK, C.V.: 'Coupled-wave theory of distributed feedback lasers', *J. Appl. Phys.*, 1972, **43**, pp. 2327–2335
47 NAKAMURA, M., YEN, H.W., YARIV, A., GARMIRE, E., SOMEKH, S. and GARVIN, H.L.: 'Laser oscillation in epitaxial GaAs waveguides with corrugation feedback', *Appl. Phys. Lett.*, 1973, **23**, pp. 224–225
48 SCIFRES, D.R., BURNHAM, R.D. and STREIFER, W.: 'Distributed feedback single heterojunction diode laser', *Appl. Phys. Lett.*, 1974, **25**, pp. 203–204
49 CASEY, H.C., SOMEKH, S. and ILEGEMS, M.: 'Room temperature operation of low-threshold separate confinement distributed feedback diode lasers', *Appl. Phys. Lett.*, 1975, **27**, pp. 487–489
50 UEMATSU, Y., OKUDA, H. and KINOSHITA, J.: 'Room-temperature cw operation of 1.3-μm distributed-feedback GaInAsP/InP lasers', *Electron. Lett.*, 1982, **18**, pp. 857–858
51 UTAKA, K., AKIBA, S., SAKAI, K. and MATSUSHIMA, Y.: 'Room-temperature cw operation of distributed-feedback buried heterostructure InGaAsP-InP lasers emitting at 1.57 μm,' *Electron. Lett.*, 1981, **17**, pp. 961–963
52 WHITEAWAY, J.E.A., GARRETT, B., THOMPSON, G.H.B., COLLAR, A.J., ARMISTEAD, C.J. and FICE, M.J.: 'The static and dynamic characteristics of single and multiple phase-shifted DFB laser structures', *IEEE J. Quantum Electron.*, 1992, **28**, pp. 1227–1293
53 GHAFOURI SHIRAZ, H. and LO, B.S.K.: 'Structural dependence of three phase shift distributed feedback semiconductor laser diodes at threshold using the transfer matrix method (TMM)', *Semicond. Sci. Technol.*, 1994, **9**, pp. 1126–1132
54 HENRY, C.H.: 'Performance of distributed feedback lasers designed to favour the energy gap mode', *IEEE J. Quantum Electron.*, 1985, **21**, pp. 1913–1918
55 NOWELL, M.C., CARROLL, J.E., PLUMB, R.G.S., MARCENAC, D.D., ROBERTSON, M.J., WICKES, H. and ZHANG, L.M.: 'Low chirp and enhanced resonant frequency by direct push-pull modulation of DFB lasers', *IEEE J. Select Topics Quantum Electron.*, 1995, **1**, pp. 433–441
56 TOHYAMA, M., ONOMURA, M., FUNEMIZU, M. and SUZUKI, N.: 'Wavelength tuning mechanism in three electrode DFB lasers', *IEEE Photonics Technol. Lett.*, 1993, **5**, pp. 616–618

57 ISHII, H., KANO, F., TOHMORI, Y., KONDO, Y., TAMAMURA, T. and YOSHIKUNI, Y. 'Broad range (34 nm) quasi continuous wavelength tuning in superstructure DBR lasers', *Electron. Lett.*, 1994, **30**, pp. 1154–1155

58 JAYARAMAN, V., MATHUR, A., COLDREN, L.A. and DAPKUS, P.D.: 'Extended tuning range in sampled grating DBR lasers', *IEEE Photonics Technol. Lett.*, 1993, **5**, pp. 489–491

59 OUGIER, C., TALNEAU, A., DELORME, F., SLEMPKES, S. and MATHOORASING, D.: 'High number of wavelength channels demonstrated by a widely tunable sampled grating DBR laser', *IEE Proc. Optoelectronics*, 1996, **143**, pp. 77–80

60 HILDEBRAND, O., SCHILLING, M., BAUMS, D., IDLER, W., DUTTING, K., LAUBE, G. and WUNSTEL, K., 'The Y-laser—a multifunctional device for optical communication-systems and switching-networks', *Lightwave Technol.*, 1993, **11**, pp. 2066–2075

61 FELLS, J.A.J., GIBBON, M.A., THOMPSON, G.H.B., WHITE, I.H., PENTY, R.V., WRIGHT, A.P., SAUNDERS, R.A., ARMISTEAD, C.J., and KIMBER, E.M.: 'Chirp and system performance of integrated laser modulators', *IEEE Photonics Technol. Lett.*, 1995, **7**, pp. 1279–1281

62 DORGEUILLE, F. and DEVAUX, F.: 'On the transmission performances and the chirp parameter of a multiple-quantum-well electroabsorption modulator', *IEEE J. Quantum Electron.*, 1994, **30**, pp. 2565–2572

63 JEWELL, J.L., HARBISON, J.P., SCHERER, A., LEE, Y.H. and FLOREZ, L.T. 'Vertical-cavity surface-emitting lasers—design, growth, fabrication, characterization', *IEEE J. Quantum Electron.*, 1991, **27**, pp. 1332–1346

64 GEELS, R.S., THIBEAULT, B.J., CORZINE, S.W., SCOTT, J.W. and COLDREN, L.A.: 'Design and characterization of $In_{0.2}Ga_{0.8}As$ MQW vertical-cavity surface-emitting lasers', *IEEE J. Quantum Electron.*, 1993, **29**, pp. 2977–2987

65 ZEEB E., MOLLER B., REINER G., RIES M., HACKBARTH T. and EBELING K.J.: 'Planar proton implanted VCSELs and fiber coupled 2-D VCSEL arrays', *IEEE J. Select. Topics Quantum Electron.*, 1995, **1**, pp. 616–623

Chapter 2
Gain, loss and spontaneous emission

2.1 Introduction

The operation of semiconductor lasers is strongly affected by the materials from which they are made. Spontaneous emission which initiates lasing, optical gain which is essential to achieve lasing, and other processes involved in lasing all use quantum processes at the level of single atoms and electrons within the lasing material. Indeed, without simplifications, the physics and mathematics necessary to describe such atomic systems fully is too complicated, certainly for the level of this book. The simplification and approximations must, however, be done in a way which is adapted to the requirements of semiconductor lasers, and the limitations must be understood. The results and implications of quantum physics are discussed here but there are only illustrative outlines of any derivations. Greater depth may be obtained from specialist material [1,2] (see Bibliography).

In the context of lasers, the atomic behaviour of semiconductors can be summarised under three headings:

(i) the distribution of energy states;
(ii) the occupation probability of those states; and
(iii) transitions between states.

The first few sections of this chapter will cover these headings at the atomic level, while the later sections will apply them at a macroscopic level to optical gain, spontaneous emission and noise.

2.2 Electronic processes in semiconductors

2.2.1 *Energy states*
In a semiconductor, the electrons which are of interest in the conduction processes are not normally bound to any single atom, but

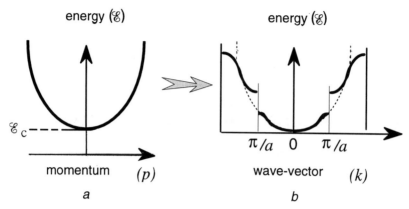

Figure 2.1 Energy-momentum diagrams
  (i) In free space
  (ii) In a periodic lattice of period a with $p=hk/2\pi$ and $k=2\pi/\lambda_{electron}$

may move around in the crystal. Simple Newtonian mechanics forms a starting point where the kinetic energy of a particle $\mathscr{E}-\mathscr{E}_c$ is measured from the conduction band energy edge $\mathscr{E}_c$, and given in terms of its momentum p and effective mass m_c from

$$\mathscr{E}-\mathscr{E}_c = p^2/(2m_c) \tag{2.1}$$

Such a free particle also has wave characteristics with a wavelength given by De Broglie's relationship $\lambda_{electron}=h/p$. Consider this particle/wave moving through a one-dimensional periodic array of atoms spaced a apart giving a periodic electric potential of period a forming an approximation to a crystalline material. As far as diffraction is concerned, the electron waves in this periodic electric potential behave analogously to electromagnetic waves in a periodic refractive index as outlined in Section 1.6. Waves, with frequencies having a corresponding wavelength $\lambda_{electron}=2a/N$ (N=1, 2, 3), are strongly reflected so that, around these critical frequencies (electron energies), there is a range of frequencies where propagation cannot occur (similar to the concept of a stop band that is met in DFB lasers in Chapters 1 and 5). In crystals, it is convenient to use the *wave vector* $|k|=2\pi/\lambda_{electron}$ and to derive an energy–momentum, or $\mathscr{E}$–k diagram, which has discontinuities or jumps in the energy whenever $|k|=2\pi/\lambda_{electron}=N\pi/a$ (Figure 2.1). The principal disallowed energy range for conduction and valence electrons forms the band gap of the semiconductor with the allowed energy ranges on either side of this band gap known as the *conduction* and *valence* bands. A complementary approach starts from the discrete energy levels of the hydrogen-like atom (for example the Bohr model) and then considers large numbers of atoms interacting to broaden the discrete energy levels into bands [3,4].

An analysis is straightforward for a one-dimensional array of atoms but becomes extremely complex [5] when modelling the $\mathcal{E}$–k diagram for a real three-dimensional crystal. In many cases, an adequate approximation for bulk material is to assume that the crystal is spherically symmetric, and to retain the one-dimensional parabolic relationship used above but now in all three dimensions so that $\mathcal{E} - \mathcal{E}_c = |\hbar k|^2/(2m_c)$. Pauli's exclusion principle then requires that no two electrons (or holes) may have exactly the same energy and this principle, combined with the fact that momentum and energy are quantised, determines an expression for density of states (per unit volume per unit energy) in the conduction band [6]

$$N_d(\mathcal{E}) = 4\pi \left(\frac{2m_c}{h^2}\right)^{3/2} (\mathcal{E} - \mathcal{E}_c)^{1/2} \qquad (2.2)$$

A real crystal is not spherically symmetric: the energy bands and effective masses can vary by ~50% as the direction of momentum changes modifying the values in eqn. 2.1 by a similar amount, but such variations are found not to be important for lasers using unstrained bulk active material. With lasers employing composite-layered materials—quantum-well material—then the departure from spherical symmetry combined with any anisotropic straining of the active quantum-well regions (Section 2.7) has significant, and often beneficial, effects on the energy bands and density of states [7,8].

The arguments used above for electrons in a crystalline semiconductor apply similarly to holes in the same material, except that a hole may be regarded as a wave packet characteristic of one missing electron in an otherwise full ensemble of electrons within the valence band and so the hole has a positive charge. This absence of an electron equivalently gives to the hole a negative mass in eqn. 2.1 ($m_c \to -m_v$) and holes therefore appear as downward parabolas on the $\mathcal{E}$–k diagram, inverting Figure 2.1 with energy now measured down from the valence-band edge as ($\mathcal{E}_v - \mathcal{E}$). Because the movement of holes really means movement of many electrons within an almost filled band, it is intuitively reasonable that the associated $\mathcal{E}$–k diagrams should be more complicated than for electrons in a sparsely filled band and that holes should be more difficult to move, or heavier than

40 Distributed feedback semiconductor lasers

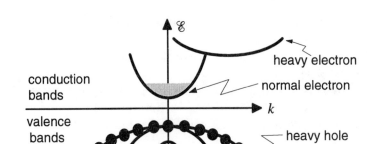

Figure 2.2 $\mathscr{E}-k$ *(energy–momentum) diagram for a direct-band-gap semiconductor*

Lowest energy in conduction band is directly above highest energy in valence band. Hole-split-off band is caused by coupling of electron spins and orbital angular momentum

electrons. A rigorous discussion is outside the scope of this book but the sketch (for one-dimensional motion) in Figure 2.2 shows the principal energy bands for a semiconductor, including the light and heavy holes and bands coupled through the effects of electron spin.

It is common to show occupation of a band as being the shaded part of the 'normal-electron' band which is convenient, because the density of states varies with $(\mathscr{E}_v - \mathscr{E})^{1/2}$, or $(\mathscr{E} - \mathscr{E}_c)^{1/2} \propto |k|$, and the shaded area is therefore proportional to the number of electrons present. However, this shading can be misleading, because electrons in the band can only exist *on the outer* $\mathscr{E}-k$ *line*. The equally spaced dots in the k direction are a schematic representation of energy states in the heavy-hole band and should be thought of as similarly existing on all the other bands.

2.2.2 Occupation probabilities

The large numbers of states indicated in Figure 2.2 for the conduction and valence bands may or may not be occupied by electrons/holes depending on the thermal excitation and the level of the energy. The *probability of occupation* for an electron state is given by the Fermi–Dirac function

$$F_N(\mathscr{E}) = 1/\{1 + \exp(\mathscr{E} - \mathscr{E}_{fN})/kT\} \tag{2.3}$$

where $\mathscr{E}_{fN}$ is referred to as the 'quasi-Fermi' level appropriate to determine the density N of electrons. In equilibrium with no applied voltage, $\mathscr{E}_{fN}$ is the Fermi level for all the electrons, but when a forward voltage is applied to a p–n junction the probability of occupation in the conduction band increases (i.e. $\mathscr{E}_{fN}$ increases) so that, for lasing, typically $\mathscr{E}_{fN} > \mathscr{E}_c$, the energy of the conduction-band edge.

Holes are absences of electrons in a state, with densities denoted by P, and at a given energy $F_P = 1 - F_N$ gives the probability of a 'hole' occupying that state but it is conventional to denote hole energies in the reverse direction to electrons, so for holes

$$F_P(\mathscr{E}) = 1/\{1 + \exp(\mathscr{E}_{fP} - \mathscr{E})/kT\} \quad (2.4)$$

The energy $\mathscr{E}_{fP}$, like $\mathscr{E}_{fN}$, is a quasi-Fermi level determining the occupation probability for holes. Increasing the applied voltage decreases this quasi-Fermi level so that, for strong lasing, $\mathscr{E}_{fP} \lesssim \mathscr{E}_v$, the energy of the valence band edge. The really important condition for lasing is that the net applied voltage V_J to the junction given from $qV_J = \mathscr{E}_{fN} - \mathscr{E}_{fP}$ is greater than the band gap [9], i.e.: $V_J > (\mathscr{E}_c - \mathscr{E}_v)/q$. Because the positions of the quasi-Fermi levels are then inside or near to the band edges, it is necessary to use the full Fermi–Dirac statistics (rather than the Boltzmann approximation as in elementary texts [10]).

2.2.3 Radiative recombination and absorption

If an electron moves from one state to another of lower energy, conservation of energy requires an emission of the energy difference: either *radiatively* involving a photon, or *nonradiatively* in which case *phonons* (which are the particles associated with lattice vibrations having quantised energies) are emitted. In optical devices it is desirable and usual for radiative transitions to dominate over nonradiative ones. A key factor is that, although the optical photons which are of interest in diode lasers have energies (~1–2 eV) which are much larger than the energies of phonons (~1–50 meV), the photon momentum is typically less than 1% of the average phonon momentum. On the scale of the $\mathscr{E}-k$ diagram, the photon momentum is negligible when compared with that of phonons.

In silicon, or other *indirect* materials, the conduction and valence bands have their extrema centred on very different momenta in the energy momentum diagram (Figure 2.3). In moving an electron from the conduction band to the valence band, a radiative recombination would need to involve a phonon with momentum but little energy and a photon with energy but little momentum so as to conserve both momentum and energy. The relatively low density of phonons combined with the complex interaction means that such radiative

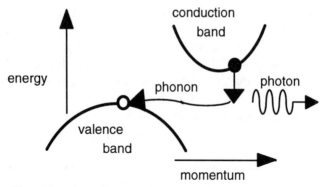

Figure 2.3 Transitions in indirect band-gap material

For an electron to move from the conduction band to the valence band requires momentum and energy to be conserved. Only a photon has enough energy to balance the energy loss and only the phonon has enough momentum to balance the momentum change

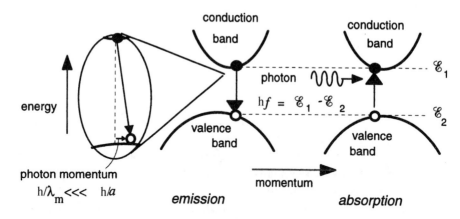

Figure 2.4 Transitions in direct-band-gap material

The photon wavelength in the medium and frequency are λ_m and ν, respectively; a is the lattice periodicity with h/a giving a rough order of a phonon's momentum in the crystal

indirect transitions are generally improbable. On the other hand, in direct-gap materials the direct radiative transitions tend to dominate over other more complex transitions, and even in materials such as $Ga_{0.5}Al_{0.5}As$, which are only 'slightly indirect' light is emitted quite well.

Figure 2.4 shows schematically the transition processes in direct-gap material of interest for lasing action and indicates that optical absorption is the reverse of radiative recombination with energy and momentum balance applied appropriately. The rate of emission of photons $(\mathrm{d}S/\mathrm{d}t)_{emission}$ through recombination of electrons and holes and the rate $(\mathrm{d}S/\mathrm{d}t)_{absorption}$ of absorption of photons generating electron hole pairs has to be closely linked as discussed next.

2.2.4 Transitions and transition rates

To calculate the transition probabilities of electrons, there are several factors to be considered. In a specialist study, the start might be the basic quantum theory of transitions between N energy levels. This requires the quantum formalism of an N×N complex density matrix M, where the probability of a spontaneous transition from state i to state j is proportional to $|M_{ij}|^2$ with the diagonal elements $|M_{nn}|^2$ giving the probability of occupation of state n. In practice, the rates of emission and absorption are primarily discovered from measurements on materials, but a reader might start by referring to a model due to Kane [11] for the quantum theory which indicates that an approximate transition probability between the conduction and valence bands is very roughly proportional to the bandgap energy $\mathcal{E}_g$ and the electronic rest mass m_o:

$$|M_{cv}|^2 \simeq \frac{m_o^2 \mathcal{E}_g [\mathcal{E}_g + \delta_s]}{12 m_c \left[\mathcal{E}_g + \frac{2\delta_s}{3} \right]} \approx \xi m_o \mathcal{E}_g \tag{2.5}$$

Here m_c is the effective mass in the conduction band and δ_s is the splitting energy caused by the electrons' spin. For GaAs, the constant of proportionality ξ is about 1.3, and other optical semiconductors are of the same order. An important conclusion of Kane's theory for laser-modelling purposes is that transition probabilities (and therefore rates) increase with the bandgap as a first-order effect, as indicated in eqn. 2.5.

The problem of the transitions which is to be solved can be expressed graphically as in Figure 2.5. Here, because of the negligible photon momentum on the scale of this diagram, radiative transitions between bands are indicated where the electrons and holes have equal k, which is referred to as the strict k-selection rule and is a useful approximation here. Total spontaneous emission etc. is determined by integrating for occupied states over all energies, discussed later in

44 *Distributed feedback semiconductor lasers*

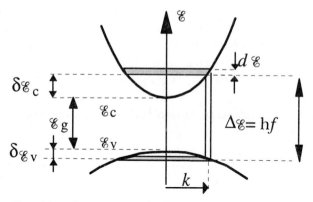

Figure 2.5 Transitions between energy levels in the conduction and valence bands

As $\Delta\mathcal{E}$ changes so $|k|$ varies. Over a band of energy $d\mathcal{E}$, there is a corresponding band or range dk in $|k|$.

Section 2.3.2. As an introduction, a simpler two-level system is considered in Section 2.3.1

Integrating over all k as in Figure 2.5 takes into account the bands of energy states but does not take into account impurities or defects where the electron states are localised spatially, typically over subnanometre distances Δx. From Heisenberg's uncertainty principle, uncertainty in momentum Δp and uncertainty in position Δx are linked by $\Delta p.\Delta x \approx h/2\pi$, so that the momentum (wave vector) becomes ill-defined for localised states, and electrons in such states can interact with particles over a wide range of k: the k-selection rule becomes relaxed, often giving better agreement between the theory of transitions and observed results [12–14].

Localised impurities can also contribute to so-called band-tail states just inside the band edges, and move lasing to longer wavelengths, particularly at low temperatures. Impurity atoms and defects with states within the bandgap also encourage recombination in multiple steps; a high proportion of these will involve phonons and no optical radiation, or optical radiation at lower frequencies than those of interest. Consequently, almost all recombination via bandgap states represents loss of charge carriers without useful radiation. Gold and copper result in such bandgap states and are carefully excluded from laser active regions during processing.

2.2.5 Auger recombination

One special set of transitions known as Auger recombination, unlike recombination via bandgap impurity states, is always of concern and

Gain, loss and spontaneous emission 45

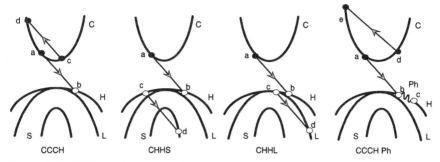

Figure 2.6 Different Auger-recombination processes

C=conduction band, H=heavy-hole band, L=light-hole band, S=hole-spin-split-off band, Ph=phonon
Three charge carriers are involved in each process so that Auger-recombination rates are proportional to (charge-carrier density)^3

this concern increases as the bandgap decreases [15,16]. Auger recombination is not a single process, but a group of different ones: depending on definitions, over 80 such processes may be identified. A key feature is that at least three particles (two electrons and one hole or two holes and one electron) and four energy states will be involved, but transitions may be band-to-band, may involve phonon assistance, or may involve impurities. A full analysis of all Auger processes in a semiconductor is difficult and only illustrative examples are given below.

Individual Auger processes are identified in Figure 2.6 by listing the bands in which states take part, using the codes C for conduction band, H for heavy-hole band, L for light-hole band and S for hole-spin-split-off band. The first diagram in Figure 2.6 shows an electron in state *a* recombining with a heavy hole at *b*, and *both the energy and momentum* are taken up by an electron which moves from state *c* to state *d*. It is readily seen that even now the process is not a single one, but might involve numerous combinations of energy and momentum. This first CCCH process is a likely one for *n*-doped semiconductor, because it needs two electrons; the next two examples require two holes, and are therefore more likely in *p*-doped material. The final diagram shows an extension to the first three processes whereby a phonon is involved as well as carriers and a photon.

In theory the rates of all the separate Auger processes could be calculated, but the following summary of measured results provides a good basis for modelling:

(i) Since Auger processes involve three particles, they increase with the cube of carrier density.
(ii) All Auger rates increase with decreasing bandgap because, in general, the range of likely transitions increases with decreasing bandgap.
(iii) Processes such as CCCH, where an increase in temperature increases the numbers of electrons available to participate (as in practical lasers), will have rates which increase with temperature.
(iv) Since numbers of dopant atoms or defects are almost constant with temperature, Auger processes involving these increase less rapidly with temperature than the processes of (iii) above.

As a direct consequence of (ii) above, it can be seen that 1.55 μm, 1.3 μm and 0.9 μm lasers are steadily less affected by Auger recombination as the wavelengths decrease, as indicated by the sensitivity of the threshold currents in lasers to temperature with Auger processes typically becoming significant at around 50°C for 1.5 μm devices, 70°C at 1.3 μm, and 120°C + for devices operating at 0.9 μm or less.

2.3 Absorption, emission rates and spectra

2.3.1 Absorption, stimulated and spontaneous emission in a semiconductor

Spontaneous emission, absorption and gain in independent atoms (as, for example, in gas lasers [17]) are more straightforward to discuss than in semiconductors where the high occupation density of electrons and holes in semiconductor lasers means that there are difficulties of using Fermi–Dirac statistics and the simplifications which arise in gas lasers do not apply. However, Fermi–Dirac statistics may be applied in a straightforward manner using a two-level model which illustrates the basic principles [18] and aids understanding of laser models.

Figure 2.7 represents an idealised unit volume of material with N_1 states at energy $\mathcal{E}_1$ of which N are occupied by electrons, and N_2 states at energy $\mathcal{E}_2$ of which N_2 are occupied. Since it is a unit volume, the number of electrons at the upper level is also the electron density N and similarly the hole density $P = N_2 - N_2$. Also within the volume there are S photons, which can be absorbed or stimulate further photons.

The first process to consider is spontaneous emission where electrons at a higher energy spontaneously fall to a lower energy giving out photons at the energy difference. The spontaneous power-output/volume is proportional to the number of high-energy electrons N and

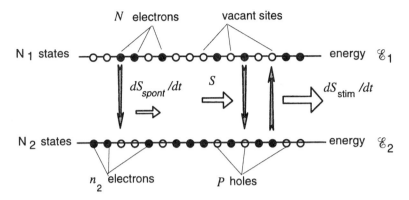

Figure 2.7 *Idealised two-energy level photonic/electronic interactions*

the number of vacant states (holes, density P) available for spontaneous recombination; hence:

$$\mathrm{d}S_{spont}/\mathrm{d}t = K_{sp}NP \cong N_{ref}/\tau_{sp} \qquad (2.6)$$

where K_{sp} is a constant of proportionality and the recombination time τ_{sp} is a parameter often measured experimentally. In a typical laser with $N_{ref} \cong N \cong P \cong 10^{18}$ cm^{-3} then $\tau_{sp} \sim 10^{-9}$ s.

The second process is stimulated emission where again electrons at a higher energy fall into vacant states at a lower energy but in this case are stimulated to make the transition by the photons already present. The rate of stimulated power within the unit volume is therefore again proportional to the same numbers of holes and electrons as above, but also now proportional to the photon number S, and with a different constant of proportionality K_{stim} and hence one obtains

$$(\mathrm{d}S_{stim}/\mathrm{d}t)_{emission} = K_{stim}SNP \qquad (2.7)$$

However, this stimulated photon rate is offset by absorption, which is proportional to the number of *holes* at the upper energy $(N_1 - N)$ and the number of *electrons* at the lower energy $N_2 = (N_2 - P)$ and also proportional to S, the photon density, so that

$$(\mathrm{d}S_{stim}/\mathrm{d}t)_{absorption} = -K_{stim}S(N_1 - N)(N_2 - P) \qquad (2.8)$$

The *net* rate of emission is the sum of these three processes above:

$$(\mathrm{d}S/\mathrm{d}t)_{net} = K_{stim}S(NN_2 + PN_1 - N_1N_2) + K_{sp}NP \qquad (2.9)$$

To obtain the net rate of generation of photons being *coupled into a lasing mode*, all of the stimulated photons S at the lasing frequency above contribute, but spontaneous photons are emitted almost isotropically and also over a much wider band of frequencies than the lasing linewidth. Consequently only a very small proportion β_{sp} (typically 10^{-6} to 10^{-2}) couple to the lasing mode and eqn. 2.9 must be modified to read

$$dS/dt = K_{stim} S(NN_2 + PN_1 - N_1 N_2) + \beta_{sp} K_{sp} NP \qquad (2.10)$$

Under lasing conditions when injected charge dominates over doping, $N \cong P$, and P may be eliminated, allowing a photon-interaction equation often used in modelling:

$$dS/dt = G'_m (N - N_{tr}) S + \beta_{sp} B N^2 \qquad (2.11)$$

One may link the parameters in eqn. 2.11 with those in eqn. 2.10 as follows:

$$G'_m = K_{stim}(N_1 + N_2); \qquad N_{tr} = N_1 N_2 / (N_1 + N_2); \qquad B = K_{sp} \qquad (2.12)$$

where G'_m is the differential power gain/unit time, and N_{tr} is the *transparency* value of electron density at which the stimulated optical gain is cancelled by the stimulated optical absorption—the material is *transparent* for the light at that frequency.

A nonzero acceptor density N_a in the gain region causing $P \cong N + N_a$ from electrical neutrality could, from the theory, reduce the *transparency electron density*. However, a high density of acceptors increases the nonradiative recombination and in practice this technique is not normally used deliberately, though *p*-dopants such as Zn may diffuse from the *p*-cladding layer during growth. Here doping in the gain region is ignored.

From eqns. 2.7 and 2.8, if gain is to exceed absorption, then one requires

$$NP - (N_1 - N)(N_2 - P) = NP\{1 - (1 - N_1/N)(1 - N_2/P)\} > 0 \qquad (2.13)$$

But N and P are related to N_1 and N_2 via the Fermi relationships of eqns. 2.3 and 2.4:

$$N_1/N = \{1 + \exp(\mathcal{E}_1 - \mathcal{E}_{fN})/kT\} = 1/F_N(\mathcal{E}_1) \qquad (2.14)$$

$$N_2/P = \{1 + \exp(\mathcal{E}_{fP} - \mathcal{E}_2)/kT\} = 1/F_P(\mathcal{E}_2) \qquad (2.15)$$

and so, for net gain (substituting the above into eqn. 2.13),

$$\exp[\{(\mathcal{E}_1 - \mathcal{E}_2) - (\mathcal{E}_{fN} - \mathcal{E}_{fP})\}/kT] < 1 \text{ or } (\mathcal{E}_{fN} - \mathcal{E}_{fP}) = qV_j > (\mathcal{E}_1 - \mathcal{E}_2) \qquad (2.16)$$

Eqn. 2.16 confirms the earlier result that the *voltage* V_j applied to the laser junction, given from the separation in quasi-Fermi levels, must be greater than the equivalent bandgap voltage to obtain lasing action[9]. A further implication is that, if the contact resistances in the laser structure (see Section 5.7) could be made very low, the carrier density in the active region would be determined solely by the applied voltage and the optical gain could then be regarded as a direct function of applied voltage, rather than the electron density.

Now in thermal equilibrium, the net photon-emission rate dS/dt is zero with no voltage applied across the junction (i.e. $\mathcal{E}_{fN}=\mathcal{E}_{fP}$). Also, select the situation where spontaneous and stimulated emission cover the same frequency ranges and angles so that the coupling factor $\beta_{sp}=1$. These conditions force the system to be in equilibrium at a temperature T with black-body thermal radiation described by Planck's formula [19] for the photon density over a range of frequencies, say $\Delta_{sp}f$, around an optical frequency f:

$$S = 8\pi f^2 n^2 n_g \Delta_{sp}f / c^3 \{\exp(hf/kT) - 1\} \quad (2.17)$$

Here n is the refractive index and n_g is the group index, i.e. $c/n_g = v_g$ is the group velocity or the velocity at which the energy propagates while $c/n = v_p$ is the phase velocity. In equilibrium, $dS/dt=0$, and eqn. 2.10 becomes

$$0 = K_{stim}SNP - K_{stim}SNP(N_1/N - 1)(N_2/P - 1) + K_{sp}NP \quad (2.18)$$

Also in equilibrium $\mathcal{E}_{fN}=\mathcal{E}_{fP}$ so that eqns. 2.14 and 2.15 yield

$$S = (K_{sp}/K_{stim}) / \{\exp(\mathcal{E}_1 - \mathcal{E}_2)/kT - 1\} \quad (2.19)$$

In the simple two-level system, $hf = \mathcal{E}_1 - \mathcal{E}_2$ so comparing eqns. 2.17 and 2.19:

$$(K_{sp}/K_{stim}) = 8\pi f^2 n^2 n_g \Delta_{sp}f / c^3 \quad (2.20)$$

A relationship having been established between spontaneous and stimulated emission coefficients, these may now be related to the differential of the (gain/unit time) with electron density given from G'_m. From eqn. 2.12, $G'_m = K_{stim}(N_1+N_2)$; and writing $(N_1+N_2) = N_{ref}$ then using this as a reference density where $N \cong P \cong N_{ref}$ in eqn. 2.6 so that $K_{sp} \cong 1/N_{ref}\tau_{sp}$, one obtains

$$G'_m = K_{stim}N_{ref} = (K_{stim}/K_{sp})/\tau_{sp} = c^3/(8\pi f^2 n^2 n_g \tau_{sp} \Delta_{sp}f) \quad (2.21)$$

Using the group velocity v_g to link the spatial distance z travelled by the light with the time t gives $z = tv_g = ct/n_g$ and the material's optical-power-gain/absorption per unit distance may be written as:

$$(1/S)\, dS/dz = c^2(N - N_{tr})/(8\pi f^2 n^2 \tau_{sp} \Delta_{sp}f) \quad (2.22)$$

showing the important link between spontaneous recombination and stimulated gain and absorption. This is related to the quantum theory of optical amplifiers as given in Appendix 9. In this book the power gain per unit time is written as $G_m(N)$ and the related field gain per unit distance is written as $g_m(N)$, and because the photon density is proportional to the square of the optical-field amplitudes,

$$(1/S)\ \mathrm{d}S/\mathrm{d}z = 2g_m(N) = G_m(N)/v_g \tag{2.23}$$

2.3.2 Stimulated-gain spectra in semiconductors

The analysis above refers to just two levels of energy when in reality there is a band of energies in both the conduction and valence band as in Figure 2.5 where one is dealing with pairs of electrons and holes just inside the conduction band and valence band, respectively. Using the notation of Figure 2.5, and assuming parabolic $\mathscr{E}/k$ relationships with k-selection of the electron/hole interactions with negligible photon momentum:

$$hf = \mathscr{E}_c - \mathscr{E}_v + \delta\mathscr{E}_c(k) + \delta\mathscr{E}_v(k) = \mathscr{E}_g + \hbar^2 k^2/2m_c + \hbar^2 k^2/2m_v = \mathscr{E}_g + \hbar^2 k^2/2m_r \tag{2.24}$$

where $1/m_r = 1/m_c + 1/m_v$ gives a combined effective mass ('reduced mass') from the effective masses in the conduction and valence bands. Now for a spread of photon energies $\mathrm{d}(hf)$ there is a spread $\mathrm{d}k$ in the magnitude of the electron momenta and a consequential spread of their energies in the conduction band given from $\mathrm{d}\mathscr{E} = (\hbar^2 k/m_c)\ \mathrm{d}k = (m_r/m_c)\ \mathrm{d}(hf)$. There then has to be a similar spread of energies $\mathrm{d}(\delta\mathscr{E}_v)$ of holes because they have the same k in the valence band, and so $\mathrm{d}(\delta\mathscr{E}_v) = (\hbar^2 k/m_v)\ \mathrm{d}k = (m_r/m_v)\mathrm{d}\mathscr{E}$.

Now consider a density of photons $S'\ \mathrm{d}(hf)$ spread over a range of frequencies $\mathrm{d}f$ around a frequency f interacting with a density of available electrons in the energy states equivalent to N_1 at an energy $\mathscr{E}_1$ in Section 2.3.1 but now $\mathscr{E}_1 = \mathscr{E} = \mathscr{E}_c + \delta\mathscr{E}_c$ and takes a spread $\mathrm{d}\mathscr{E}$. This *density* of available electrons is found by first finding the density of electron states (given by eqn. 2.2) with the range of energies $\mathrm{d}\mathscr{E}$ associated with the spectral spread $\mathrm{d}f$:

$$N_1 \text{ becomes} \quad 4\pi\left(\frac{2m_c}{h^2}\right)^{3/2} (\mathscr{E} - \mathscr{E}_c)^{1/2}\ \mathrm{d}\mathscr{E} \tag{2.25}$$

Replacing $\mathscr{E}_1$ in eqn. 2.14 with $\mathscr{E}$, the occupation probability is $F_N(\mathscr{E})$ giving the effective density of electrons available to aid the stimulated emission as

$$N_1 F_N(\mathcal{E}_1) \text{ becomes} \quad 4\pi \left(\frac{2m_c}{h^2}\right)^{3/2} (\mathcal{E} - \mathcal{E}_c)^{1/2} F_N(\mathcal{E}) \, d\mathcal{E} \quad (2.26)$$

Using a similar argument, the density of vacant sites or holes available to receive the electrons for stimulated recombination is the equivalent of $N_2 F_P(\mathcal{E}_2)$ in Section 2.3.1 where the energy $\mathcal{E}_2 = (\mathcal{E} - hf)$ will be below the valence-band edge. Bearing in mind that the spread of energies $(m_c/m_v) \, d\mathcal{E}$ in the valence band corresponds to the spread $d\mathcal{E}$ in the conduction band for the same k-values, then $N_2 F_P(\mathcal{E}_2)$ becomes

$$4\pi \left(\frac{2m_v}{h^2}\right)^{3/2} (\mathcal{E}_v - \mathcal{E} + hf)^{1/2} F_P(\mathcal{E} - hf)(m_c/m_v) \, d\mathcal{E} \quad (2.27)$$

Now hf along with the spread $d(hf)$ will be held constant but the whole range of $\mathcal{E}$ can be permitted, consistent with $\mathcal{E} \geq \mathcal{E}_c$ and $\mathcal{E} - hf \leq \mathcal{E}_v$, i.e.

$$0 \leq \mathcal{E} - \mathcal{E}_c \leq (hf - \mathcal{E}_g) \quad (2.28)$$

Then on summing or integrating over the whole energy range of permitted $\mathcal{E}$, the equivalent *stimulated emission* rate given from eqn. 2.7 now becomes

$$(dS'/dt)_{\text{stim emission}} \, d(hf)$$
$$= A_{\text{stim}} S' \{ \int (\mathcal{E} - \mathcal{E}_c)^{1/2} (\mathcal{E}_v - \mathcal{E} + hf)^{1/2} F_P(\mathcal{E} - hf) F_N(\mathcal{E}) \, d\mathcal{E} \} \, d(hf) \quad (2.29)$$

where $A_{\text{stim}} = K_{\text{stim}} 16\pi^2 \left(\frac{4 m_v m_c}{h^4}\right)^{3/2} (m_c/m_v)$ with K_{stim} as in eqn. 2.7 and the range of the integration is limited by eqn. 2.28.

Still following the methodology of Section 2.3.1, it may be seen that the equivalent expression for *stimulated absorption* is the same expression as for the *stimulated emission* except that the occupation probabilities F_N and F_P of electrons and holes, respectively, are replaced by

$$(1 - F_N) = F_N \exp\{(\mathcal{E} - \mathcal{E}_{fN})/kT\} \text{ and } (1 - F_P) = F_P \exp\{(\mathcal{E}_{fP} - \mathcal{E} + hf)/kT\} \quad (2.30)$$

giving the result that

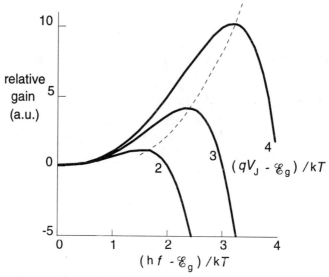

Figure 2.8 Optical gain against frequency

Simplified calculation of the approximate shape of the gain spectrum, with k-selection and no band tails

$$(1 - F_N)(1 - F_P) = F_N F_P \exp\{(hf - qV_J)/kT\} \quad (2.31)$$

Hence one finds for the absorption

$$(dS'/dt)_{\text{stim absorption}} \, d(hf) = A_{\text{stim}} S' \exp\{(hf - qV_J)/kT\} \times$$
$$[\int (\mathscr{E} - \mathscr{E}_c)^{1/2} \{(hf - \mathscr{E}_g) - (\mathscr{E} - \mathscr{E}_c)\}^{1/2} F_P(\mathscr{E} - hf) F_N(\mathscr{E}) \, d\mathscr{E}] \, d(hf) \quad (2.32)$$

with the same range of $\mathscr{E}$ as in eqn. 2.28.

Now the gain spectrum is given from the net stimulated (emission − absorption) so that approximating the Fermi–Dirac distributions to unity over the limited range of energies, one can determine the gain spectrum as approximately

$$(dS'/dz)/S' = 2g_m \approx (n_g/c) A_{\text{stim}} \times$$
$$\{1 - \exp\{(hf - qV_J)/kT\}\} [\int (\mathscr{E} - \mathscr{E}_c)^{1/2} \{\mathscr{E}_c - \mathscr{E} + (hf - \mathscr{E}_g)\}^{1/2} d\mathscr{E}] \quad (2.33)$$

The integral may be evaluated by putting $y^2 = (\mathscr{E} - \mathscr{E}_c)$ to find a rough but explicit expression for the gain spectrum with k-selection as a function of the junction voltage:

$$g_m(hf, V_J) = (\pi n_g A_{\text{stim}}/8c)(hf - \mathscr{E}_g)^2 [1 - \exp\{(hf - qV_J)/kT\}] \quad (2.34)$$

Figure 2.8 sketches this dependence of gain on the optical frequency and the junction voltage. As the junction voltage increases, so the peak optical gain increases in frequency. One can also see that as kT

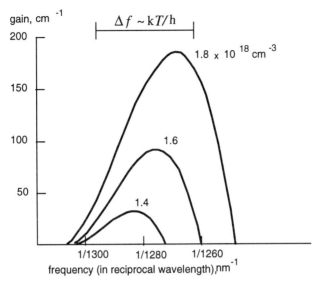

Figure 2.9 Estimates of gain against frequency (parameter is carrier density)

For a quaternary $In_{0.75}Ga_{0.25}As_{0.55}P_{0.45}$ undoped material: (parameter=carrier density). This is the material gain for the optical field and the net effect in a laser would be reduced by the confinement factor. The bar at the top indicates the spectral range corresponding to a change of photon energy of kT at room temperature of 300 K. Data taken from Osinski and Adams [21]. The general shape is common to most gain spectra with the gain covering a range of frequencies Δf typically, where $h\Delta f \sim kT$. The theory agrees reasonably with experiment [22].

decreases so the gain bandwidth decreases. This simplified analysis then demonstrates many key features of optical gain.

The results may be converted to gain as a function of electron density by estimating the electron and hole density for a given junction voltage. Gowar [20] gives an explicit example of how this can be done. However, obtaining detailed calculations of the gain, as opposed to having an understanding of the gain processes, really requires a more detailed study of the k-selection processes, band tailing and other physics of optical transitions and is left to specialist references, as for example the work shown in Figure 2.9. For many modelling purposes, expressions for gain etc. rely on measurements on relevant material rather than theory.

2.3.3 Homogeneous and inhomogeneous broadening

The actual linewidth $\Delta_{LW}f$ of a single laser mode is orders of magnitude smaller than $\Delta_{sp}f$, the width of the spontaneous spectrum. While the main discussion concerning linewidth is left until Section 6.2, a brief discussion is essential here. Ideally if there were only transitions between a single pair of energy levels, as in an ideal gas or solid-state laser where the active atoms are sparsely distributed and essentially independent of each other then this lasing linewidth $\Delta_{LW}f$ would be very small. However, even when all the photon emissions are created from identical energy transitions within the independent atoms or molecules, the lasing linewidth is *homogeneously* broadened because the photons have random phases and are emitted at random times. The effect on the linewidth $\Delta_{LW}f$ of the spontaneous lifetime τ_{sp} in such a laser is found through straightforward Fourier analysis to give

$$\Delta_{LW}f \simeq 1/(2\pi\tau_{sp}) \qquad (2.35)$$

The associated line shape is known as Lorentzian (Section 6.2).

This type of line broadening still takes place in semiconductor lasers but the energy levels involved in the photon emissions are *not* all identical so that strong *inhomogeneous* broadening should lead, through a superposition of different random processes, to a Gaussian line shape. However, the experimentally observed results give mixed conclusions [23] but for many purposes the Lorentzian line shape is still adequate with other linewidth-broadening processes dominating as discussed in Sections 2.5, 4.2, 6.2 and 8.2.5

2.3.4 Spontaneous-emission spectra from semiconductors

The close connection between spontaneous emission and stimulated emission was clearly seen in the two-level model and this link still holds in the more realistic theoretical models [24]. Spontaneous-emission spectra may be measured from both FP lasers or LEDs, but the measurement is not made at the point where the photons leave the atom; the measured light has to pass through further material where it may be absorbed and/or re-emitted, depending on the material gain and the length traversed. This is very dependent on the device type and structure, but measurements from surface-emitting LEDs give a reasonable direct estimate of true material-spontaneous spectra while those from FP lasers will need more careful interpretation because of the feedback.

The link demonstrated between spontaneous emission and stimulated emission means that the spontaneous-emission data can be processed to give gain–frequency information at different electron densities, provided that enough details are known of the laser

Gain, loss and spontaneous emission 55

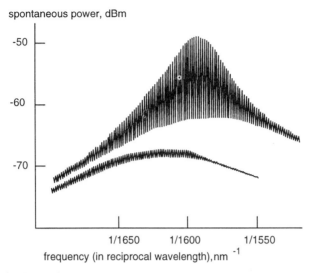

Figure 2.10 Typical measured spontaneous spectra using a Fabry–Perot laser

> Data taken from Morton *et al.* [25] of the emission from a FP laser with one AR-coated facet using 7 quantum-well laser material: top spectra at ~85% of threshold, bottom spectra at ~25% of threshold. The spectral-line structure is created by the many Fabry–Perot modes but the envelope gives the spontaneous profile

structure. Figure 2.11 indicates processed results from the data of Figure 2.10 based on Reference 25.

2.4 Semiconductor interactions with the lasing mode

2.4.1 Spontaneous-coupling factor

In Chapter 1, the lasing condition required a round-trip complex gain of unity, thereby neglecting the existence of spontaneous emission. This is unrealistic, particularly for describing a laser around threshold where the spontaneous emission plays a key role. Spontaneous emission is essentially random and, on the scale of the modelling, the emission at a position x is uncorrelated with the emission at x' in space. To a good first approximation, k-vectors of electrons in a bulk semiconductor are isotropically distributed and the band structure does not vary sufficiently with direction to invalidate this discussion. Spontaneous emission is therefore uniformly and randomly distributed in both direction and polarisation, and in any one direction is uncorrelated with any other direction.

56 *Distributed feedback semiconductor lasers*

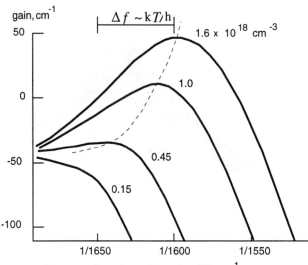

Figure 2.11 Estimates of gain-frequency against carrier density

Material with 7 QWs of 80Å width separated by 80Å barriers (parameter = carrier density). Such quantum-well material gain includes the net effect of the confinement factor which would be the order of 0.15 depending in detail on the surrounding guide structure. (Data compiled from P.A. Morton *et al.* [25].) The bar at the top indicates the spectral range corresponding to a change of photon energy of kT at room temperature. Note how the gain peak changes with electron density

A guided lasing mode may be regarded as a coupled group of rays propagating with the same velocity component in the mode direction, but at a range of different angles. The greater this range of angles, the greater is the range of spontaneous emissions that will couple to the mode. The coupling for the forward and backward directions will be identical, giving a factor of 2 in spontaneous power when both couple to a mode. When discussing forward and reverse waves separately, note that the forward and reverse spontaneous components are uncorrelated. The spontaneous-emission components associated with any two polarisation directions are similarly uncorrelated, but a single guided mode will have only one polarisation, and only one resolved component will be coupled to the mode, thereby reducing the coupling of the total spontaneous emission by another factor of two. Mathematically, one has to estimate the overlap integral between a guided mode and spontaneous emission, in both the spatial and spectral domains. Spectrally, the spontaneous emission covers a broad range ($\Delta_{sp}f \sim 10\,000$ GHz) but the lasing-mode linewidth

Gain, loss and spontaneous emission 57

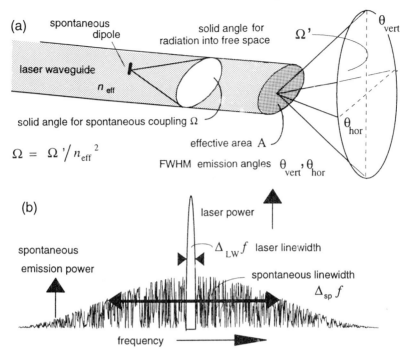

Figure 2.12 Schematic of the principal elements forming the spontaneous emission-coupling coefficient

$\Delta_{LW}f \sim 0.1$–100 MHz. To a first order, the spectral coupling can be taken as the ratio of the modal linewidth to the spontaneous spectral width. Figure 2.12 indicates these features schematically.

It is helpful to discuss orders of magnitude. The principal method of guiding considered for lasers here is index guiding, where the internal angles are a few degrees (see Chapter 3). A typical far field for a buried-heterostructure laser might have 25° full-width half-maximum (FWHM) horizontal (i.e. parallel to the junction) and 40° FWHM vertical (i.e. perpendicular to the junction) as indicated in Table A11.1. The half-cone angles in the air, $\frac{1}{2}\theta_{hor}$ and $\frac{1}{2}\theta_{vert}$, are reduced inside the waveguide through Snell's law (see Figure 2.12a) where approximating $\sin(\frac{1}{2}\theta) \sim \frac{1}{2}\theta$ the internal angles are simply reduced by the ratio n_{air}/n_{eff}, where $n_{eff} \sim 3.3$ (an effective or average value for the lasing material waveguide as in Table A11.1). Then the solid-angle emission angle $\Omega' \sim 0.21$ outside the laser is reduced inside the laser to a value $\Omega = \Omega'/n_{eff}^2 \sim 0.02$ steradians. Noting that the full solid angle is 4π, the angle-coupling factor is therefore $\sim 1.6 \times 10^{-3}$. Taking the mode spectral linewidth, for example, as 0.15 nm, with the total spontaneous bandwidth as 100 nm the spectral-coupling factor $\sim 1.5 \times 10^{-3}$ and

58 *Distributed feedback semiconductor lasers*

applying a polarisation factor of 0.5 gives the spontaneous-coupling factor:

$$\beta_{sp} \sim 0.5 \times 1.6 \times 1.5 \times 10^{-3} \times 10^{-3} \sim 1.2 \times 10^{-6} \qquad (2.36)$$

This is an appropriate estimate for travelling-wave rate equations, but is doubled in conventional rate equations which consider both forward and reverse optical waves together in a single rate equation.

The value of spontaneous-emission coupling can be evaluated in a number of ways; the method given above is straightforward, but other methods are examined and reconciled in Appendix 9 which also examines the numerical modelling of spontaneous emission.

2.4.2 Petermann's 'K factor'

In some early laser devices, the guiding of the optical waves was accomplished through the active gain. This is not studied in this book, where index guiding is assumed to dominate. However with gain guiding the propagating wave fronts have a significant curvature allowing for much greater (up to 50 times) acceptance angles for spontaneous emission. The increase in the acceptance angle is known as the Petermann transverse K factor, variously referred to as K_p or K_{tr}. The reader is referred to Petermann's work [26, 27] for further study. In DFB lasers, where optical fields may be strongly localised longitudinally, an equivalent longitudinal K factor [28] also referred to as K_z exists giving a total change in the spontaneous emission of $K_{tot} = K_z K_{tr}$. However, with the numerical modelling here, the longitudinal (or z) interactions and nonuniformities are included and so K_z is automatically taken into account and should not be inserted as an addition.

2.4.3 Gain saturation in semiconductors

Gain saturation may be regarded as arising from three different mechanisms. If no current is driven into the diode, the semiconductor laser is said to be unpumped, few conduction band states are occupied, and most valence band states are full, so the optical absorption per unit length is essentially limited by the strength of the interaction (|matrix element|^2, related to spontaneous lifetime) and the volumetric density of states. Even with very strong pumping or full inversion, the maximum gain per unit length can only approach the maximum absorption per unit length. This 'gain saturation' means that the gain is nonlinear with electron density in the laser and this applies to all materials, but the effect is especially noticeable in single-quantum-well material where the small density of states can saturate rapidly.

A similar but distinct cause of saturation is optically induced. In this latter case, for a given drive current at high optical powers, the high rate of stimulated transitions reduces the carrier density from the value at low optical powers. This saturation is naturally taken into account through the gain decreasing with electron density.

Finally, even if the average carrier density were maintained by extra drive current at high optical powers, the high optical power at a particular photon energy causes the distribution of electrons and holes to depart from their equilibrium spread of energies within the conduction and valence bands, and the gain is reduced *at the lasing frequency* as discussed further in Section 2.4.4 below. This type of saturation is dealt with by introducing a first-order 'gain-saturation parameter' ϵ, where if the gain as a function of electron density in the semiconductor is $G_m(N)_{(0)}$ at zero optical power, and $G_m(N)_{(S)}$ is the gain at an optical photon density S then

$$G_m(N)_{(S)} = G_m(N)_{(0)} / (1 + \epsilon S) \tag{2.37}$$

where $\epsilon \sim 1$ to 3×10^{-17} cm^3 is a useful guide to the accepted strength of gain saturation with S then measured in numbers per cubic centimetre to ensure that ϵS is dimensionless.

2.4.4 Spectral hole burning and carrier heating

The precise reasons for gain saturation at high optical powers are still a cause for discussion [29], but two candidates for contributing to the parameter ϵ are mentioned briefly. The first main effect is the inability of the charge carriers to redistribute themselves sufficiently quickly within the conduction or valence bands so that, at high optical powers, the electrons and holes in those particular energy levels (i.e. the shaded regions shown in Figure 2.5) which are contributing to the photon energy of highest intensity become depleted from their conventional equilibrium values, i.e. there is 'spectral hole burning' in the gain spectrum. Under normal conditions, charge carriers within a band (i.e. at energies just above and below the shaded regions in Figure 2.5) can redistribute themselves to help fill the spectral hole towards a thermal-equilibrium state on the time scale of the intraband relaxation time $\tau_{intra\ band} \sim 10^{-12}$ s. Because radiative-carrier lifetimes (or *inter*-band relaxation time) are around 10^{-9} s the effects of $\tau_{intra\ band}$ are usually negligible on this time scale. However, at sufficiently high optical-power densities, the radiative lifetime is reduced by an order of magnitude so that a 'hole' or notch in the gain occurs at the lasing wavelength and the carrier distribution in the bands is no longer in thermal equilibrium. The inclusion of the parameter ϵ in eqn. 2.37 is one method of accounting for this effect. Spectral-hole burning can have noticeable effects on the linewidth of some DFB lasers [30].

There is a related effect which occurs when charge carriers are injected from the contacts into the material confining the light. These electrons are necessarily injected at higher than average energy, and so need a little time to redistribute into the energies characteristic of the band. This effect is particularly important in quantum-well lasers where, as these carriers thermalise into the quantum wells, they arrive at a higher than average temperature and such 'hot' carriers can then transport out of a well and not contribute so effectively to the gain. All this significantly affects the details of transport from contacts into quantum wells [30] (see Section 6.6.2).

2.4.5 Scattering losses

Any slight inhomogeneities in the laser's material or waveguide will cause optical scattering [31]. Perpendicular to a p–n junction, even good epitaxial layers will have roughnesses of the order of one atom step, and parallel to the junction the photolithographically defined and etched sidewalls of a buried heterostructure mesa are likely to have RMS roughnesses of the order of 100 nm. The refractive-index steps, though small, can cause some light to be 'scattered' out of the guide and in effect give attenuation as if there were absorption but, unlike absorption, scattering does not generate heat (unless the light is subsequently reabsorbed).

2.4.6 Free-carrier absorption

An electron within the conduction band may interact with a photon and so gain the photon's energy, but because the range of energies in the conduction band is wide some such electrons can stay within the same energy band. This type of free-carrier photon absorption is almost independent of the frequency because the higher energy levels are empty and there is no equivalent bandgap to control the process. As is normal for any interaction, the higher the density of electrons, the greater the absorption, and in heavily doped ($\sim 10^{19}$ cm^{-3}) n-type substrates an absorption coefficient ~ 10 cm^{-1} can be measured.

The same occurs in principle with valence-band electrons, but because such bands are nearly full of electrons the process statistically involves more 'shuffling' of full and empty states to take place, and so is less significant. When a low-optical loss substrate is required, it can sometimes be advantageous to use *p*-type material.

2.5 Henry's α factor (or linewidth enhancement factor)

Until now, optical emission and absorption have been considered in terms of particles or energies. Advanced laser modelling requires that one considers the wave nature of light. The net stimulated gain for a propagating electric field of amplitude $E_f(z)$ has to be given from:

$$E_f(z) = E_0 \exp(-j\beta z) \exp(gz) \qquad (2.38)$$

- There has to be both propagation with a phase change as well as gain. For a plane wave in a uniform (nonmagnetic) material with gain, there are real and imaginary parts of a complex relative permittivity linked to the real and imaginary parts of a complex refractive index:

$$\beta + jg = \omega\{(\varepsilon_{rr} + j\varepsilon_{ri})\varepsilon_o \mu_o\}^{1/2} = \omega(n_r + jn_i)/c \qquad (2.39)$$

If E_0 is regarded as the input and $E_f(z)$ the output of a linear system, it can be shown (Appendix 7) that the 'gain response' is uniquely defined by the 'phase response' so that β and g are linked and consequently the real part of the complex refractive index n_r is linked to the imaginary part of the refractive index n_i. The processes of calculating the real response from the imaginary, and vice versa, are known as Hilbert transforms or Kramers–Krönig transforms. These relationships apply strictly to any linear causal system where there is an input which causes an output, but they may be extended to any piecewise-linear or continuous system. In any case, departures from linearity in semiconductor properties are not sufficiently great to invalidate the application of the Kramers–Krönig relationships for the work discussed here. The term refractive index, when used by itself, will always mean the real part of the complex refractive index but, when there could be some doubt, the notation n_r will be used to indicate the real part with n_i the imaginary part corresponding to gain.

Implicit in the above is that any change in gain with frequency *must* be associated with a refractive-index change in frequency and vice versa. However, a change in gain at a particular frequency does not *necessarily* result in a refractive-index change *at the same* frequency; the Kramers–Krönig transforms determine how the two changes are linked. Figure 2.13 illustrates schematically this link between gain and refractive index changes with frequency, basing the work on eqn.

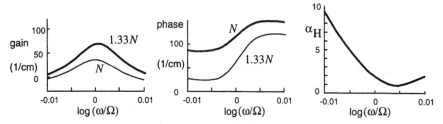

Figure 2.13 Schematic links between gain and refractive index changes with frequency caused by Kramers–Krönig relationships

The frequency Ω is the frequency of the gain peak. The real refractive index is given by the phase, measured in inverse distance to have the same dimensions as the gain (imaginary refractive index). The real index level can be set at an arbitrary level without altering the frequency dependence; only the relative changes of phase are of concern. As the electron density increases, the gain (imaginary refractive index) increases and the real refractive index decreases. The parameter α_H can then be estimated

A7.16 and limiting the spectral range to values around the gain peak. Even at the peak of the optical-gain mechanism, there is a separate physical effect caused by a plasma of nearly free electrons which adds a component into the relative permittivity, *reducing* the real refractive index with increasing electron density (Appendix 7). As the frequency moves away from the peak gain, this change in the (real) refractive index is modified by the gain curvature through the Kramers–Krönig relationship. At any particular wavelength, one can define the ratio known as 'Henry's α factor' or α_H:

$$\alpha_H = -\{(dn_r/dN)/(dn_i/dN)\} \qquad (2.40)$$

where the negative sign is inserted to make α_H positive in this work.

When amplitude modulating a laser, by turning it on and off, there have to be changes in the gain. From eqn. 2.39, the field gain/unit distance $g = \omega n_i/c = 2\pi n_i/\lambda_{fs}$, so that from eqn. 2.40 changes Δg in the gain give changes in the refractive index from

$$\Delta n_r = -\alpha_H \lambda_{fs} \Delta g / 2\pi \qquad (2.41)$$

The laser frequency is usually controlled by an optical length such as the optical length of the Bragg period or the Fabry–Perot cavity, and is therefore directly proportional to the average refractive index. Consequently the magnitude of α_H directly affects the change of frequency $\Delta_g f$ with a change in gain Δg in a laser by approximately

$$\Delta_g f / f \sim -\overline{\Delta n_r / n_r} \sim (\lambda_{fs}/2\pi)\overline{(\alpha_H \Delta g / n_r)} \qquad (2.42)$$

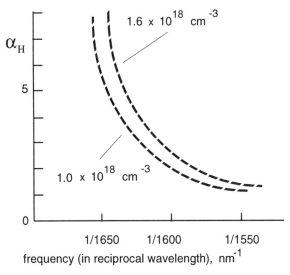

Figure 2.14 Estimates of linewidth enhancement factor α_H for quantum-well material

QW material with seven 80Å wells and 80Å barriers
Parameter = carrier density
Data compiled from P.A. Morton *et al.* [25]

where the bar denotes some appropriate average over the laser. These dynamic wavelength shifts are known as (dynamic) chirp with dynamic changes in frequency following dynamic changes in the gain. Different laser structures using the same material will have different chirp because of structure-related effects, and especially if the operating frequency can be controlled in relation to the gain peak. Figure 2.13 shows a rough estimate of the changes in α_H with operating frequency, indicating that α_H rises markedly at the lower frequencies and reaches a minimum above the gain peak.

This effect can be understood by splitting the refractive-index dependence into the component caused by free electrons (also referred to as the plasma effect) where increasing the electron density reduces the real refractive index over a broad range of frequencies (Appendix 7, Figure A7.3). The localisation of the gain gives rise to an associated ⌢-shaped change in the real refractive index, centred about the gain peak. Above the frequency of this gain peak, the net real refractive-index change with electron-density change reaches a minimum. Figure 2.14 shows measured data indicating the reduction in α_H at high frequencies (short wavelength), as also shown schematically in Figure 2.13.

In Fabry–Perot lasers, the laser must operate close to the gain peak so that major variations in α_H are not possible. In DFB lasers, the choice of operating frequency is determined more by the grating, so that the choice of the wavelength for peak gain gives an extra degree of freedom and allows one to operate at frequencies where there is a lower value of α_H than for a Fabry–Perot laser, and so in principle one can reduce the chirp. At 1550 nm it is common to offset the lasing wavelength by around 20 nm down in wavelength from the gain peak to take advantage of this lower value of chirp.

Quantum-well material allows one to engineer the gain profile to a much greater extent than merely changing the composition in bulk materials [8]. If the gain peak could be made narrower about its centre frequency, then the ⌣-shaped change in the real refractive index becomes stronger and gives a bigger offset for the plasma effect [33] (see Figure A7.3). At the correct wavelength, it may be possible to design quantum-well material to approach a 'chirpless' laser where $\alpha_H \to 0$. A detailed study of Henry's alpha factor has been undertaken by several authors, with a notable discussion by Osinksi and Buus [34].

2.6 Temperature-induced variations in semiconductor lasers

There is considerable interest in understanding the changes in the performance of lasers as temperature changes [35–37] where the temperature of importance is that of the active area. Because of the heating from injected carriers, this active region temperature may be different from that of the heat sink by 2–10°C, but this difference will be strongly affected by the technology of packaging. Even with excellent heat sinking using Peltier coolers, local changes in temperature can still alter important material parameters significantly. A full account is beyond the scope of this book which only outlines the effects of temperature on refractive index and wavelength and then briefly considers the effects of temperature increases on gain and threshold in bulk materials. Lasers using quantum-well material are no less sensitive to temperature. Indeed temperature induced effects can be used to measure material parameters [38].

From eqn. 1.13 for a first-order grating, the physical periodicity of the Bragg grating Λ is $\Lambda = \frac{1}{2}\lambda_b / n_{eff}$ where n_{eff} is the effective value of the refractive index for the guided wave and λ_b gives the Bragg wavelength in free space for the maximum reflection. An estimate of temperature coefficient of wavelength may then be obtained from

$$(\partial \lambda_b/\partial T)/\lambda_b = (\partial \Lambda/\partial T)/\Lambda + (\partial n_{eff}/\partial T)/n_{eff} \qquad (2.43)$$

The material coefficient of expansion, $\alpha = (\partial \Lambda/\partial T)/\Lambda$, for InP between 0°C and 100°C is $\alpha \approx 4.5 \times 10^{-6}$ K^{-1} ($\alpha \approx 6 \times 10^{-6}$ K^{-1} for GaAs), and is often ignored because it is an order of magnitude smaller than the coefficient of refractive-index change with temperature $(\partial n/\partial T)/n$. The guide structure may have an effect in changing n_{eff} with temperature but here only the change of the underlying value n is considered. Measurements have been discussed for this coefficient [39], which is difficult to measure and compute close to the bandgap where there are strong lattice resonance effects. Estimates of $(\partial n/\partial T)/n$ for InP (GaAs) are the order of $6(7) \times 10^{-5}$ K^{-1} around 1.53 μm (1.1 μm) leading to the conclusion that the operating wavelength increases with temperature because the refractive index becomes larger. One notices that increasing the injection of charge carriers into a laser therefore has two effects: a decrease of wavelength as the carrier density rises and reduces the refractive index (Section 2.5), but on a longer time scale the wavelength increases because of the rise in temperature induced by the additional current.

Some features of the variation of the gain spectrum with temperature can be appreciated from Figure 2.8 which plots a normalised gain against $(hf - \mathcal{E}_g)/kT$. An important effect, not included in that previous discussion, is that the bandgap $\mathcal{E}_g$ decreases in energy with rises in temperature, thereby increasing the wavelength for the peak gain by approximately 0.35 nm/K for both GaAs and InP. The *width* of the gain spectrum (and also the width of spontaneous emission) increases with kT but to keep the peak gain constant the applied voltage has to increase linearly with kT thereby requiring the electron density to increase with temperature if the gain is to be kept constant. Menzel et al. [40] have modelled these effects in some detail with empirical curve fitting.

From the practical view of stabilising the laser's optical output, the effect of temperature on the threshold is most important (Figure 1.3). Although reduction in gain with temperature increases the threshold, the more significant increase with temperature arises because of:

(i) carrier leakage over the heterojunction barriers;
(ii) increased recombination at material interfaces, for example in BH lasers;
(iii) increased non-radiative recombination; but especially
(iv) increased Auger recombination which rises more rapidly with

temperature than any of the other effects [39] and is notably significant for 1.3 and 1.55 μm lasers.

The effects of (i) should be small in all well designed lasers [41]. One measure often used with threshold current is the *characteristic temperature* T_0. This empirically relates the threshold currents at temperatures T_1 and T_2 by

$$I_{thresh}(T_2) = I_{thresh}(T_1)\ \exp\{(T_2 - T_1)/T_0\} \quad (2.44)$$

As a rough guide,

for 0.9 μm lasers $T_0 \sim 120\text{–}200$ K
for 1.3 μm lasers $T_0 \sim 60\text{–}100$ K
for 1.55 μm lasers $T_0 \sim 40\text{–}70$ K

increasing with temperature as Auger recombination starts to be significant.

2.7 Properties of quantum-well-laser active regions

2.7.1 Introduction to quantum wells

The physics of quantum-confined states and quantum-well lasers [7] is a major subject but, in this book on DFB lasers, quantum-well material is regarded as essentially a material with special parameters, particularly high differential gain, caused by the quantum-well layers. This section limits the discussion to outlining how these special parameters arise.

Electron-wave functions in bulk semiconductors, found through solving Schrödinger's equation, spread over the dimensions of many atoms. This permits average relative permittivities to be used. Equally, conventional device dimensions are much larger than the electron wavelengths which then allows the discrete energy levels to be approximated by a continuum of electron energies and wavelengths. With the dimensions of conventional heterostructure band diagrams (Figure 2.15a), holes and electrons may exist at energies between the active-region bandgap energy and the confining layer bandgap energies.

Once the active region width falls below about 20 nm, and more markedly at smaller dimensions, then quantum effects emerge and the permitted energies are quantised between $\mathscr{E}_{act}$ and $\mathscr{E}_{clad}$ (Figure 2.15b). Transitions are only possible at energies $\mathscr{E}_{Q1}$, $\mathscr{E}_{Q2}$ etc. with only the lowest states occupied except at high carrier-injection levels. The

Gain, loss and spontaneous emission 67

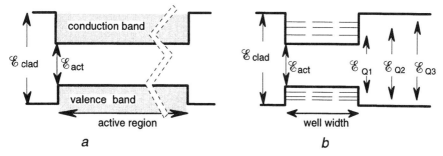

Figure 2.15 Energy-band/distance diagrams

     a Bulk material with active region widths over 20 nm so that a continuum of energy bands is the appropriate approximation rather than discrete levels
     b Quantum-well material with active region widths in 1–10 nm range for which quantised levels have to be considered

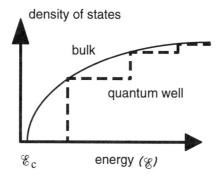

Figure 2.16 Schematic comparison of density of states in bulk material and quantum-well material

density-of-states function is proportional to $\sqrt{(\mathcal{E}-\mathcal{E}_c)}$ for bulk-material conduction electrons, but for quantum wells it becomes a 'staircase' touching the original parabola (as in Figure 2.16).

Several advantages of quantum-well material, compared with bulk material, can be deduced from this diagram; in particular, fewer states have to be filled to reach inversion, so threshold currents should be reduced. Further consideration shows that if only one energy level is filled, the spectral width of spontaneous emission and gain should be much smaller than for bulk material, though electron–electron collisions still broaden the energy bands. This advantage may be lost at higher drive currents, as additional energy levels are filled.

Transitions in unstrained bulk material are essentially isotropic, i.e. on average do not depend on the electron's direction of travel. In such lasers, the excitation of a TE or TM optical mode is partly determined by the interaction, or mode overlap as it is called, of the particular mode with the gain region. Mode overlap is still important in QW lasers, but now the transition probabilities, which give gain, depend on the direction as well as the magnitude of the momentum k, and so the polarisation of the E field and its orientation with respect to the quantum wells also help to determine whether TE or TM modes are favoured. In fact, any process which breaks the lattice symmetry, such as mechanical strain, can also differentiate TE and TM gain. Higher strain levels are possible in quantum wells without dislocations being generated, and hence strain effects are usually larger than in bulk materials.

Finally, it must be said that some of the improvements claimed for lasers using quantum wells may be caused by the advances in fabrication technology which are steadily being made over the years, so that improvements would also be seen on bulk lasers using such advanced fabrication.

2.7.2 Gain saturation and the need for multiple quantum wells

Many of the benefits of quantum-well materials result from the lower density of states and the lower carrier densities needed to reach transparency; however, these features can also be detrimental. A separate (optical) confinement structure is essential in quantum-well lasers to ensure that the optical filament is concentrated around the quantum well(s); the quantum well does give some positive waveguiding, but it is far too weak to confine the mode. However, because of the nanometre scale of the quantum regions, confinement factors $\Gamma < 0.01$ are typical, in contrast to $\Gamma \sim 0.5$ for bulk lasers. Hence the required gain for quantum wells may need to be nearly two orders of magnitude larger than the gain in similar lasers using bulk material. This higher gain leads to a more rapid filling of the available states at a particular energy, making any increase in gain at that desired energy (i.e. at a desired wavelength λ_o) impossible. As the drive current increases steadily, yet higher energy levels in the wells may be filled, increasing the gain at shorter and shorter wavelengths, but this will not increase the gain at λ_o. The gain therefore saturates at a given λ_o with increasing electron density, as was pointed out in Section 2.4.3. These serious effects of gain saturation, especially with 1.5 μm material, are mitigated by stacking, say, three to seven quantum wells in an active

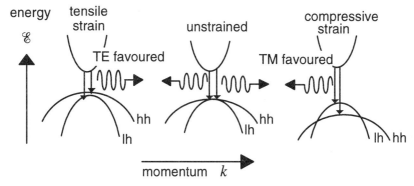

Figure 2.17   Schematic effects of strain

layer and so engineering a material which has a much higher differential gain than bulk material but has saturation and other properties which are intermediate between those of a single quantum-well and bulk material.

2.7.3 Strained quantum wells

Strain in the active regions of lasers has been regarded as deleterious, leading to the generation and migration of defects, and causing premature failure. However, because quantum-well layers are so thin, remarkably large strains are possible; in fact there is a 'critical thickness' for any level of strain, compressive or tensile. A useful guide is to keep the material thickness for each quantum well below a critical value which is around 20 nm per well for 1% strain [42]. If this critical thickness is exceeded, subsequent epitaxial growth will usually have a poor morphology, creating large numbers of defects which can move during subsequent processing of material and devices. Up to these critical levels, the laser designer is then able deliberately to build strain into quantum-well structures, thereby achieving several advantages.

One of the most direct advantages is an increase in the range of wavelengths of materials grown on a given substrate because lattice matching is relaxed. A well studied example [43] of this is InGaAs grown on GaAs with about 1% compressive strain in the active region, which allows an operating wavelength of 980 nm, which is needed for pumping erbium amplifiers and is difficult to achieve using other materials systems. Figure 2.17 indicates that, for a given material, the bandgap increases under compressive strain, but more importantly the heavy-hole band moves away faster than the light-hole band, so the degeneracy of these bands is lost. Furthermore, it is found that transitions to the light-hole band couple better to the TM than TE mode, providing better polarisation mode selectivity.

The degeneracy of light- and heavy-hole bands causes problems via *intervalence-band absorption* (IVBA), particularly in long-wavelength lasers [44]. The effect here is that photons can (and do) cause hole transitions from one valence band to the other, and the resulting energy change is dissipated in the form of multiple phonons as the excited hole relaxes back to the centre of the band. Such a process is a source of excess optical loss. This is also one of the major causes of poor temperature sensitivity of long-wavelength lasers, since IVBA becomes much more prominent at the high injected carrier densities which are needed for lasing at high temperatures. This problem is relieved by strain, which shifts the bands apart, and reduces the probability of transitions between them.

In longer-wavelength lasers, based on InP, both compressive and tensile strains have been employed to reduce thresholds and IVBA, with some success. In DFB lasers the removal of the competition between transverse-electric-field (TE) and transverse-magnetic-field (TM) mode is also important, but in semiconductor optical amplifiers (SOA), by contrast, the gains for TE and TM should be closely matched to ensure independence of the output power of the polarisation at the input.

2.7.4 Carrier transport

When quantum-well lasers were first introduced, the expected increase in frequency response for optical output with direct-current modulation was not realised, and research showed that one contribution to the problem [30] was transporting the holes (and electrons) across waveguide layers into the active region of the quantum wells. The quantum wells, by themselves, often provide an inadequate change in the refractive index to guide the light effectively, and also any holes/ electrons captured by the wells tend to escape by thermal emission over the well barriers. To combat both of these difficulties, separate (optical) confinement-heterostructure (SCH) layers have to be introduced (Figures 2.18 and A11.1) which help to guide the light and also keep any escaping carriers to the physical region where they can be recaptured by the quantum wells.

The time taken by the carriers to transport from the contacts into the wells is the sum of a characteristic diffusion time across the separate confinement regions and a characteristic capture time (τ_{cap})

Gain, loss and spontaneous emission 71

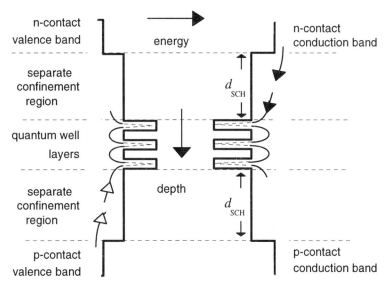

Figure 2.18 Transport into quantum wells

Energy/depth band diagram showing the lengths d_{SCH} of the separate confinement zones, designed to confine the photons close to the quantum wells. These must be kept small to allow electrons and holes to move rapidly into the quantum wells. Graded impurity composition of the confining regions can be used to assist in the carrier transport

into the wells. The electric field is particularly low in these separate-confinement regions, so that carrier transport is often dominated by diffusion with a limiting time constant $\tau_{diff} \sim q\, d_{SCH}^2 / \mu_h kT$ which is of the order of the time taken for holes (with their lower mobility than electrons) to diffuse across their separate confinement region. The problem is reduced if the waveguide layers are thin (d_{SCH} is small), and reduced still further if a graded-composition separate confinement heterostructure (GRIN-SCH) is used. Here either a gradient or a series of steps in the composition of the SCH layers can change the conduction-band edge with depth, thereby assisting the carriers to transport from the contacts to the wells in much the same way that varying composition in the base of a transistor can help the transport of charge carriers through the base. The small-signal dynamics of carrier transport are discussed in Appendix 4, along with implications for changes in emission frequency caused by changes of the free carrier density in the confinement regions. Once this effect is understood and the confinement lengths are reduced appropriately, then transport effects can often be neglected, at least for modulation frequencies up to tens of gigahertz [45].

2.8 Summary

Starting from a background of degree-level knowledge about semiconductors and band diagrams, this chapter provides key tutorial background for understanding photon–electron interactions in semiconductors so that one can appreciate how laser materials provide optical gain and loss, spontaneous and stimulated emission, and how these quantities change with electron density and with wavelength. Important discussions include a two-level model for a semiconductor which gives the key rate equations for photon and electron interactions and includes the full Fermi–Dirac statistics; this progresses to an illustrative account of determining the gain as a function of frequency and excitation.

The various material factors which are important in laser models are discussed. Among these effects are the coupling factor for the spontaneous emission to the lasing mode and the changes that occur with gain and nonuniformity (the Petermann factors); and Henry's spectral linewidth-broadening factor α_H, which describes the changes of refractive index and gain with electron density. Important effects such as Auger recombination, temperature dependence, strain in material, quantum-well material and carrier transport are all touched on to provide the required background for later chapters.

2.9 Bibliography

Some helpful books on materials:

ADACHI, S.: 'Physical properties of III–V semiconductor compounds' (Wiley 1992)

RAZEGHI, M.: 'The MOCVD challenge' (Adam Hilger–IOP Publishing, 1989), vol. 1

MURARKA, S.P., and PERKERAR, M.C.: 'Electronic materials: science and technology' (Academic Press, Boston 1989) (although this is not about opto-electronic materials, it contains helpful accounts of a wide range of materials technology)

PEARSALL, T.P.: 'GaInAsP alloy semiconductors' (Wiley, Chichester, 1982)

IEE EMIS Series (Electronic Materials Information Service)

BROZEL, M.R., and STILLMAN, G.E. (Eds.): 'Properties of gallium arsenide' (EMIS series, IEE, London, 1996), 3rd ed.

BHATTACHARYA, P. (Ed.): 'Properties of III–V quantum wells and superlattices' (EMIS series, IEE, London, 1996)

BHATTACHARYA, P. (Ed.): 'Properties of lattice matched and strained indium gallium arsenide' (EMIS series, IEE, London, 1993)

ADACHI, S. (Ed): 'Properties of aluminium gallium arsenide' (EMIS series, IEE, London, 1993)
'Properties of indium phosphide' (EMIS series, IEE, London, 1991)

2.10 References

1 SALE, T.E.: 'Gain calculations for strained InGaAs/GaAs quantum wells' *in* 'Vertical cavity surface emitting lasers' (Research Studies Press, Taunton; Wiley, New York, 1995), chap. 3
2 CORZINE, S.W., YAN, R.-H., and COLDREN, L.A.: 'Optical gain in III-V bulk and quantum well semiconductors' *in* ZORY, P.S. (Ed.): 'Quantum well lasers' (Academic Press, San Diego, 1993)
3 COLES, B.R., and CAPLIN, A.D.: 'The electronic structure of solids' (Arnold, London, 1976), p. 88
4 SHOCKLEY, W.: 'Electrons and holes in semiconductors' (Van Nostrand, New York, 1950)
5 KITTEL, C.: 'Introduction to solid state physics' (J. Wiley, New York, 1966), 1st ed., pp. 316–317
6 CARROLL, J.E.: 'Physical models of semiconductor devices' (Arnold, London, 1974), Section 6.5
7 ZORY, P.S. (Ed.): 'Quantum well lasers' (Academic Press, San Diego, 1993)
8 O'REILLY, E.P.: 'Valence band engineering in strained layer structures', *Semicond. Sci. Technol.* 1989, **4**, pp. 121–137
9 BERNARD, M.G.A., and DURAFFOURG, G.: 'Lasers conditions in semiconductors', *Phys. Status Solid.*, 1961, **1**, pp. 699-703
10 CARROLL, J.E.: 'Physical models of semiconductor devices' (Arnold, London, 1974), Section 3.5
11 KANE, E.O.: 'Thomas–Fermi approach to impure semiconductor band structure', *Phys. Rev.*, 1963, **131**, pp. 79–88
12 STERN, F.: 'Calculated spectral dependence in gain in excited GaAs', *J. Appl. Phys.*, 1976, **47**, pp. 5382–5386
13 HALPERIN, B.I., and LAX, M.: 'Impurity band tails in the high density limit. 1: Minimum counting methods', *Phys. Rev.*, 1966, **148**, pp. 722–740
14 BRINKMAN, W.F., and LEE, P.A.: 'Coulomb effects on the gain spectrum of semiconductors', *Phys. Rev. Lett.*, 1973, **33**, pp. 237-240
15 BEATTIE, A.R., and LANDSBERG, P.T.: 'Auger effect in semiconductors', *Proc. Roy. Soc. London*, 1959, **249**, pp. 16-29
16 DUTTA, N.K., and NELSON, R.J.: 'The case for Auger recombination in $In_{1-x}Ga_xAs_yP_{1-y}$', *J. Appl. Phys.*, 1982, **53**, pp. 74-92
17 BECK, A.H., and AHMED, H.: 'An introduction to physical electronics' (Arnold, London, 1968), chap. 12
18 CARROLL, J.E.: 'Rate equations in semiconductor electronics' (Cambridge University Press, Cambridge, 1985).
19 HECHT, E.: 'Optics' (Addison–Wesley, 1987), 2nd ed., Section 13.2
20 GOWAR, J.: 'Optical communication systems' (Prentice–Hall, Hemel Hempstead, 1993), Section 18.2

21 OSINSKI, M., and ADAMS, M.J.: 'Gain spectra of quaternary semiconductors', *IEE Proc. I*, 1982, **129**, pp. 229-236
22 WESTBROOK, L.D.: 'Measurement of dg/dN and dn/dN and their dependence on photon energy in $\lambda = 1.5$ μm InGaAsP laser diodes', *IEE Proc. J.*, 1986, **133**, pp. 135-142
23 COLDREN, L.A., and CORZINE, S.W.: 'Diode lasers and photonic integrated circuits' (Wiley, New York, 1995), Section 4.3.2
24 PEES, P., and BLOOD, P.: 'Derivation of gain spectra of laser-diodes from spontaneous emission measurements', *IEEE J. Quantum Electron.*, 1995, **31**, pp. 1047-1050
25 MORTON, P.A., ACKERMAN, D.A., SHTENGEL, G.E., KAZARINOV, R.F., HYBERTSEN, M.S., TANBUN-EK, T., LOGAN, R.A., and SERGENT, A.M.: 'Gain characteristics of 1.55-μm high-speed multiple quantum well lasers', *IEEE Photonic Technol. Lett.*, 1995, **7**, pp. 833-835
26 PETERMANN, K.: 'Calculated spontaneous emission factor for double heterostructure injection lasers with gain induced waveguiding', *IEEE J. Quantum Electron.*, 1979, **15**, pp. 566-570
27 ELSÄBER, W., and GÖBEL, E.O.: 'Einstein relations for gain guided semiconductor lasers', *Electron. Lett.*, 1983, **19**, pp. 335-336
28 WANG, J., SCHUNK, N., and PETERMANN, K.: 'Linewidth enhanced for DFB lasers due to longitudinal field dependence in the laser cavity', *Electron. Lett.*, 1987, **23**, pp. 715-716
29 AGRAWAL, G.P.: 'Effect of gain and index nonlinearities on single-mode dynamics in semiconductor-lasers', *IEEE J. Quantum Electron.*, 1990, **26**, pp. 1901-1909
30 WILLIAMS, K.A., GRIFFIN, P.S., WHITE, I.H., GARRETT, B., WHITEAWAY, J.E.A., and THOMPSON, G.H.B.: 'Carrier transport effects in long-wavelength multi-quantum-well lasers under large-signal modulation', *IEEE J. Quantum Electron.*, 1994, **30**, pp. 1355-1357
31 THOMPSON, G.H.B., KIRKBY, P.A., and WHITEAWAY, J.E.A.: 'The analysis of scattering in double-heterostructure GaAlAs/GaAs injection lasers', *IEEE J. Quantum Electron.*, 1975, **11**, pp. 481-488
32 PAN, X., OLESEN, H., and TROMBORG, B.: 'Influence of nonlinear gain on DFB laser linewidth', *Electron. Lett.*, 1990, **26**, pp. 1074-1076
33 YAMANAKA, T., YOSHIKUNI, Y., YOKOYAMA, K., and LUI, W.: 'Theoretical study on enhanced differential gain and extremely reduced linewidth enhancement factor in quantum well lasers', *IEEE J. Quantum Electron.*, 1993, **29**, pp. 1609-1616
34 OSINKSI, M., and BUUS, J.: 'Linewidth broadening factor in semiconductor lasers—an overview', *IEEE J. Quantum Electron.*, 1987, **23**, pp. 9-28
35 OE, K., and ASAI. H.: 'Proposal on a temperature insensitive wavelength semiconductor laser', *IEICE Trans. Electron.*, 1996, **E-79C**, pp. 1751-1754
36 PASCHOS, V., and SPHICOPOULOS, T.: 'Influence of thermal effects on the tunability of three-electrode DFB lasers', *IEEE J. Quantum Electron.*, 1994, **30**, pp. 660-667
37 LI, X., and HUANG, W.P.: 'Simulation of DFB semiconductor lasers incorporating thermal effects', *IEEE J. Quantum Electron.*, 1995, **31**, pp. 1848-1855

38 TALGHADER, J., and SMITH, J.S.: 'Thermal dependence of the refractive index of GaAs and AlAs measured using semiconductor multilayer optical cavities', *Appl. Phys. Lett.*, 1995, **66**, pp. 335–337
39 MCCAULLEY, J.A., DONELLEY, V.M., VERNON, M., and TAHA I.: 'Temperature dependence of the near infrared refractive index of silicon, gallium arsenide and indium phosphide', *Phys. Rev. B*, 1994, **49**, pp. 7408–7417
40 MENZEL, U., BÄRWOLFF, A., ENDERS, P., ACKERMANN, D., PUCHERT, R., and VOSS, M.: 'Modelling the temperature dependence of threshold current, external differential efficiency and lasing wavelength in QW laser diodes', *Semiconductor Sci. Technol.*, **10**, pp. 1382–1392
41 THOMPSON, G.H.B.: 'Physics of semiconductor laser devices' (Wiley, Chichester, 1980), section 3.2.4
42 MATTHEWS, J.W., and BLAKESLEE, A.E.: 'Defects in epitaxial multilayers. I: Misfit dislocations', *J. Cryst. Growth*, 1974, **27**, pp. 118–125
43 FALLAHI, M., DION, M., CHATENOUD, F., TEMPLETON, I.M., MCGREER, K.A., CHAMPION, G., and BARBER, R.: 'Low divergence electrically pumped circular grating surface emitting DBR laser on an InGaAs/GaAs structure', *Electronics Lett.*, 1993, **29**, pp. 1412–1414
44 FUCHS, G., HORER, J., HANGLEITER, A., HARLE, V., SCHOLZ, F., GLEW, R.W. and GOLDSTEIN, L. : 'Intervalence band absorption in strained and unstrained InGaAs multiple quantum well structures', *Appl. Phy. Lett.*, 1992, **60**, pp. 231–233
45 WRIGHT, A.P., GARRETT, B., THOMPSON, G.H.B., and WHITEAWAY J.E.A.: 'Influence of carrier transport on wavelength chirp of InGaAs/InGaAsP MQW lasers', *Electron. Lett.*, 1992, **28**, pp. 191–212

Chapter 3
Principles of modelling guided waves

3.1 Introduction

Without well controlled waveguiding to confine the light to the region where the electrons and holes recombine most strongly, diode lasers, and Bragg lasers in particular, would not have achieved their pre-eminence as signal sources for communication systems. The waveguide provides a stable platform for the electronic interaction even with changes in optical power. Understanding waveguiding, then, is an essential part of the study of DFB lasers. There are many fine texts on guiding of electromagnetic waves [1–3] and many methods for solving the problems of waveguiding ranging from finite-element and finite-difference to integral equations, series expansions, separation of variables, effective refractive index, beam-propagation methods and slab-guide methods. Chiang [4] has provided a review of many methods, giving over 200 references. This chapter therefore limits itself to the basic concepts of optical waveguiding appropriate for semiconductor lasers using slab guides which provides, in the authors' view, one of the most helpful and readily accessible techniques. This chapter, together with Appendixes 1 and 2, therefore provides only the basic theory, but unlike many previous texts it provides the reader, through the World Wide Web, with a versatile numerical package (using MATLAB) which can illustrate the principles of guiding in complex slab waveguides with both gain and loss and many layers, and so gives the reader a powerful tool to explore some of the effects of waveguiding. The reader who wishes to advance to discussions specifically on DFB devices can move rapidly through this chapter, referring back to it later as necessary.

Principles of modelling guided waves 77

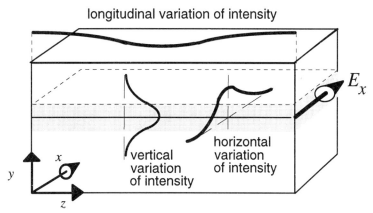

Figure 3.1 Vertical and horizontal variations

The terms transverse and lateral variations are also used in the literature but with insufficient consensus

3.1.1 Vertical and horizontal guiding

It is helpful to make a formal distinction between vertical and horizontal guiding, as indicated in Figure 3.1. The light is normally propagating along the 'longitudinal' Oz direction and the terms 'vertical' and 'horizontal' refer here to the two directions which are perpendicular to the Oz direction. Taking the normal orientation of a laser on a test bench, these terms avoid the confusion which can arise over the terms 'lateral' and 'transverse' which are also used. Horizontal variations will refer to changes in the Ox direction, taken as parallel to the junction or layer interfaces, while vertical variations will refer to changes in the Oy direction, taken as perpendicular to the junction or layer interfaces; Ox, Oy, Oz as usual form a right-handed set of axes. The term transverse is occasionally used here, but will in general mean both x and y directions, as for example in the transverse-field pattern $u(x, y)$. Longitudinal variations are for the present taken to be simple forward and reverse travelling waves of constant amplitude. These forward and reverse waves beat together to form standing waves with a period of $\lambda_m/2$, but their average intensity is taken to be uniform until later on in this work, when 'spatial hole burning' is considered, and the average longitudinal variations can be as shown in Figure 3.1.

3.1.2 Index and gain guiding

The reader is reminded of the principles of guiding light in a layer of slightly higher refractive index than its surrounds. Figure 1.5 summarised the concepts where one views the light, propagating along the layers of dielectric, as plane electromagnetic waves (see Appendix 1) with a cone of directions determining a pencil of optical rays. Each ray experiences total internal reflection (TIR) at shallow ray angles of incidence when reaching the outer boundaries of the guide. It is this TIR which confines the light to one main region which has an incrementally larger refractive index than the other layers. To keep the principles clear, only nonmagnetic material, with permeability μ_r sufficiently close to unity, is considered. The complex refractive index $n = n_r + jn_i$ and the complex relative permittivity $\varepsilon_r = \varepsilon_{rr} + j\varepsilon_{ri}$ of the material are found to be related by $n^2 = \varepsilon_r$.

For a plane wave in a uniform slab of such a material, the wave propagates as $\exp\{j(\omega t - \beta z)\} \exp gz$ where

$$(\beta + jg)^2 = \omega^2 \{\mu_0 \varepsilon_r(\omega) \varepsilon_0\} \tag{3.1}$$

so that $\beta = n_r \omega / c$ and $g = n_i \omega / c$.

There is a unique value of β and g for each ω given $\varepsilon_r(\omega)$. For 'index-guided' lasers there are several layers of different materials with a range of real optical refractive indices n_r so that the result of eqn. 3.1 has to be modified. The whole concept of a mode within a guide is that there is a single axial propagation coefficient β over the whole cross-section of the guide and one has to solve the guiding problem matching the field boundary conditions between each layer. Nevertheless, eqn. 3.1 is still most helpful provided that it is written in the form

$$(\beta + jg)^2 = \omega^2 \{\mu_0 \varepsilon_{r\,eff}(\omega) \varepsilon_0\} \tag{3.2}$$

with an effective value of complex relative permittivity representing the average effects of the guide through a complex effective refractive index $n_{eff} = (n_{r\,eff} + jn_{i\,eff}) = (\varepsilon_{r\,eff})^{1/2}$. For guides where there is no gain or loss, n_{eff} is purely real and $g = 0$.

In general, throughout this book, the gain is considered as a perturbation in the refractive index so that

$$n_{r\,eff} \gg n_{i\,eff} \quad \text{with} \quad n_{r\,eff} \simeq (\varepsilon_{rr\,eff})^{1/2} \text{ and } n_{i\,eff} \simeq \varepsilon_{ri\,eff} / 2n_{r\,eff} \tag{3.3}$$

Unless stated otherwise this is the usual assumption. For a 'good' laser, the guided transverse profile of the laser's optical fields should not significantly alter as the gain increases. The interaction between fields and charge carriers should then simply be proportional to gain without serious nonlinearities over a useful range of power levels, say up to around 5–10 mW in a laser for communication.

In early lasers, the horizontal guiding was often provided primarily by the gain [5]: 'gain-guided' lasers. From the point of view of the mathematical analysis, layers of complex permittivity with different gains provide reflections of the waves in a similar way to the reflections provided by layers of different real permittivity, but the physics of gain and index guiding rely on fundamentally different principles. Whenever a beam of light is confined to a finite area, the beam diffracts as it propagates and so expands after a few wavelengths unless there is some compensating mechanism to maintain the confinement of the optical energy to a constant area. In index guiding, the diffracted light is reflected back into the main stream in the correct phase through TIR, but in gain guiding this diffraction loss is compensated by the generation of light as it propagates through an active medium with gain. The narrower the guide, the greater the diffraction loss and the greater has to be the gain in the material. It follows that layers which only have changes in the optical loss cannot guide stable propagating light because there is no gain to compensate for the diffractive loss.

In gain guiding, the gain changes as the electron density changes and the complex interactions between light, current drive and electron density mean that lasers, which rely on weak gain guides, tend to be less stable under changes of operating conditions. Indeed, a laser can be particularly sensitive to varying combinations of index and gain guiding and in early lasers this led to instabilities or 'filamentation' in the transverse structure [6]. This chapter presupposes that good modern lasers are designed with strong dominant index guiding.

3.1.3 Effective area and confinement factor

The importance of this chapter comes in designing appropriate real refractive-index changes in layers of material so that the optical mode remains close to the region where the optical gain (i.e. where electrons and holes recombine most) is strongest. Significant parameters for laser design are the effective area $\mathcal{A}$ occupied by the optical fields and the confinement factor Γ. The discussion of Section 1.3.4 can be placed on a firmer footing as soon as one has discovered how the electric field propagates, found the effective refractive index $\varepsilon_{r\,eff}$ and found the transverse modal pattern $u(x, y)$. For example, if there are forward and reverse propagating fields with negligible gain,

$$E = F u(x, y) \exp\{j(\omega_0 t - \beta_0 z)\} + R u(x, y) \exp\{j(\omega_0 t + \beta_0 z)\} \quad (3.4)$$

then β_0 is determined for a given mode from eqn. 3.2. The phase difference between the forward and reverse waves is, say, $\arg(R/F) = \phi$.

80 Distributed feedback semiconductor lasers

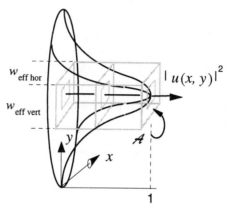

Figure 3.2 Effective-area schematic

Relationship between effective area $\mathcal{A}$ and field profile $u(x, y)$ using an equivalent 'rectangular' distribution (see also Figure 1.9). $\mathcal{A} \sim w_{\it{eff\,hor}} \times w_{\it{eff\,vert}}$. Appendix 11 gives estimates of these values. (An elliptical effective area might be used instead of a rectangular equivalent area.)

Then the local field intensity exhibits standing waves with a periodicity $\lambda_m/2$ of the form

$$E^*E = u(x, y)^* u(x, y) \{F^*F + R^*R + 2|FR|\cos(2\beta_0 z + 2\phi)\} \quad (3.5)$$

At first sight the periodic changes of intensity might be thought to react with the charge carriers and set up periodic changes to the optical gain in the active region. However, in general, these standing waves have little effect because diffusion of the charge carriers along the length of the guide smooths out and greatly reduces any impressed periodicity in the electron density. There are exceptions when the standing waves do have some effect, and these are discussed briefly at appropriate times. Over one whole optical wavelength the product term $|FR|\cos(2\beta_0 z + 2\phi)$ normally averages to have zero effect (but see Section 6.4.3).

The modal pattern $u(x, y)$ may be normalised to have a peak value $|u(x, y)|_{peak} = 1$ permitting definition of an effective area for the mode in the guide, see Figure 3.2, to be given from

$$\mathcal{A} = \int\int |u(x, y)|^2 \, dx \, dy \quad (3.6)$$

The equivalent average optical density S of the photons in this area may then be given from the stored electromagnetic energy:

$$(F^*F+R^*R)\iint \varepsilon_{rr\,eff}(\omega_0)\varepsilon_0\,|\,u(x,y)\,|^2\,dx\,dy = \hbar\omega_0\mathcal{A}S \quad (3.7)$$

The optical gain of the material may now be introduced from the imaginary part $n_i(x, y)$ of the refractive index. The gain is confined to the active region which occupies an area $\Gamma\mathcal{A}$ where the electrons and holes recombine. The density of photons S interacts with the material gain per unit distance g_m only within the active region and Γ is known as the confinement factor, given from

$$(F^*F+R^*R)\iint_{active\ area}\{\omega n_i(x,y)/c\}\varepsilon_0\,|\,u(x,y)\,|^2\,dx\,dy = \hbar\omega_0\Gamma g_m\mathcal{A}S$$

$$(3.8)$$

A specific value for the confinement factor Γ is determined by taking g_m to be the peak value of $\{\omega\,n_i(x, y)/c\}$ and taking $\mathcal{A}S$ as determined from eqn. 3.7.

A *slab* waveguide, which is used to confine plane electromagnetic waves in one dimension, provides an instructive starting point for optical waveguiding to determine $u(y)$ and $\varepsilon_{r\,eff}$ in one plane for a particular mode. MATLAB programs are used to discover the main effects of guiding and then an effective refractive-index model is used for estimating $u(x, y)$ in two dimensions.

3.2 The slab guide

3.2.1 TE and TM guided waves

In its simplest version, a slab guide can be formed from three slabs where material of lower refractive index surrounds material with the highest refractive index (Figure 3.3). Consider then a plane wave propagating in one direction where the E and H fields, together with the direction of propagation, form a right-handed set with $|E/H| = (\mu_r\mu_0/\varepsilon_r\varepsilon_0)^{1/2}$. The E field is assumed to be oriented along the plane of the junctions as in Figure 3.3a so that the E field has a component only in the horizontal direction, i.e. transverse to the junction; hence the nomenclature TE mode. It is found that the guide will only act as a guide when this plane wave is launched into it at some

82 *Distributed feedback semiconductor lasers*

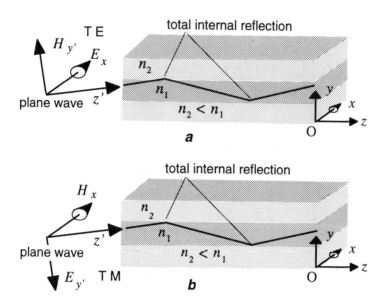

Figure 3.3 TE and TM modes

To guide plane electromagnetic waves, the slab guides are theoretically infinite in the *x* dimension

angle to the Oz axis so that Oz', the direction of launch of the plane wave, is not exactly parallel to the Oz axis. This means that the H_y field is not exactly parallel to the Oy axis and there is therefore a small component of longitudinal or H_z field (see also Figure A1.1). Provided that the angle of launching into the slab guide is shallow enough, then the plane wave is reflected through TIR at the boundaries. Here any one optical ray can be thought of as performing a slalom down the guide with components of E_x, H_y, and H_z defined by the right-handed set of axes shown in Figure 3.3. In general, there is a distribution, or *pencil*, of rays at a number of different shallow angles at any one position z. Appendix A1 discusses plane waves further.

The TM mode is very similar to the TE mode except that it is now the H_x field which is purely transverse to the junction (i.e. in the horizontal direction as in Figure 3.3*b*). The angle of launching of the plane wave means that there has to be an E_z field as well as an E_y field so that there is a longitudinal component of the E field.

3.2.2 Multilayer slab guides

Although a ray model can, in principle, give information about propagation velocities through finding the angle of the ray to the axis,

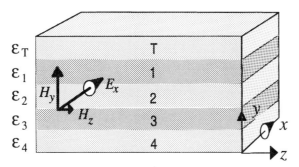

Figure 3.4 Multilayer slab guide (notation for TE excitation)

In the ideal theory the slab is infinite in the x direction with no x variations; in practice the x variations are usually over a significantly longer spatial distance than the y variations

it cannot give field or intensity distributions, which require solutions of Maxwell's equations. In the initial cases the 'horizontal' variations of the fields in the Ox direction are assumed to be sufficiently slow that one can approximate with variations only in the 'vertical' (or y) direction. The three-slab guide is a well studied problem and the reader can refer with confidence to alternative texts to obtain analytic solutions for symmetric and asymmetric three-layer guides with TE or TM modes [2,7]. With modern lasers, one has to have general methods which aid designing slab guides with many more layers. Figure 3.4 makes a start with five layers, labelling the top layer T, and this notation is used in Appendix 2 to discuss the formalisms for a computer program.

3.3 Wave equations for the TE and TM guided waves

The TE and TM classification shown in Figure 3.3 is discussed further in Appendix 1 but the outcome for either mode is that, within each uniform layer, the E field which is the main driving field for interaction with the electrons is given from the wave equation

$$\nabla^2 E - \mu_r \mu_0 \varepsilon_r \varepsilon_0 (\partial^2 E / \partial t^2) = 0 \qquad (3.9)$$

with $\quad H_{TE} = (j/\omega\mu_0\mu_r) \nabla \times E_{TE} \quad$ and $\quad E_{TM} = (-j/\omega\varepsilon_0\varepsilon_r) \nabla \times H_{TM}$

With the axes as shown in the figures, E_x is the key field component for the TE mode while H_x is the key field component for the TM mode. The boundary conditions are determined by seeking the field components which are tangential to the dielectric interfaces, as these components are always continuous across the interface:

$$E_x \text{ and } H_z = \frac{1}{j\omega\mu_r\mu_0}\frac{\partial E_x}{\partial y} \text{ are continuous for TE fields}$$

$$H_x \text{ and } E_z = -\frac{1}{j\omega\varepsilon_r\varepsilon_0}\frac{\partial H_x}{\partial y} \text{ are continuous for TM fields}$$

One notices that, for the TE fields with nonmagnetic material, where $\mu_r=1$ everywhere, the E_x field and its derivative are continuous across the boundaries. For the TM modes there will be slight discontinuities in E_y $[=-\beta H_x/(\omega\varepsilon_0\varepsilon_r)]$ at boundaries between layers of different permittivity whereas $\partial E_y/\partial y$ $[=j\beta E_z]$ is continuous.

An important caveat is required. Although interchanging E and H can be helpful to keep a similar mathematical formalism for the calculations of field profiles for TE and TM fields, care is needed when considering energy or power interactions. Optical gain, in either TE or TM waves, relies on the *electric field* interacting with those oscillating electronic dipoles formed by the recombining holes and electrons which can radiate into the waveguide in the semiconductor material. When it comes to the exchange of power and energy between electrons and photons, it is therefore essential to consider the interaction of the appropriate transverse E fields taken here as E_x for the TE mode and $E_y=(\beta H_x/\omega\varepsilon)$ for the TM mode in the important gain region.

3.4 Solving multislab guides

3.4.1 Effective refractive index

Appendix 2 discusses how to construct a numerical program to solve the propagation for multilayers, taking five layers as an example which can be extended to many more (e.g. 30) layers. The first numerical illustrations in this chapter are simply for a three-slab guide, as in Figure 3.5. The layers are lossless and only one wavelength (1.55 μm) is considered. The main guiding layer will be chosen to be sufficiently thick so that three modes can appear with different vertical patterns sandwiched between the slabs.

The method of solution is best appreciated by noting that the axial-propagation coefficient β and the resulting growth rate g for the wave have to be the same in all the different layers, and it is possible to find

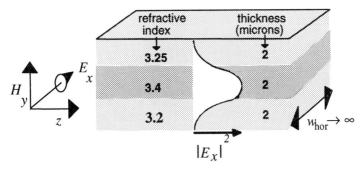

Figure 3.5 Three-layer guide

Although mathematically the slab has the dimension $w_{hor} \to \infty$, it is useful for the purposes of calculating a finite power to have an effective and finite horizontal width w_h

an effective permittivity ε_{eff} as in eqns. 3.2 and 3.3 with the fields propagating over the whole vertical and horizontal extent of the guide as in eqn. 3.4. The concept of an effective permittivity also can be approached through a weighted mean relative permittivity. By the definition of 'guiding', the optical fields have to vanish far enough away from the main guiding layer. Now consider a solution $E_x(y, z)$ to the wave equation

$$\{\partial^2 E_x(y, z)/\partial z^2\} + \{\partial^2 E_x(y, z)/\partial y^2\} + k_0^2 \varepsilon_r(y) E_x(y, z) = 0 \qquad (3.10)$$

with $k_0^2 = \omega^2 \mu_0 \varepsilon_0$.

$$H_y(y, z) = (j/\omega\mu_0\mu_r)\partial E_x(y, z)/\partial z \qquad (3.11)$$

Then one may define

$$E_{Ix}(z) = \int_{-\infty}^{\infty} \{E_x(y, z)\}\, dy \qquad (3.12)$$

Using

$$\int_{-\infty}^{\infty} \{\partial^2 E_x(y, z)/\partial y^2\}\, dy = 0 \qquad (3.13)$$

$$\{\partial^2 E_{Ix}(z)/\partial z^2\} + k_0^2 \varepsilon_{r\,eff} E_{Ix}(z) = 0 \qquad (3.14)$$

The weighted real effective relative permittivity is then given as

$$\varepsilon_{r\,eff} = \int_{-\infty}^{\infty} \varepsilon_r(y) E_x(y, z)\, dy \bigg/ \int_{-\infty}^{\infty} E_x(y, z)\, dy \qquad (3.15)$$

If the mode is asymmetric such that $\int_{-\infty}^{\infty} E_x(y, z)\, dy = 0$, then one has to consider different limits between, say, y_1 and y_2, values of y for which $\partial E_x(y, z)/\partial y$ vanishes so that with the new limits eqn. 3.13 still holds. The value of eqn. 3.15 lies in showing that the effective refractive index is a weighted mean, rather than any utility in computation.

For guides with no loss or gain, the real effective refractive index $n_{eff} = \sqrt{\varepsilon_{r\,eff}}$ always lies between the maximum refractive index of all the layers and the maximum refractive index of the two outermost layers. This gives a clear limited area of search for appropriate numerical solutions. It is instructive to consider the physical reasons for these limits. Vertical variations $\exp(-\gamma_y y)$ or $\exp(\gamma_y y)$ occur with $\gamma_y = k_0 \sqrt{(\varepsilon_{r\,eff} - \varepsilon_r)}$ when $\varepsilon_{r\,eff} > \varepsilon_r$ and such fields are said to be *evanescent*. To confine the optical power, the fields in both outer layers have to decay as the magnitude of y increases (i.e. be evanescent). However, when $\varepsilon_{r\,eff} < \varepsilon_r$ the vertical variations are of the form $\cos(\beta_y y + \phi)$ where $\beta_y = k_0 \sqrt{(\varepsilon_r - \varepsilon_{r\,eff})}$ and can be said to give vertical propagation. If the fields were to propagate vertically across the whole set of layers, the power would dissipate away from the guiding region. Equally, the modes cannot evanesce everywhere. There has to be at least one layer where vertical 'propagation' occurs with both outer layers giving an evanescent vertical variation with negligible energy at the far edges of the guide. These requirements then force the effective real relative permittivity to lie between the extreme actual values of the real relative permittivity.

3.4.2 Reflection coefficient calculation

The detailed algebra and mathematics for multislab guides is left to Appendix 2. The method of solution starts with a trial value of an effective refractive index. Then one considers a single frequency and the appropriate component of an appropriately evanescing electric field, varying vertically as $E_{top} \exp(-\gamma_y y)$, which is 'fed' into the top of the guide (i.e. decaying as $y \to \infty$ away from the guide region). All the fields, with the correct vertical propagation coefficients in each layer, are matched for the required field-continuity conditions at the boundaries. Finally, at the bottom of the guide one assumes a form of the field which has to evanesce as $\exp(+\gamma_y y)$ (i.e. decaying as $y \to -\infty$)

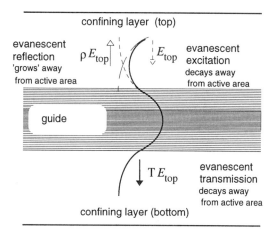

Figure 3.6 Evanescent excitation, reflection, transmission

> The transversely (vertical) propagating wave in the guide area is excited by the evanescent wave E_{top} leading to an evanescent wave T E_{top} at the bottom, but also a reflected wave ρE_{top} with the incorrect form of evanescence growing as one moves away from the guide, i.e. decaying towards the guide. The correct solution has $\rho = 0$

with a magnitude TE_{top} where T is a transmission factor. In general, with an arbitrary choice of the trial effective refractive index, all these boundary conditions will only match when there is a field $\rho E_{top} \exp(+\gamma_y y)$ 'scattered' back at the top of the surface where ρ is a reflection coefficient. Given that ρ is not zero, the solution is not correct and the fields are not confined (Figure 3.6).

By scanning (both real and imaginary parts if required) the complex relative permittivity $\varepsilon_{r\,eff}$ numerically, one finds distinct values where $|\rho|$ is a minimum and, for precisely the correct complex $\varepsilon_{r\,eff}$, one can, in principle, find $|\rho|=0$. At this value there is the correct form of evanescence of the field at the top and similarly the appropriate form of evanescence for the 'transmitted' field TE_{top} at the bottom. The field amplitudes decay to negligible values away from the guide, confirming the process of guiding. Looking for a minimum in the modulus $|\rho|$ of the reflection coefficient is found to be a more stable mathematical process (especially for lossy/gain guides with complex refractive indices) than looking for a maximum transmission or an impedance match at the bottom boundary [8]. Figure 3.7 shows typical results with a simple three-layer guide found using a MATLAB program **slabexec** in the directory **slab**. Note that the lowest-order mode has the highest value of effective refractive index (largest value of $\varepsilon_{r\,eff}$ in the scanning process) and the effective refractive index decreases as the number of nodes increases.

88 Distributed feedback semiconductor lasers

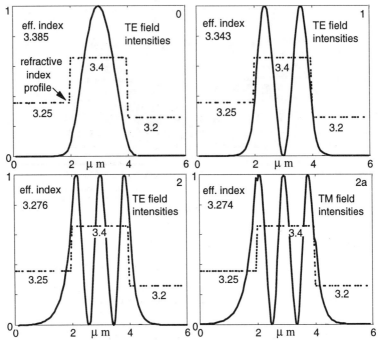

Figure 3.7 Three-layer TE/TM-mode intensities

Index changes here support three modes for both polarisations at 1.55 μm: note how the modal index is steadily decreasing from just below 3.4 to just above 3.25 as mode number increases. Only one TM mode shown. Note the field intensities are the transverse $|E|^2$ fields for both TE and TM modes because it is the E fields which interact with the electrons and holes to give the optical gain

The example shown in Figure 3.7 makes the layer sufficiently wide so that, at 1.55 μm wavelength, it supports three modes. The zero-order mode has no nulls in the E field, while the first and second order modes have one and two null intensities, respectively. The E-field intensities in the TM mode are very similar in this instance to those for the TE mode but, because of the different boundary conditions, there are small discontinuities in the field intensities at the dielectric interfaces. Only the second-order TM mode has been illustrated because the other orders show hardly any significant difference from the TE mode in their field profiles for this geometry. The modal indices are very similar for both TE and TM modes. One notices, particularly for the higher-order modes, how the fields evanesce most strongly in the layer of lowest refractive index so that the optical intensities do not penetrate as far into the bottom layer (index 3.2) as

Principles of modelling guided waves 89

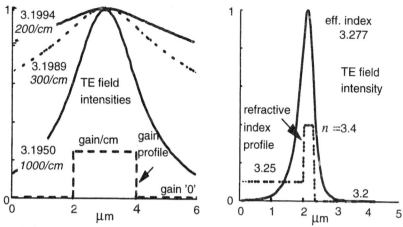

Figure 3.8 More three-layer examples: gain/index modes

Left-hand graph shows pure *gain* guiding with no index change ($n=3.2$). Field-intensity profile depends on gain. Figure in italics shows field gain per centimetre while the upper figure shows the effective refractive index. Note poor optical confinement compared with index guiding. On the right, a narrow *index* guide supports only one mode

they penetrate into the top layer (index 3.25). With loss/gain in the guide, the fields do not have strict nulls although the strong minima/maxima remain.

3.4.3 Gain-guiding example

With a 2 μm-thick layer (Figure 3.8) it is possible to obtain gain guiding, but one may notice how the optical fields are poorly confined, compared with the confinement that can be achieved with index guiding. Practical values of gains over large volumes of material are typically measured in the hundreds of inverse centimetres. Only with unusually strong gains can the confinement start to match that of the index guide. The change of field shape with increasing gain is a problem and can lead to unstable emission patterns with pure gain guiding. This instability of the intensity profile is made worse by photon/electron interactions with regions of strong optical intensity stimulating recombination of electrons and holes and reducing the gain locally, but this has not been illustrated here.

In general, a good guide should support only one mode over the range of frequencies of interest. To achieve this with the index guide of Figure 3.7 one needs to make the high-index region considerably narrower, as on the right-hand side of Figure 3.8. A comparable gain guide with a 0.3 μm-thick gain region would require a much larger value of gain than for the 2 μm guide because of the high diffraction losses. Trying out numerical examples on a multipurpose slab-layer program can rapidly give a sense of what can and cannot be achieved in terms of guiding.

3.5 Scaling

Classic early work [9] on three-slab guides showed how to formulate the guide in terms of normalised parameters which can be useful once experience has been gained in the required orders of magnitude for these normalised parameters. Here the required scaling laws are discussed briefly to show how to change a successful design of a guide to another wavelength.

Taking some appropriate vertical layer thickness d, a normalised parameter can be defined:

$$V(y) = k_0 d \sqrt{\{\varepsilon_r(y) - \varepsilon_{r\,eff}\}} \tag{3.16}$$

This enables eqn. 3.10 to be written in the form

$$\{\partial^2 E_x(y, z) / \partial (y/d)^2\} + V(y)^2 E_x(y, z) = 0 \tag{3.17}$$

For a symmetrical three-slab guide where the central thickness is $w_1 = d$ with relative permittivity ε_{r1}, then a V *number* can be defined from $k_0 d \sqrt{(\varepsilon_{r1} - \varepsilon_{r0})}$ and this permits scaling of the guide with straightforward rules. Typical V numbers lie within the range 1–3 to ensure a clear single 'zero'-order mode. With multilayers, one can see that, provided that $V(y)$ is maintained with y scaled with d, then the shape of the field profile has the same shape, at least on the scale of d. Narrowing the guide by a factor of two requires a doubling of $[\sqrt{\{\varepsilon_r(y) - \varepsilon_{r\,eff}\}}]$ which is determined by the differences of the relative permittivity.

3.6 Horizontal guiding: effective-index method

In taking into account the vertical guiding, the effects of any horizontal guiding have so far been ignored. Here the principles of the 'effective refractive index' method [10] are outlined to show one useful way to estimate for horizontal guiding without having to use a full three-dimensional numerical simulation. A convenient example for horizontal guiding combined with vertical guiding is given from the ridge waveguide shown schematically in Figure 3.9.

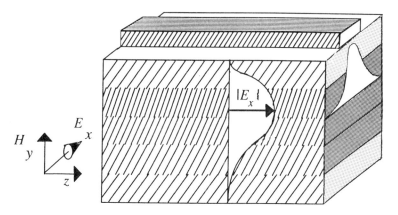

Figure 3.9   Vertical and horizontal guiding under a ridge guide

An approximate solution assumes that there is a slow variation in the Ox direction so that approximate fields $E_x(x, y, z)$ can be found at a succession of x values taking into account the different set of vertical layers at each x. Then return to the wave equation

$$\{\partial^2 E_x(x, y, z)/\partial x^2\} + \{\partial^2 E_x(x, y, z)/\partial y^2\} + \{\partial^2 E_x(x, y, z)/\partial z^2\} \\ + k_0^2 \varepsilon_r(x, y) E_x(x, y, z) = 0 \quad (3.18)$$

By integrating over the y-direction one reduces the problem to an equivalent slab guide in the x-direction. Define then:

$$E_{\mathit{eff}\,x}(x, z) = \int_{-\infty}^{\infty} E_x(x, y, z) \, dy \quad (3.19)$$

$$\varepsilon_{r\,\mathit{eff}}(x) E_{\mathit{eff}\,x}(x, z) = \int_{-\infty}^{\infty} \varepsilon_r(x, y) E_x(x, y, z) \, dy \quad (3.20)$$

These three equations then lead to

$$\{\partial^2 E_{\mathit{eff}\,x}(x, z)/\partial x^2\} + \{\partial^2 E_{\mathit{eff}\,x}(x, z)/\partial z^2\} + k_0^2 \varepsilon_{r\,\mathit{eff}}(x) E_{\mathit{eff}\,x}(x, z) = 0$$

One then is back again at a one-dimensional problem with horizontal changes of refractive index now dominating the calculation. The program for solving slab guides can be invoked again, using sufficient layers to approximate to the variations in $\varepsilon_{r\,\mathit{eff}}(x)$. The process could be iterated. In slab guides, effective-refractive-index methods for rectangular-shaped guides give varying accuracy [11,12]. In some sophisticated guides where the vertical and horizontal geometries are complex and similar, more sophisticated techniques may be required such as finite elements [13,14] or beam propagation [15,16].

3.7 Orthogonality of fields

An important point about the optical fields is that the modes are *orthogonal*. Here the distinction has to be made between guides which are lossless (or nearly lossless) and guides with gain. The lossless guide has modal field patterns $u_n(x, y)$ where (again normalising the pattern so that the peak values are unity)

$$\int\int u_n(x, y)^* u_m(x, y) \, dx \, dy = \delta_{mn} \mathcal{A}_n \quad \text{where} \quad \delta_{mn} = \begin{cases} 1 \text{ for } m=n \\ 0 \text{ for } m \neq n \end{cases} \quad (3.21)$$

The value of $\mathcal{A}_n$ gives the modal equivalent area. The conjugation is significant because it leads to what Marcuse refers to as being 'energy orthogonal' [17]. The energy storage can be split into separate modes with all the consequences of equipartition of modal energy for quantum statistics which is significant in mode counting for spontaneous emission (Appendix 9). There is an orthogonality (without the conjugation) for guides with strong loss or gain [1] but these are not true modes when it comes to counting modes for spontaneous emission, as is highlighted in [18].

3.8 Far fields

The spatial distribution within the semiconductor guide is only half of the problem of the design of a waveguide. It is often necessary to be concerned with the far-field pattern. The conventional far-field optical intensity is found to be proportional to the square of the modulus of the Fourier transform of the near field just in front of the laser. In general, the larger the radiating aperture, the narrower is the angle of the beam. The principles of finding the far field are well recorded in most textbooks on electromagnetic theory which have sections on aerials [18]. However, less well recorded in texts are the effects of the semiconductor surface which give rise to a special *obliquity* factor modifying this far-field pattern. The origins of this obliquity factor are given by the Huygen's radiation pattern, whereby any aperture, no matter how small, has a radiation pattern which is dependent on the obliquity and the simple far-field transform has to be multiplied by the obliquity factor, as summarised in Appendix 2. However, for a laser aperture this factor depends on whether the fields are TE or TM modes. These different factors come about because of the slight

Principles of modelling guided waves 93

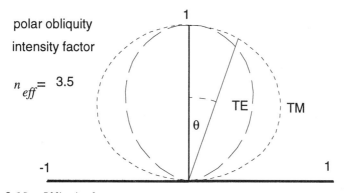

Figure 3.10 Obliquity factors
  Correction factors to far field dependent on direction of emission

differences of reflection at the surface of a dielectric, as shown in Figure 3.10, and also in Figures A2.4 and A2.5.

Figure 3.10 shows that there is a slight broadening of the far-field pattern for the TM mode in relation to the TE mode. The composite slab structure of the semiconductor guide is taken into account in estimating these obliquity factors through the effective refractive index, which is taken as the appropriate mean value of n to use in such calculations.

This section is concluded with two examples of the normalised far-field (TE) patterns for a 'narrow' (0.5 μm-wide) and a 'wide' (2 μm-wide) guide so as to indicate the orders of magnitude of the effects (Figure 3.11). One can see how the 0.5 μm-wide guide gives a far field with a half angle around 30° while the 2 μm guide gives a half angle around 15°. The confinement factor was briefly outlined in Chapter 1 and one may observe that, in the wider guide, the confinement factor is close to 1 and the effective refractive index is almost as if the wave was entirely within the central layer of highest-index material.

3.9 Waveguiding with quantum-well materials

Before leaving this chapter on waveguiding, a short word is needed about waveguides using material with quantum wells. In much of this book, such material is treated as forming an equivalent bulk material which has an enhanced gain but a reduced confinement factor and increased optical-gain saturation. Although totally ignoring the marvellous technology required to make these materials, this is

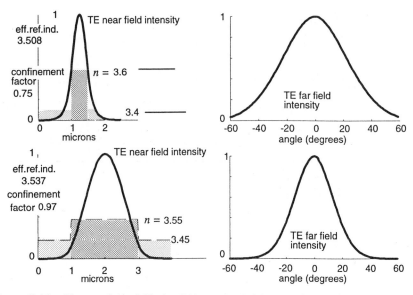

Figure 3.11 Near and far fields for different guided layers

adequate for understanding much of the performance of quantum-well lasers. In terms of designing waveguides to guide the light, this assumption is inadequate. In bulk-material lasers, the active region usually forms the main guiding layer within which the light rays are totally internally reflected. However, with quantum-well material a single well by itself would be inadequate to play the same role and even with multiquantum wells one is too concerned with getting their electrical composition right to have yet another parameter of confining the light to consider in the optimisation problem. The answer is to have separate-confinement heterolayers (SCH layers) which guide the light and also have the role of keeping any escaping electrons or holes close to their respective quantum wells. The equivalent slab-waveguide then has many layers (Figure 3.12). Because the quantum-well layers and barriers are so thin compared with a transverse wavelength, one can estimate appropriate averages of the refractive indices, but an automated slab-waveguide program can solve the problem rigorously. The index in the SCH layers may be graded because the composition of the separate confinement layers may be graded, as discussed in Section 2.7.4, and this could be modelled using the slab-waveguide techniques by a discrete step change but in many thin layers.

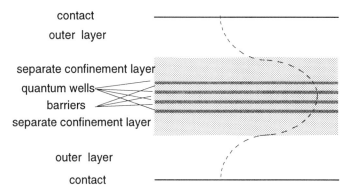

Figure 3.12 Schematic of multilayer guide

Outer layer, separate confinement layer, quantum wells and barriers, all with different refractive indices and gains/losses

3.10 Summary and conclusions

The chapter and its associated appendixes has discussed the mechanisms and mathematics required for optical guiding in semiconductor-slab guides so that one appreciates the importance of confining the light appropriately to the active region and can in Chapter 4 use the transverse-field patterns to estimate the effect of the grating on reflecting the optical fields. The MATLAB programs in the directory **slab** available through the IEE Web site are started by running **slabexec** . These give calculations for field profiles in slab layers along with the far-field patterns. The default settings in the program are for a five-layer slab guide. Trying out a few examples will provide some insight and sense for the guiding processes. One may observe how, with a thick guide, many modes are permitted and note how the effective refractive index changes with increasing mode number from a value close to the largest value of refractive index down to a value close to the smallest value of refractive index of the different layers. Occasionally, a result will come up where the field profiles show clearly that the guide is failing to guide. These results happen if the guide parameters are close to their cutoff values (i.e. the condition where the guide is only just guiding, or not guiding, as the case may be).

For the remainder of the book, it is assumed that the design values of the transverse-field patterns $u(x, y)$ are known and are sufficiently stable with respect to changes in gain and also, because the modal patterns are strongly gain guided, satisfy the orthogonality relationship (eqn. 3.21) with respect to any other modal pattern. The MATLAB programs provided permit the user to explore the validity of this for any particular slab-guide design by simply changing the material gain within the guide.

3.11 References

1 COLLIN, R.E.: 'Field theory of guided waves' (McGraw Hill, 1960)
2 ADAMS, M.J.: 'An introduction to optical waveguides' (Wiley, 1981)
3 SNYDER, A.W., and LOVE, J.D.: 'Optical waveguide theory' (Chapman and Hall, 1983)
4 CHIANG, K.S.: 'Review of numerical and approximate methods for the modal-analysis of general optical dielectric wave-guides', *Opt. Quantum Electron.*, 1994, **26**, pp. s113–s134
5 THOMPSON, G.H.B.: 'Physics of semiconductor laser devices' (Wiley, Chichester, 1980), Section 6.4
6 HESS, O., KOCH, S.W., and MOLONEY, J.V.: 'Filamentation and beam-propagation in broad-area semiconductor-lasers', *IEEE J. Quantum Electron.*, 1995, **31**, pp. 35–43
7 HAUSS, H.A.: 'Waves and fields in optoelectronics' (Prentice Hall, Englewood Cliffs, 1984), p. 175
8 ROZZI, T.E., and IN'TVELD, G.H.: 'Field and network analysis of interacting step discontinuities in planar dielectric waveguides', *IEEE Trans.*, 1979, **MTT-27**, pp. 303–309
9 KOGELNIK, H., and RAMASWAMY, V.: 'Scaling rules for thin fibre optical waveguides', *Appl. Opt.*, 1974, **8**, p. 1857
10 MUNOWITZ, M., and VEZZETTI, D.J.: 'Beam-propagation computations in one and 2 transverse dimensions', *Opt. Comm.*, 1993, **100**, pp. 43–47
11 LEE, J.S., and SHIN, S.Y.: 'On the validity of the effective-index method for rectangular dielectric wave-guides', *J. Lightwave Technol.*, 1993, **11**, pp. 1320–1324
12 CHIANG, K. S.: 'Performance of the effective-index method for the analysis of dielectric wave-guides', *Opt. Lett.*, 1991, **16**, pp. 714–716
13 SILVESTER, P.P., and FERRARI, R.L.: 'Finite elements for electrical engineers' (Cambridge University Press, 1996), 3rd ed.
14 KOSHIBA, M., and INOUE, K.: 'Simple and efficient finite-element analysis of microwave and optical wave-guides', *IEEE Trans.*, 1992, **MTT-40**, pp. 371–377
15 YOUNG, T.P.: 'CAD tools for optoelectronic subsystems', *GEC J. Res.*, 1994, **11**, pp. 110–121
16 HOEKSTRA, H.J.W.M.: 'On beam propagation methods for modelling in integrated optics', *Opt. Quantum Electron.*, 1997, **29**, pp. 157–171 (over 70 references)
17 MARCUSE, D.: ' Quantum mechanical explanation of spontaneous emission K-factor', *Electron. Lett.*, 1982, **18**, pp. 820–822
18 RAMO, S., WHINNERY, J.R., and VAN DUZER, T.: Fields and waves in communication electronics' (Wiley, New York, 1994), 3rd. ed., Section 12.4

Chapter 4
Optical energy exchange in guides

4.1 The classic rate equations

4.1.1 Introduction

There are two basic classical methods of modelling mathematically the operation of semiconductor lasers. The first method applies the concepts of photon/electron particle exchange, outlined in Chapter 2, where one discusses the rate of absorption and emission of photons along with the rate of recombination of holes and electrons, ensuring at each stage that there is a detailed balance between photon generation and electron/hole recombination leading to particle conservation and energy conservation. This is the standard rate-equation approach which is robust and well researched but can be difficult to apply when there are strong nonuniformities, and even more difficult when the phase of the electromagnetic field is important. For distributed-feedback lasers, both the phase of the field and nonuniformities are important and so one has to abandon the photon-rate equation in favour of an approach based on interactions between electromagnetic fields and the electric dipoles in an active optical medium. This chapter presents both methods and shows (mainly in Appendix 5) that they are, indeed, just different aspects of the same physics of energy conservation and are wholly compatible with one another.

The electromagnetic-field analysis is essential when the refractive index/permittivity changes periodically inside the laser. The chapter then concludes with an analysis of the coupled-mode equations which determine how a fraction of the forward-travelling field is coupled into the reverse-travelling field with a medium which has a periodic permittivity, i.e. the waveguide contains a Bragg grating.

98 Distributed feedback semiconductor lasers

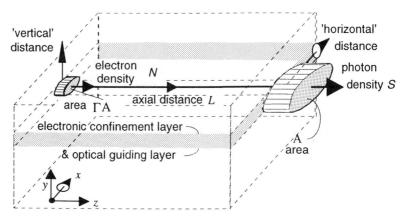

Figure 4.1 'Ideal' laser densities

Schematic uniform photon density S over an area $\mathcal{A}$ along whole length L of laser giving a volume $\mathcal{V} = L\mathcal{A}$. Similarly, electrons of uniform density N but occupying area $\Gamma \mathcal{A}$ giving a volume $\Gamma \mathcal{V}$

A natural starting point uses the time-honoured rate-equation approach. Understanding the technique is simplest with an ideal uniform laser where there is a waveguide which determines a fixed transverse pattern of the optical mode. In this model there is an approximately uniform density S of photons being guided along an area $\mathcal{A}$ (typically the order of a square micrometre or so) and length L (typically a few hundred micrometres), giving a total active volume $\mathcal{V} = L\mathcal{A}$. These photons S interact with a uniform density of electrons N in an active volume $\Gamma \mathcal{V} = L\Gamma \mathcal{A}$ as in Figure 4.1. As indicated in previous chapters, the photons are not as easy to confine to a sharply defined region in the way that the electrons and holes are confined to the active region through double heterojunctions. The concept of confinement factor along with effective area $\mathcal{A}$ was discussed in Section 3.1.3 and $\Gamma(<1)$ is then the ratio of the effective areas of equivalent uniform distributions of gain and photon density.

4.1.2 Rate of change of photon density

The rate of change of the photon number within such a laser cavity is given from an equation of the form

$$\frac{d(\mathcal{V}S)}{dt} = \Gamma \mathcal{V} G_m(N) S - \mathcal{V} A_m S - \frac{\mathcal{V}S}{\tau_p} + \beta_{sp}\Gamma \frac{\mathcal{V}N}{\tau_{sp}} \quad (4.1)$$

Here the concepts outlined in Section 2.3 are followed and that section should be read first. The first term on the right-hand side in eqn. 4.1 gives the rate of stimulated emission of photons by the electrons within the laser's active region, as determined by the material's optical power-gain coefficient $G_m(N)$ per unit time. The photon density is taken to be uniform so that any standing-wave effects have been neglected, as in eqn. 3.7. Only those photons within the region of electronic gain, of volume $\Gamma \mathcal{V}$, experience gain so the confinement factor Γ appears here. The second term is the optical-power absorption per unit time as determined by a material/guide parameter A_m. This term accounts for absorption of photons outside the gain region. [There can be an additional loss term ($-\Gamma b N S \mathcal{V}$) caused by IVBA, as discussed in Section 2.7.3, but it is omitted in this first analysis.]

The photon lifetime τ_p gives an estimate for the rate of escape of photons out of the cavity and can be determined in a very rough-and-ready manner by assuming that a proportion T of the photons get transmitted at each facet on each pass down the lasing cavity so that a photon has an approximate effective lifetime $\tau_p \sim L/(T v_g)$ within the cavity. More accurate estimates require the use of field calculations (see eqn. 4.10). In the final term, N/τ_{sp} gives the spontaneous emission caused by the electrons spontaneously recombining with holes, giving out a random emission of light over a range of frequencies and all directions. Only those frequencies and directions which can couple to the lasing mode can participate in the stimulated interaction (see Section 2.4.1) so that typically there is a coupling factor $\beta_{sp} \sim 10^{-5} \ll 1$. When the device is lasing strongly, one can often ignore this spontaneous contribution but it is essential around threshold to excite lasing action.

Obviously in eqn. 4.1 the volume $\mathcal{V}$ cancels but its inclusion at this stage makes it more clear as to where and why the confinement factor Γ occurs. The electron–hole recombination provides the stimulated emission over the smaller volume $\Gamma \mathcal{V}$ than the volume $\mathcal{V}$ occupied by the photons as a whole. The confinement factor is then a convenient 'single-parameter' method of modelling complicated transverse distributions of photons and electrons on the assumption that these transverse patterns are fixed by the guiding mechanisms and do not change significantly with optical intensity or carrier density. That assumption is not always true and then serious modifications have to be made. It is usually a good approximation when the photons are guided through well designed changes of the refractive index in the material structure rather than being guided by the electronically induced optical gain. Spontaneous emission could have a slightly different confinement factor from that for the stimulated gain because of the different distributions, but this complication is ignored here.

Later the net power gain $G(N)$ per unit time and net field gain $g(N)$ per unit distance along with their differentials with electron density need to be distinguished and related. Table 4.1 lists these forms as used in this book. The field gain $g(N)$ is typically measured in units of inverse distance (with modern laser material $g \sim 100\text{–}1000$ cm^{-1}—the lower values for bulk material and higher values for quantum well material) and will be zero when the electron density N is at a value known as the *transparency density* N_{tr}. Around the transparency value, it is often adequate to consider a linear change of gain with electron density and the differential of the gain with electron density $g'(N)$ becomes the important parameter.

4.1.3 Rate of change of electron density

The equivalent equation for the rate of change of electrons within the interaction region $\Gamma \mathcal{V}$ is next considered, and this is given from

$$\frac{\mathrm{d}(\Gamma \mathcal{V} N)}{\mathrm{d}t} = -\Gamma \mathcal{V} G_m(N) S - \frac{\Gamma \mathcal{V} N}{\tau_r} + \frac{I}{q} \tag{4.2}$$

where the first term on the right-hand side is the counterpart of the similarly placed term in eqn. 4.1 and gives the rate of removal of electrons from the interaction region $\Gamma \mathcal{V}$ caused by stimulated recombination associated with the material optical power gain. In a unit volume, N/τ_r gives the net spontaneous rate of recombination, representing this for simplicity by a single (carrier-dependent) rate constant $1/\tau_r$. The difference between the time constants τ_r and τ_{sp} is that the latter relates only to radiative recombination giving out photons that can interact with the lasing field while the former covers all recombinations including nonradiative recombinations.

This recombination in detail is made up from the following main factors:

$$N/\tau_r = AN + B(NP) + C'(N^2 P) + C''(NP^2) \tag{4.3}$$

The first term on the right-hand side is recombination proportional to N and is usually nonradiative; the second term is the bipolar radiative recombination of electrons, density N, and holes, density P. The last pair of terms in eqn. 4.3 represent Auger recombination where there is no radiation given out, as discussed in Chapter 2. The fact that,

Table 4.1   Relationships between different forms of gain

Subscript m for material power gain/loss per unit time

$G_m(N)$ Material optical power gain per unit time as a function of electron density

$G_m(N) = G'_m(N - N_{tr})$

           G'_m is differential of photon G_m with electron density

A_m Optical-material power absorption per unit time

Upper-case G for net power gains per unit time

$G = \Gamma G_m(N) - A_m$ Net power gain per unit time including confinement factor and loss

$G(N) = G'(N - N_{tr})$

           G' is differential of G with electron density

Lower case g for field gain/loss per unit distance

$g_m(N)$ Material optical-field gain per unit distance for fields $(dE/dz = g_m E)$

$g_m(N) = g'_m(N - N_{tr})$

           g'_m is differential of photon g_m with electron density

a_m Material optical-field-absorption coefficient per unit distance

Lower case g for net field gains per unit distance

$g = \Gamma g_m - a_m$ Net field gain per unit distance including confinement factor and loss

$g(N) = g'(N - N_{tr})$ g' is differential of g with electron density

Link between power gain/loss per unit time and field gain/loss per unit distance

$G = 2gv_g$ $G_m = 2g_m v_g$ $A_m = 2a_m v_m$

where v_g is group (energy) velocity in material; the factor of 2 arises because $S \propto E^2$

$dS/dt = GS = v_g\, dS/dz = 2v_g E\, dE/dz = (2v_g g) E^2$

---

under operational conditions, $N \sim P$ within the laser means that one can approximately simplify these terms into three major terms:

$$N/\tau_r \sim AN + BN^2 + CN^3 \tag{4.4}$$

The analysis will show shortly that once the laser is lasing, N remains roughly constant or clamped and a linear approximation (N/τ_r) with τ_r in the nanosecond range is a useful first approximation.

The final term in eqn. 4.2 is determined by the driving current I which pushes electrons at a rate I/q into the conduction band within the interaction region. This drive current is the source of electrons and, under equilibrium conditions, must balance the net rate of loss of electrons caused by all forms of recombination. In interpreting this 'ideal' rate equation, the current I which is referred to here is the current which may usefully increase the electron density in the active region and may be some 40% less than the external current being driven into the local contact. The reason for this loss is the sideways diffusion and spreading of the charge carriers as, for example, indicated in Figure 1.6. For simplicity this loss has been ignored in this analysis or alternatively lumped into an 'equivalent' nonradiative recombination, a term which also leads to loss of carriers without producing any useful photons. The convention here then is that I is the useful driving current into the active region. It can also be the case that, if the processing technology is exceptionally good, the net impedance of the diode's contacts is less than the intrinsic junction impedance so that the injected charge carriers are determined mainly by the external voltage along with the appropriate diode equation (Section 5.7), with the result that the electron density is more uniform.

In more sophisticated analyses, as in Appendix 4, it is also recognised that the electrons and holes, once they have been injected into the laser contacts, take time to reach the interaction region, especially with quantum-well material. This carrier-transport problem is of real significance when one is dealing with such materials in order to gain a high gigabit-per-second modulation (see Section 6.6). If the delay or loss of charge carriers is significant (Section 2.7.4), then there has to be another rate equation linking the actual external current I_{ext} with the driving current I.

Yet another complication arises because all the experimental evidence points to the optical gain reducing as the local optical intensity increases at a fixed carrier density. As discussed in Chapter 2, a useful practical approximation is to take the optical gain $G(N)$ to be represented by $G'(N - N_{tr})/(1 + \epsilon S)$, so that the gain saturates at high photon densities but is approximately linear in the electron density around the transparency value N_{tr}. It has been argued that this saturation of the gain arises because the electrons in the conduction band and the holes in the valence band take time (on the subpicosecond scale) to rethermalise as their densities within these bands are depleted locally around those energy levels which are participating most in the lasing process. This leads to spectral-hole burning of the gain when the laser is lasing strongly. To a first order, the redistribution of gain may be assumed to be immediate with the effect approximated simply by the gain-saturation parameter.

It has already been mentioned that, for a laser, $N \sim P$, an assumption which arises through the requirement of electrical neutrality. If the charge carriers did not neutralise one another in the gain region, then their high densities would give rise to such strong fields that other charge carriers would be forced into the region to ensure neutralisation. The time scale on which this effect occurs is also on the subpicosecond scale and is ignored in this work with no extra rate equation being included.

Eqns. 4.1 and 4.2 are the classic rate equations which, in various forms, have been discussed in great detail by many authors ([1–8] are a small selection). The variations that are possible include nonlinear recombination, multimodes with different thresholds and coupled lasers, along with the other factors already mentioned such as internal impedance and charge transport from contact to quantum well.

4.2 Some basic results from rate-equation analysis

4.2.1 Simplifying the rate equations

It can be helpful, whenever possible, to group key parameters together. For example, by definition of the 'gain region', the photons absorbed within the gain region can create electrons within the conduction band and these electrons are immediately then available to participate again in stimulated photon emission. The electrons can be 'recycled'. Photons absorbed outside the gain region do not create electrons that are available to participate in stimulated emission by the guided photons. One may then recognise that photons which escape from the facets and photons absorbed outside the gain region can essentially be grouped to give a net rate of loss of photons from the interaction process and this requires only one parameter:

$$\frac{S}{\tau'_p} = A_m S + \frac{S}{\tau_p} \tag{4.5}$$

$$\frac{dS}{dt} = \Gamma G_m(N) S - \frac{S}{\tau'_p} + \beta_{sp} \Gamma \frac{N}{\tau_{sp}} \tag{4.6}$$

$$\frac{dN}{dt} = -\frac{1}{\Gamma}\left(\frac{dS}{dt} + \frac{S}{\tau'_p} - \beta_{sp}\Gamma\frac{N}{\tau_{sp}}\right) - \frac{N}{\tau_r} + \frac{I}{q\Gamma\mathcal{V}} \tag{4.7}$$

There are two important ways of studying the results of these equations:

(i) small-signal analytic studies, and
(ii) large-signal dynamic computations.

Appendix 4 covers both features in more detail with MATLAB programs available to demonstrate some main features from normalised equations. A few of the main results are reported in this section along with a small-signal analytic examination of the equations in order to highlight some features that are either common to Fabry–Perot lasers and distributed-feedback lasers or can be quite different.

4.2.2 Steady-state results

In the steady lasing state with negligible spontaneous emission ($\beta_{sp} \to 0$), it is found from eqn. 4.6 that the electron density $N_0 \approx N_{th}$ ($>N_{tr}$) where

$$S_0 = \left(\beta_{sp}\Gamma \frac{N_0}{\tau_{sp}}\right) \bigg/ \left\{\frac{1}{\tau_p'} - \Gamma G_m(N_0)\right\} \tag{4.8}$$

with

$$\Gamma G_m(N_{th}) = \frac{1}{\tau_p'} \tag{4.9}$$

The light output is proportional to the internal photon density S_0, so that eqn. 4.8 shows that the light output is amplified spontaneous emission with the amplification factor determined by (loss − gain)$^{-1}$ = $\{(1/\tau_p') - \Gamma G_m(N_0)\}^{-1}$ which is equivalent to the round-trip gain referred to in Section 1.2. With small enough spontaneous emission, it becomes clear that the steady-state electron density $N_0 \approx N_{th}$, a constant value known as the threshold electron density where the round-trip gain becomes infinite and eqn. 4.9 holds.

As an aside, eqn. 4.9 is equivalent to eqn. 1.5 with $\Gamma = 1$ and $G_m = 2v_g g_m$, and provides a better estimate of the effective photon lifetime for the FP laser (than in Section 4.1.2) from

$$1/\tau_p' = 2v_g a_m + (v_g/L) \ln(1/\rho_{left}\rho_{right}) \tag{4.10}$$

With $N_0 = N_{th}$, the net steady-state photon output rate $\Phi_0 = \mathcal{V} S_0/\tau_p$ is also determined from

$$\Phi_0 = \frac{\tau'_p}{\tau_p}\left(\frac{I}{q} - \frac{\Gamma \mathcal{V} N_{th}}{\tau_r}\right) \quad (4.11)$$

If the gain saturation is included through a nonzero parameter ϵ, then its main effect on the static characteristics is to cause the electron density to rise slightly with increasing drive rather than be clamped at a fixed level once the lasing starts. The concept of a fixed photon lifetime in eqn. 4.10, while satisfactory for many FP lasers, is not so useful for distributed-feedback (DFB) lasers where the photon lifetime τ_p arises from distributed reflectivities so that a different effective photon lifetime has to be estimated. The electron density changes with distance and power level—so-called spatial-hole burning—and this in turn can alter the effective photon lifetime.

From eqn. 4.11 there is a threshold current $q\Gamma \mathcal{V} N_{th}/\tau_r$ required to create a threshold gain to start lasing. The concept of a threshold gain is common to all forms of oscillator, and lasers are no different in this respect. The analysis here assumes that there is no change in confinement factor with power level so this is equivalent to assuming stable optical-mode patterns. Above the threshold, in this ideal uniform laser, the electron density changes little provided that one does not wish to change the light output rapidly. The steady light output increases linearly with an increase in the steady current (Figure 4.2). If the Auger recombination, photon absorption and optical-scattering loss are all zero ($\tau_p = \tau'_p$) and there is no leakage or horizontal diffusion of the electrons, then all the electrons driven into the laser by the current above the threshold value are turned into photons. The internal differential efficiency of the laser in such circumstances can approach 100% and this forms one useful check on any numerical analysis.

If the spontaneous coupling factor β_{sp} is no longer negligible, then there is significant spontaneous light output below threshold where the device acts like a light-emitting diode and the threshold current for lasing is no longer sharply defined. Some research work envisages the concept of a sufficiently high value of β_{sp} that one might talk about the 'thresholdless' laser [9,10].

As noted in Section 1.2.3, a knowledge about only the magnitude of the gain cannot by itself determine the frequency of lasing; conse-

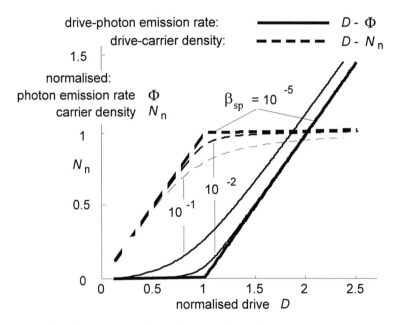

Figure 4.2 Idealised normalised light/current characteristics

Predicted from single-mode rate equations with uniform fields. The spontaneous coupling factor has a marked effect on removing a sharp threshold when it reaches values of a few percent. The normalised electron density N_n is almost constant once the stimulated emission dominates, though with gain saturation N_n will rise slightly with increasing drive

quently steady-state rate equations give no phase information and so cannot determine the precise optical frequency. However, dynamic rate equations will shortly be shown to give an indication of the *changes* of frequency when the laser is modulated.

4.2.3 Dynamic analysis

With sufficiently small dynamic changes S_1 in the photon density from a steady-state value S_0 and equivalently small changes N_1, I_1, G_1 also from steady-state values, the preceding results may be rearranged, ignoring products of any two small quantities such as $S_1 N_1$, to give

$$\frac{dS_1}{dt} = \Gamma G_{m1} S_0 \qquad (4.12)$$

$$\frac{dN_1}{dt} = -\frac{1}{\Gamma}\left(\frac{dS_1}{dt} + \frac{S_1}{\tau'_p}\right) - \frac{N_1}{\tau_r} + \frac{I_1}{q\Gamma\mathcal{V}} \qquad (4.13)$$

where $G_{m1} = G'_m N_1$ with G'_m giving the differential material (power) gain/unit time (Table 4.1). This pair of coupled equations becomes a classic second-order equation

$$\frac{d^2 S_1}{dt^2} + \frac{1}{\tau'_r}\frac{dS_1}{dt} + \omega_r^2 S_1 = G'_m S_0 \frac{I_1}{q\mathcal{V}} \qquad (4.14)$$

where

$$\frac{1}{\tau'_r} = G'_m S_0 + \frac{1}{\tau_r} \quad \text{and} \quad \omega_r^2 = \frac{G'_m S_0}{\tau'_p}$$

Note that the rates of change with time refer to modulation rates (typically microwave rates up to several gigahertz) and not to rates of change at optical frequencies (terahertz).

It is now helpful to list the basic results augmented by calculations from a large-signal analysis outlined at the end of Appendix 4.

(i) Eqn. 4.14 shows that there is a damped resonant response of the light output to the changes in the electron drive. Figure 4.3 demonstrates that gain saturation at the higher optical powers [putting $G_m \to G_m/(1+\epsilon S)$] has a strong dynamic effect on the damping of the resonance, and indeed without it one obtains an unrealistically low damping of the transient. Typically, the damping rate is measured on the nanosecond scale. The resonance frequency, referred to as the *photon–electron resonant frequency*, increases as the square root of the steady-state light output and takes values measured in the gigahertz range. This resonance, which provides an upper limit to the useful bandwidth of direct modulation, can be pushed up into the tens of gigahertz by using material with high optical gain.

Horizontal carrier diffusion in the plane of the layers, referred to in Section 1.3.3 but not considered in detail in this work, adds an additional rate of loss of electrons and changes the details of the damping of the electron–photon resonance. In general, these carrier losses increase the damping and reduce the peak magnification in the AM response (shown in Figure 4.5) with the maximum effect when the width of the optical mode is approximately the same as the diffusion length [11,12].

108   *Distributed feedback semiconductor lasers*

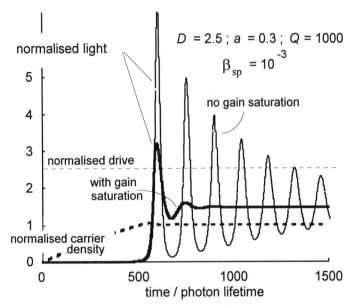

Figure 4.3   Normalised dynamic light/current characteristics

The electron–photon resonance is shown from a large-signal dynamic model with and without gain saturation: $Q=\tau_r/\tau_p$; D is normalised drive; $a=N_{tr}/N_{th}$ (Appendix 4). The drive current is switched on at $T=0$. The parameters are normalised as in Appendix 4. The normalising power might be ~10 mW with the photon lifetime measured in picoseconds. Because the light has to grow from statistically varying spontaneous emission, this emission will alter the turn-on delay of the stimulated output

(ii) The light output (or rate of photon emission) can increase rapidly only if the gain rises significantly above the threshold value. Similarly, the light output can decrease only if the gain falls below the threshold value so that dynamic changes of the light must be accompanied by changes in the gain. For small signals, these changes of gain and light are in quadrature (eqns. 4.12 and 4.13 and Figure 4.4) where the peak gain (peak electron density) gives the greatest rate of increase of light while the minimum gain gives the greatest rate of decrease of light.

Figure 4.4 also holds the key to the explanation of significant changes in frequency and spectral-line broadening when one turns lasers on and off by changing the gain. Recall eqn. 2.42, repeated here for convenience, relating the changes in the frequency $\Delta_g f$ with changes in that gain:

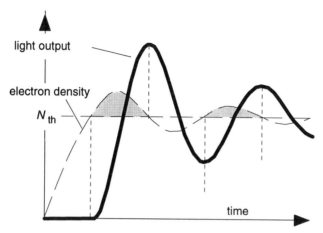

Figure 4.4 Light/electron-density/time

The figure shows the shaded regions where the electron density (and hence the gain) are above the threshold value and so the light output can increase with time. Because of a direct link between increases of electron density and associated decreases of refractive index, the electron-density curve also represents a dynamic shift of the central lasing frequency, typically measured in tens of gigahertz for gigahertz modulation rates

$$\Delta_g f/f \sim -\overline{\Delta n_r/n_r} \sim (\lambda_{fs}/2\pi)\, \overline{(\alpha_H \Delta g/n_r)} \qquad (4.15)$$

The average changes in gain with time, as predicted by the dynamics of a rate-equation analysis and eqn. 4.15, can give an estimate, at least for a Fabry–Perot laser, of the change in the frequency [13] with time. This dynamic change of frequency is called frequency 'chirp' or linewidth broadening [14] so that α_H has the obvious name of *linewidth broadening factor* or *linewidth-enhancement factor*. In communication lasers, this chirp can readily give more than 0.1 nm change in wavelength, thereby broadening the spectrum of the laser many times over the fundamental broadening required from Fourier analysis of a modulated optical carrier.

(iii) The significance of the resonance frequency is demonstrated in Figure 4.5 by a typical Bode plot of the power in the photodetector's load resistor ($\propto |\Phi_0|^2$) for a given modulation power ($\propto |I_1|^2$) with increasing modulation frequency ω_M. Above the photon-electron resonance frequency, this Bode plot shows a fall in output power at 40 dB/decade even with an idealised package with negligible parasitic capacitances, resistances or inductances and no effects from the internal charge-carrier transport within the laser.

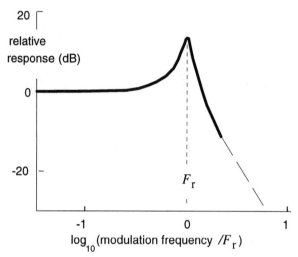

Figure 4.5   *Normalised Bode response*

Showing photon–electron resonance for small-signal AM response with $F_r \sim 3$ gigahertz and 0.25 ns effective recombination time. Fall off of response above the resonance is at 40 dB per decade, indicated by broken line

4.2.4 Problems of particle balance

The rate equations developed here do not require either detailed quantum theory or electromagnetic theory. They rely on the concepts of detailed balance in the numbers of particles being generated, absorbed or lost from a given region. They cannot give explicit information about the frequency of oscillation, even if one uses specific information about gain as a function of frequency, as in Figure 2.8. In the form presented here, the rate equations neither make allowance for nonuniformities in the field nor can they allow for the optical field's phase which creates important new effects in DFB lasers. Distributed-feedback lasers are sensitive to the nonuniformities of the electrical fields as well as relying on careful phasing of the fields with respect to the gratings, and so the simple rate equations cannot succeed. Further equations are also needed if there are significant delays or loss of charge carriers between their injection into the laser structure at the contacts and those carriers reaching the actual recombination regions. Despite their deficiencies, the rate equations (eqns. 4.1 and 4.2) are key foundations on which to build the relationships between the particle approach of the rate equations and electromagnetic theory.

4.3 Field equations and rate equations

4.3.1 Introduction

Whenever the phase of the optical field becomes an important effect in the physics, straightforward rate equations have to be augmented. For example, the modes of a Fabry–Perot laser have to be determined from arguments about the field and the round-trip phase. In a DFB laser the phase of the field, which is reflected back from the grating in relation to the phase of the forward-travelling field, is a crucial factor. A field analysis is therefore essential deriving, from Maxwell's equations, the propagation of the forward field combined with some distributed reflection, caused by the grating, of the reverse field into a forward component. The propagation of the reverse field and the coupling of the forward field will follow. It is also important to understand that energy conservation as exhibited by the photon rate equation of Section 4.1 is also demonstrable from Maxwell's classical equations. Appendix 5 does this necessary detailed work developing:

(i) the role of electronic polarisation in the optical medium as the key method of energy exchange between electrons and photons;
(ii) the connection between 'particle' exchange and electromagnetic energy conservation; and
(iii) the detailed calculations for forward and reverse waves in a guided medium with gain and periodic refractive indices.

This second part of the chapter presents a more tutorial account referring, for detail, to Appendix 5.

4.3.2 Wave propagation

From Chapter 3, one can assume that within a lossless waveguide of a laser there are forward- and reverse-travelling electric-field components at an (angular) frequency ω_0:

$$E = F u(x, y) \exp\{j(\omega_0 t - \beta_0 z)\} + R u(x, y) \exp\{j(\omega_0 t + \beta_0 z)\} \quad (4.16)$$

where the propagation coefficient β_0 is identical over the whole (x, y) cross-section of the guide. Within the permitted propagation ranges for a single mode, there is a unique β_0 for each ω_0. The transverse variation which gives the guided field pattern is given from a distribution $u(x, y)$ (as in Figure 4.6) normalised so that the modulus of the peak value is unity to define an effective area $\mathcal{A}$ as in eqn. 3.6.

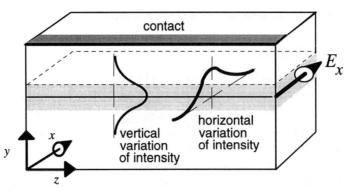

Figure 4.6 **TE mode**

Schematic field patterns guided by a laser structure with the E field mainly in the Ox direction. In many semiconductor diode lasers, the field pattern in the x-direction is broader than that in the y-direction.

The frequency and propagation coefficient are related by solving the guided-wave problem (see Chapter 3) to give a unique β for every ω from the equation:

$$\beta^2 = \omega^2 \{\mu_0 \varepsilon_{r\,eff}(\omega) \varepsilon_0\} \qquad (4.17)$$

In a uniform medium with a plane wave throughout, the effective permittivity $\varepsilon_{r\,eff} = \varepsilon_r$, the real relative permittivity of the medium. In a waveguide, the effective permittivity $\varepsilon_{r\,eff}$ also gives the value of the mean stored optical energy (electric and magnetic) per unit length (associated with the forward wave) from eqn. 3.7.

Now consider only a forward wave and assume that this varies in amplitude as

$$E_f = F(t, z)\, u(x, y)\, \exp\{j(\omega_0 t - \beta_0 z)\} \qquad (4.18)$$

Provided that the modulation frequencies within this amplitude change are small compared with the optical frequency, the transverse-field variation does not change markedly but there will be a 'wave packet', as in Figure 4.7, and the envelope of the fields associated with this wave packet within the laser will travel not at the phase velocity $v_p = \omega_0/\beta_0$ but at the group velocity $v_g = d\omega_0/d\beta_0$.

A straightforward mathematical description of the way in which these forward waves move in space and time is then given from the 'advection' equation:

$$\frac{\partial F}{\partial z} + \frac{1}{v_g}\left(\frac{\partial F}{\partial t}\right) = (\Gamma g_m - a_m) F \qquad (4.19)$$

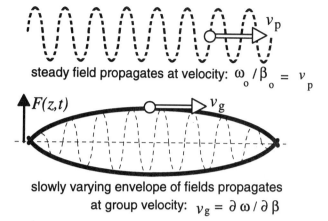

Figure 4.7 Waves and wave-packets

A steady wave pattern varies as $\exp(j\omega_0 t - j\beta_0 z)$ but a wave packet may be defined with the wave pattern inside the envelope varying as $\exp(j\omega_0 t - j\beta_0 z)$ and moving at the velocity ω_0/β_0 while the envelope $F(t, z)$ (thick line) moves at the group velocity $\partial\omega_0/\partial\beta_0$ not in general equal to ω_0/β_0

where the first two terms ensure that, with no gain or loss, the fields propagate with a fixed wave-packet envelope at the group velocity which is the velocity at which energy propagates and is slightly lower than the phase velocity in semiconductor materials. The terms on the right-hand side of eqn. 4.19 give the net optical *field* gain using the notation of Table 4.1. Here the field gain g_m is a function of the electron density N and may also saturate as the photon density S increases. This field gain is reduced by the confinement factor Γ and also by any absorption of the fields outside the gain region denoted by a_m. The term $(\Gamma g_m - a_m)$ appears regularly and, as shown in Table 4.1, will be replaced with the net field gain per unit distance g.

Next one has to include the excitation of the fields by spontaneous emission so that the advection equation for the forward wave becomes, for example:

$$\frac{\partial F}{\partial z} + \frac{1}{v_g}\frac{\partial F}{\partial t} = gF + i_{spf} \qquad (4.20)$$

The detailed evaluation of the spontaneous excitation, written here as i_{spf}, is discussed in Chapters 2 and 5 and Appendix 9. This emission is emitted over a broad range of frequencies and angles so that only a small fraction β_{sp} couples to the lasing mode in the narrow waveguide, with β_{sp} taking values $\sim 10^{-6}$. The spontaneous emission is strongly linked to the electron density; consequently there is a confinement factor Γ_{sp} defining the effective area $\Gamma_{sp}\mathcal{A}$ for the spontaneous emission over the guide with Γ_{sp} similar in magnitude and concept to the confinement factor Γ for the gain. Both β_{sp} and Γ_{sp} can be included in i_{spf} so as to avoid additional cumbersome terms. Semiconductor lasers operate typically with electron densities $\sim 10^{18}$ cm^{-3} and recombination rates $\sim 10^9/$s so that spontaneous recombinations emitting photons which couple to the lasing field occur at $\sim \beta_{sp}\ 10^{27}$ events per cubic centimetre second. As a rough example, in a 1.55 μm-wavelength laser with a volume of the order of 400 μm × 5 μm × 0.5 μm there are $\sim 10^{12}$ photons/s emitted spontaneously which will then give ~ 100 nW of spontaneous emission.

4.3.3 Decoupling of frequency and propagation coefficient

When it comes to Bragg gratings with a periodicity determined from $\cos(2\beta_b z)$, it will be found to be analytically helpful to consider waves varying as $E(t, z) \exp(j\beta_b z)$. However, at an arbitrary desired central frequency ω_0, the wave will not in general have a propagation coefficient of β_b to couple with the Bragg grating. This problem is overcome by 'removing' the link between the central frequency ω_0 and its associated propagation coefficient β_0 by writing $\beta_0 = \delta + \beta_b$ or defining $\delta = \beta_0 - \beta_b$. Now write eqn. 4.18

$$E_f = F(t, z) \exp(-j\delta z) u(x, y) \exp\{j(\omega_0 t - \beta_b z)\} \tag{4.21}$$

Assume that δ is sufficiently small so that all the previous assumptions leading to the advection equation are still valid. Now *redefine* $F(t, z)$ so that the new F is equal to the old value $F(t, z) \exp(-j\delta z)$. Similarly, *redefine* a new value of i_{spf} to be equal to the old value $i_{spf} \exp(-j\delta z)$. The phase factor $\exp(-j\delta z)$ cancels, to leave eqn. 4.20 as:

$$\frac{\partial F}{\partial z} + \frac{1}{v_g}\frac{\partial F}{\partial t} = gF - j\delta F + i_{spf} \tag{4.22}$$

with $E_f = F(t, z)\ u(x, y) \exp\{j(\omega_0 t - \beta_b z)\}$. This has now allowed the fields to have a central frequency ω_0 and at the same time to appear to have a central propagation coefficient β_b as if the wave operated at the Bragg frequency.

Because the spontaneous emission is random, the phase factor $\exp(-j\delta z)$ multiplying this emission is found to be of no consequence. The key random properties of spontaneous emission are described through complex autocorrelations (see Section 5.2) where terms like the time average of $|<i_{sp\,f}(t_1, z_1)^* \, i_{sp\,f}(t_2, z_2)>|$ determine the magnitude and bandwidth of the spontaneous emission. It can be seen that phase factors do not affect the value of such complex correlations.

The photon density per unit guide length associated with the forward power flow is referred to here as S_f and can, by normalising the field values appropriately, be related to $F^*F = S_f$ so that eqn. 4.22 is first multiplied by F^* and then has the complex conjugate of the result added to arrive at a travelling-wave rate equation for the photon density per unit length associated with the forward field:

$$\frac{\partial S_f}{\partial z} + \frac{1}{v_g}\frac{\partial S_f}{\partial t} = 2gS_f + i_{sp\,f} \qquad (4.23)$$

The spontaneous term is given from: $s_{sp\,f} = i_{sp\,f}F^* + i_{sp\,f}^*F$.

The reverse amplitude R may also be redefined (as F was redefined) to obtain a reverse-propagating wave varying as $E_r = R(t, z) \, u(x, y) \, \exp\{j(\omega_0 t + \beta_b z)\}$ where

$$-\frac{\partial R}{\partial z} + \frac{1}{v_g}\frac{\partial R}{\partial t} = gR - j\delta R + i_{sp\,r} \qquad (4.24)$$

paired with eqn. 4.22. Similarly, one can associate a photon density per unit guide length S_r with this reverse-travelling field and obtain the equation comparable to eqn. 4.23:

$$-\frac{\partial S_r}{\partial z} + \frac{1}{v_g}\left(\frac{\partial S_r}{\partial t}\right) = 2gS_r + i_{sp\,r} \qquad (4.25)$$

with $s_{sp\,r} = i_{sp\,r}R^* + i_{sp\,r}^*R$. The forward and reverse spontaneous emissions are uncorrelated and so, if there is no coupling of the forward and reverse waves through reflections or scattering, these two waves are independent and uncorrelated. Nonlinear mixing inside the laser can, of course, alter this result.

In some papers one may find that the advection equations are written in terms of the power flow $P_{f(r)}$ rather than in terms of $S_{f(r)}$, the photon density per unit length. Provided that the group velocity is sufficiently constant over the packet of frequencies, power and photon density per unit length are linearly related by the group velocity so that

$$P_{f(r)} = \hbar \omega v_g S_{f(r)} \mathcal{A} \tag{4.26}$$

It follows that, apart from the dimensions and calculation of the spontaneous excitation, the advection equations for $P_{f(r)}$ are the same as for $S_{f(r)}$.

4.4 Field equations with a grating

4.4.1 The periodic permittivity

The periodic change of refractive index or Bragg grating within a laser has been met in Chapter 1 and it is convenient at this point, while discussing field equations, to develop the coupled-mode field equations for a laser containing a Bragg grating. The whole point of finding the travelling-wave fields with a central frequency ω_0 and an independent propagation coefficient β_b will now become apparent.

Suppose that inside the laser material there is a periodic permittivity

$$\varepsilon = \{\varepsilon_{rr}(x, y) + \varepsilon_{rfz}(x, y, z)_{\cos(2\beta_b z)}\} \tag{4.27}$$

The subscript *rfz* on the permittivity component indicates that it will couple the reverse and forward waves in the z direction through its periodicity denoted by the subscript $\cos(2\beta_b z)$ denoting that its lowest-order spatial period is $\Lambda = \pi/\beta_b$, where Λ is a half wavelength in the medium at the Bragg frequency. It is convenient to refer to Λ as the Bragg period. In principle, the periodic term ε_{rfz} may be real or complex, with the imaginary part depending on whether there is any periodic component of the gain.

Historically, Bragg gratings grown within diode lasers had the periodic part of their permittivity as essentially real because the grating is most easily created by etching the cladding layers above or below the gain region followed by regrowth of materials with different permittivity, as outlined in Chapter 1. In practice, the axial variation is not sinusoidal as possibly suggested from eqn. 4.27 but is determined by the etch pit that is defined to make the grating, as indicated schematically in Figure 4.8. The magnitude of the step change of refractive index depends on the regrowth material. With sufficiently deep etching down into the gain region (Figure 4.8c), it is possible to generate a periodic gain as well as a periodic permittivity, so that ε_{rfz} has an imaginary component. The reader may recall that in Section 3.1 it was argued that diffusion of the charge carriers smoothed out variations in the gain which could be caused by optical standing waves

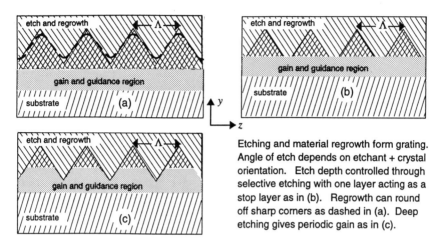

Figure 4.8 *First-order grating inside laser (side-view cross-section)*

with a similar periodicity. However, here the etching of the material and the refill process limit the carrier diffusion so that the rapid variations of gain can be embedded into the laser. Such technology has only recently become available [15].

At each y value, Fourier analysis over one period of length Λ finds the Fourier amplitude of the permittivity component varying as $\cos(2\beta_b z)$, where the component is given from

$$\varepsilon_{rf1}(x, y) = (2/\Lambda)\left\{\int_{\text{one period }\Lambda} \varepsilon_{rfz}(x, y, z)_{\cos(2\beta_b z)} \cos(2\beta_b z) \, dz\right\} \quad (4.28)$$

so that the relative permittivity variations of interest can be written as

$$\varepsilon_r = \varepsilon_{rr}(x, y) + \varepsilon_{rf1}(x, y) \cos(2\beta_b z) \quad (4.29)$$

4.4.2 Phase matching

As indicated in Section 4.3, wave packets of fields travelling in the $+Oz$ direction vary as $F(t, z)\, u(x, y)\, \exp\{j(\omega_0 t - \beta_b z)\}$ and when travelling in the $-Oz$ direction vary as $R(t, z)\, u(x, y)\, \exp\{j(\omega_0 t + \beta_b z)\}$. The net electric field E is then proportional to

$$E = F(t, z)\, u(x, y)\, \exp\{j(\omega_0 t - \beta_b z)\} + R(t, z)\, u(x, y)\, \exp\{j(\omega_0 t + \beta_b z)\} \quad (4.30)$$

To understand the key features of the wave coupling, take the full wave equation in one dimension, without gain or loss or spontaneous emission, and for simplicity ignore any dispersion, i.e. take the permittivity to be independent of frequency so that phase and group velocity are the same and take the parameter $\delta=0$ so that $\beta_b^2 = \omega_0^2\,\varepsilon_{rr}\varepsilon_0\mu_0$, to give:

$$\frac{\partial^2 E}{\partial z^2} - \{\varepsilon_{rr} + \varepsilon_{rf1}(x,y)\cos(2\beta_b z)\}\varepsilon_0\mu_0 \frac{\partial^2 E}{\partial t^2} = 0 \qquad (4.31)$$

Insert the fields of eqn. 4.30 into eqn. 4.31, assuming slowly varying changes relative to the propagation coefficient or frequency so that only first-order derivatives in time and space are retained (Appendix 5 shows how this can be done in more detail). The 'slow' variations in the forward wave F are placed on the left-hand side to give the *first-order* (i.e. ignoring terms such as $\varepsilon_{rf1}\,\partial F/\partial t$) rates of change of field:

$$2j\beta_b u(x,y)\frac{\partial F}{\partial z} + 2j\omega_0\varepsilon_{rr}\varepsilon_0\mu_0 u(x,y)\frac{\partial F}{\partial t}$$
$$\simeq -\omega_0^2\varepsilon_{rf1}(x,y)\cos(2\beta_b z)\varepsilon_0\mu_0 E \exp(-j\omega_0 t + j\beta_b z) \qquad (4.32)$$

The terms on the right-hand side couple strongly to the terms on the left-hand side only if both sides vary with the same spatial frequency. This is called the *phase-matching condition*. Any higher-order Fourier components (HOFCs) such as $\varepsilon_0\mu_0 E\exp(-j\omega_0 t + j3\beta_b z)$ change so rapidly in space that any terms they excite over the first half of a wavelength will be excited in the opposite phase over the next half wavelength and the excitation effectively cancels. It is found that elimination of HOFCs varying as $\sim\exp(jN\beta_b z)$, $N=\pm 1; \pm 2$ etc., is a very good approximation, when integrating eqn. 4.31 over a few wavelengths. Taking E as in eqn. 4.29 and writing

$$\cos(2\beta_b z) = \tfrac{1}{2}\{\exp(j2\beta_b z) + \exp(-j2\beta_b z)\} \qquad (4.33)$$

the elimination process discussed above selects the slowly varying envelope of the reverse fields $R(z,t)$ and allows the variations $\exp(j\omega_0 t + j\beta_b z)$ to be cancelled to give

$$\omega_0^2 \varepsilon_{rf1}(x,y)\cos(2\beta_b z)\varepsilon_0\mu_0 E\exp(-j\omega_0 t + j\beta_b z)$$
$$\rightarrow \tfrac{1}{2}\omega_0^2\varepsilon_{rf1}(x,y)\varepsilon_0\mu_0 R(t,z)u(x,y) \qquad (4.34)$$

Eqn. 4.32 then simplifies into

$$u(x,y)\frac{\partial F}{\partial z} + u(x,y)\frac{1}{v_g}\frac{\partial F}{\partial t} = -j\tfrac{1}{4}(\omega_0/v_g)\{\varepsilon_{rf1}(x,y)/\varepsilon_{rr}\}R(t,z)u(x,y) \qquad (4.35)$$

The transverse variation $u(x, y)$ has been left in place deliberately to emphasise that the only components which couple to the normal mode $F(t, z)\, u(x, y)$ also have to have the same normal mode field pattern $u(x, y)$. As stated earlier, the transverse-field pattern is unique to a guided mode and orthogonal to any other modal pattern $u_n(x, y)$ (eqn. 3.21). To find the magnitude of the correct modal interaction, start by writing

$$\tfrac{1}{4}(\omega/v_g)\{\varepsilon_{rf1}(x,y)/\varepsilon_{rr}\}u(x,y) = \kappa_{rf}u(x,y) + \sum_n \kappa_n u_n(x,y) \quad (4.36)$$

and then using modal orthogonality as in eqn. 3.21 and approximating $(\omega/v_g) \sim \beta_b$:

$$\kappa_{rf}\mathcal{A} \approx \iint \tfrac{1}{4}\beta_b\{\varepsilon_{rf1}(x,y)/\varepsilon_{rr}\}\,|u(x,y)|^2\,dx\,dy \quad (4.37)$$

where

$$\mathcal{A} = \iint |u(x,y)|^2\,dx\,dy$$

The reader is reminded that the periodic component $\varepsilon_{rf1}(x, y)\cos(2\beta_b z)$ is found from Fourier analysis of the actual patterns of etching and regrowth and eqn. 4.37 is then the second part of a calculation to find a coupling coefficient κ_{rf} describing how forward and reverse waves couple. This is the calculation which is performed approximately with the program **slabexec** in directory **slab** on selecting a Bragg grating while calculating the waveguide's field patterns.

Now use modal orthogonality as above to remove the transverse variation $u(x, y)$, and make the same assumptions as in eqn. 4.22 to add in spontaneous emission and to include the confinement factor and attenuation in the gain. Include also the change between the Bragg frequency and the operating frequency determined by the detuning parameter $\delta = \beta_0 - \beta_b$; then the periodic permittivity changes eqn. 4.22 to read

$$\frac{\partial F}{\partial z} + \frac{1}{v_g}\frac{\partial F}{\partial t} - gF + j\delta F = j\kappa_{rf}R + i_{spf} \quad (4.38)$$

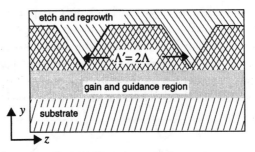

Figure 4.9 Second-order grating (side-view cross-section)

It is sometimes technologically easier to place the etch pits at twice the first-order distance shown in Figure 4.8

The equivalent equation for the reverse wave is found by reversing the sign of z:

$$-\frac{\partial R}{\partial z}+\frac{1}{v_g}\frac{\partial R}{\partial t} - gR+j\delta R = j\kappa_{fr} F + i_{sp\,r} \qquad (4.39)$$

With the phase of the grating chosen relative to the spatial origin as in eqns. 4.29 and 4.31, $\kappa_{fr}=\kappa_{rf}=\kappa$. Chapter 5 revisits these equations, gives an alternative perspective on their physics and highlights also the meaning of the factor j in front of the coupling coefficient κ.

4.4.3 Second-order gratings

Lithographic constraints on defining the etch pits means that it is sometimes technologically easier to space these pits at twice the first-order distance apart (Figure 4.9), giving a grating period $\Lambda'=2\Lambda$ known as a *second-order* grating (Section 1.6). Then the permittivity is expandable in the form

$$\varepsilon(x,y,z)=\varepsilon_{rr}(x,y)+\{\varepsilon_{rf1}(x,y)\cos(\beta_b z)+\varepsilon_{rf2}(x,y)\cos(2\beta_b z)+\cdots\} \qquad (4.40)$$

where $\beta_b=\pi/\Lambda$ as before. However, $\varepsilon_{rf1}(x,y)\cos(\beta_b z)$ does not have the correct period to cause direct coupling between the forward and reverse waves. The direct coupling requires the second-order term with a periodic permittivity varying as $\cos(2\beta_b z)$ extracted through spatial Fourier analysis of the full variation of permittivity:

$$\varepsilon_{rf2}(x,y)=(2/\Lambda')\left\{\int_{\text{one period }\Lambda'} \varepsilon_{rfz}(x,y,z)\cos(2\beta_b z)\,\mathrm{d}z\right\} \qquad (4.41)$$

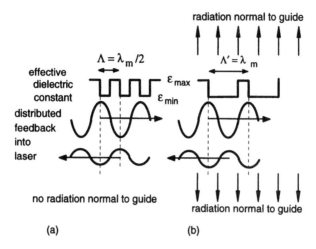

Figure 4.10 Feedback and scattering properties of first- and second-order gratings

    a First-order grating
    b Second-order grating
    Symmetrical radiation is assumed for second-order gratings here but reflective layers would modify this assumption

Given appropriate designs of the grating, this component in the spatial Fourier analysis of the grating can still have a significant amplitude. Higher-order gratings where the etch pits are spaced at $p\Lambda$ ($p=3$, etc.; see Table 1.1) are only occasionally used.

The mathematics of the coupling for second-order gratings becomes more complicated and the reader may be advised on a first reading to skip lightly over the ideas rather than burrow into the details. The key results are sketched in Figure 4.10, indicating sideways radiation occurring as well as the forward and reverse wave coupling. The key equations will be eqns. 4.48 and 4.49 which will contain the additional loss and coupling terms caused by the radiation.

Assuming only vertical (y) and axial (z) variations, the required wave equation is

$$\frac{\partial^2 E}{\partial z^2}+\frac{\partial^2 E}{\partial y^2} - [\varepsilon_{rr}(x, y) +\{\varepsilon_{fr1}(x, y) \cos(\beta_b z)$$

$$+\varepsilon_{fr2}(x, y) \cos(2\beta_b z)\}]\varepsilon_0\mu_0 \frac{\partial^2 E}{\partial t^2}=0 \quad (4.42)$$

As for the first-order grating, the terms in each of $\exp(j\omega_0 t \pm j\beta_b z)$ are isolated and equated to zero, but because of the term $\varepsilon_{fr1}(x, y) \cos(\beta_b z)$

interacting with the E fields, there also has to be an additional radiation-field term $E_{rad}(t, z)\, v(x, y)\, \exp(j\omega_0 t)$ having an axial propagation coefficient $\beta_z \sim 0$. To fit Maxwell's equations, this field has a 'vertical' propagation coefficient $\beta_y \sim \beta_b$ away from the interaction region and thereby radiates some optical power at approximately 90° to the structure [16,17]. The required E fields are now written as

$$E = F(t, z)\, u(x, y)\, \exp(j\omega_0 t - j\beta_b z) + R(t, z)\, u(x, y)$$
$$\times \exp(j\omega_0 t + j\beta_b z) + E_{rad}(t, z)\, v(x, y)\, \exp(j\omega_0 t) \qquad (4.43)$$

where $u(x, y)$ and $v(x, y)$ are approximately real. The magnitude of $E_{rad}(t, z)$ is estimated directly by putting $\partial^2/\partial z^2 \to 0$ with $\partial^2/\partial t^2 \to -\omega_0^2$. For compatibility with Maxwell's wave equation (eqn. 4.42) and the phase-matching condition around $\beta \sim 0$, one has approximately:

$$\left\{ \frac{\partial^2 v(x, y)}{\partial y^2} + \omega_0^2 \varepsilon_{rr} \varepsilon_0 \mu_0 v(x, y) \right\} E_{rad} \simeq$$

$$-\tfrac{1}{2} \omega_0^2 \varepsilon_0 \mu_0 \{F(t, z) + R(t, z)\} u(x, y)\, \varepsilon_{rf1}(x, y) \qquad (4.44)$$

The small variations in the permittivity have relatively little effect on the transverse propagation so that the approximation is made that $\omega_0^2 \varepsilon_{rr} \varepsilon_0 \mu_0 \sim \beta_b^2$. The physical clue to an approximate solution for eqn. 4.44 comes from considering an infinitesimally thin-layer excitation $\delta(y - y')$ so that the fields radiate symmetrically and vertically (y direction) on either side as plane waves with fields $\propto \exp\{-j\beta_b |(y - y')|\}$ as they propagate away from the layer. The x variation is ignored and, on integrating over the whole range of layers at each y',

$$v(y) E_{rad} =$$

$$(j/4)\beta_b \left(\left[\int_{-\infty}^{y \geq y'} \frac{\varepsilon_{rf1}(y')\, u(y')}{\varepsilon_{rr}} \exp\{-j\beta_b(y - y')\}\, dy' \right] \right.$$

$$\left. + \left[\int_{y \leq y'}^{\infty} \frac{\varepsilon_{rf1}(y')\, u(y')}{\varepsilon_{rr}} \exp\{j\beta_b(y - y')\}\, dy' \right] \right) \{F(t, z) + R(t, z)\} \qquad (4.45)$$

The formal mathematics uses Green's functions [16,18] but the physical concepts of a solution like a radiating plane wave, checked through differentiation, show that eqn. 4.45 is a solution. The

Optical energy exchange in guides 123

relationship between E_{rad} and $F+R$ can be written more compactly as

$$v(y)E_{rad} = (j/4)\beta_b \left\{ \int_{-\infty}^{\infty} \frac{\varepsilon_{rf1}(y')u(y')}{\varepsilon_{rr}} \right.$$
$$\left. \times \exp(-j\beta_b|y-y'|)\,dy' \right\} \{F(t,z)+R(t,z)\} \qquad (4.46)$$

The modified eqn. 4.32 now has an extra term including E_{rad}:

$$2j\beta_b u(y)\frac{\partial F}{\partial z} + 2j\omega_0 \varepsilon_{rr}\varepsilon_0\mu_0 u(y)\frac{\partial F}{\partial t}$$
$$\simeq -\omega_0^2\varepsilon_{rf2}(y)\cos(2\beta_b z)\varepsilon_0\mu_0 Ru(y)\exp\{j(\omega_0 t+\beta_b z)\}\exp\{-j(\omega_0 t-\beta_b z)\}$$
$$-\omega_0^2\varepsilon_{rf1}(y)\cos(\beta_b z)\varepsilon_0\mu_0 E_{rad}v(y)\exp(j\omega_0 t)\exp\{-j(\omega_0 t-\beta_b z)\} \qquad (4.47)$$

Note that the radiating fields have a variation $v(y)$ with y that is not the same as $u(y)$. Ignoring rapidly changing terms in z, as explained previously, but equating the slowly varying terms on both sides which have the same spatial periodicity and selecting the correct modal variation $u(x,y)$ through the orthogonality relation, then further, on adding in the detuning parameter δ, one finally obtains

$$\frac{\partial F}{\partial z} + \frac{1}{v_g}\frac{\partial F}{\partial t} - gF + j\delta F = j\kappa_{rf2}R + i_{spf} - \kappa_{rf1}(F+R) \qquad (4.48)$$

where

$$\kappa_{rf2}\mathcal{A} \approx \iint \tfrac{1}{4}\beta_b\{\varepsilon_{rf2}(y)/\varepsilon_{rr}\}u(x,y)^2\,dx\,dy$$

and

$$\mathcal{A} = \iint u(x,y)^2\,dx\,dy$$

with

$$\kappa_{rf1}\mathcal{A}(F+R) \simeq -j\iint \tfrac{1}{4}\beta_b\{\varepsilon_{rf1}(y)/\varepsilon_{rr}\}u(x,y)v(y)E_{rad}\,dx\,dy \qquad (4.49)$$

Substituting for E_{rad} leads to

$$\kappa_{rf1}\mathcal{A} \approx$$

$$\iint \tfrac{1}{4}\beta_b^2 \frac{\varepsilon_{rf1}(y)u(x,y)}{\varepsilon_{rr}} \int_{-\infty}^{\infty} \frac{\varepsilon_{rf1}(y')u(x,y')}{\varepsilon_{rr}} \exp(-j\beta_b|y-y'|)\,dy'\,dy\,dx$$

For symmetrical guides,

$$\kappa_{rf1} \simeq \int_{-\infty}^{\infty} \left| \int_{-\infty}^{\infty} \tfrac{1}{2}\beta_b \frac{\varepsilon_{rf1}(x,y)u(x,y)}{\varepsilon_{rr}} \exp(-j\beta_b y)\,dy \right|^2 dx/\mathcal{A} \qquad (4.50)$$

Eqn. 4.48 has important additional terms $-\kappa_{rf1}(F+R)$ when compared with the first-order grating. It can be seen, by taking the term $\kappa_{rf1}F$ over to the left-hand side, that this term is simply loss which reduces the net gain. The term $-\kappa_{rf1}R$ gives complex coupling which again adds to loss and it is found that if this term is large enough then one can help to select the modes which lase within a Bragg laser [11], as will be touched on later.

The correct evaluation of the radiation-loss coefficient κ_{rf1} is more complicated than indicated above because it depends on the relative phasing of the fields F and R (e.g. $F+R=0$ gives no radiation) which in turn depends on conditions at the laser's facets. The radiation also need not be symmetrical above and below the laser as assumed here but would depend on reflective (e.g. metallic) or attenuating layers around the waveguide. Experimentally, one will find widely varying results depending on the detailed fabrication methods. The values of κ_{rf1} for second-order gratings which are approximately evaluated by the program slabexec therefore have to be used as indicative rather than definitive parameters, and more detailed modelling/measurement is required.

The corresponding equation to eqn. 4.48 for the reverse wave is

$$-\frac{\partial R}{\partial z} + \frac{1}{v_g}\frac{\partial R}{\partial t} - gR + j\delta R = j\kappa_{rf2}F + i_{sp\,r} - \kappa_{rf1}(F+R) \qquad (4.51)$$

4.4.4 Shape of grating

Considerable work has been done in demonstrating how the shape of the grating alters the coupling coefficient [19,20]. A square-tooth shape $\lambda_m/4$ in length is very satisfactory but the etching processes often prefer triangular teeth, as in Figures 4.8 and 4.9, when the top of the etch pit should be greater than $\lambda_m/4$ (but less than $\lambda_m/2$) by an amount that depends on the etch depth and angle. To obtain the largest values of κ, which is the general thrust of this design work, requires maximising the appropriate spatial Fourier component of the effective permittivity by the choice of the width and depth of the etching within the technological constraints of the etching processes. The algebra is 'serious' and in general requires computation to reach a solution. The MATLAB program slabexec in directory slab permits relatively simple shapes to be explored in a way which is appropriate to the limited shaping capabilities for grating teeth caused by typical etching processes.

4.5 Summary

Starting with the classic rate equations which approximate the physics of a uniform Fabry–Perot laser, several key features about lasing were discovered:

(i) The round-trip complex gain including the feedback has to be unity so that both the phase and magnitude of the field gain play a significant role in the mode selection.

(ii) There is a resonance between the electron and photon energy exchange which limits the capabilities for modulating a laser directly.

(iii) There are increases/decreases in the electron density which decrease/increase the refractive index and increase/decrease the lasing frequency as the laser is turned on/off. The magnitude of this significant effect is determined by α_H, the linewidth-broadening factor.

(iv) In a Fabry–Perot laser there are very few mechanisms to keep the laser lasing in just one single mode, which is the most desirable condition for modern communication systems.

(v) The necessity to consider both the phase and the gain suggests that one should consider travelling-wave field equations for a laser, and the necessary mathematics was set up to permit wave packets to be considered with independent frequency and propagation coefficients centred about relatively arbitrary choices so as to prepare the ground for the final main section.

(vi) The introduction of a periodic grating into the waveguide which forms the laser introduces a frequency-selective element that is the main feature of this book—the Bragg grating inside the laser diode.

(vii) The wave packets are centred about a frequency close to, but not necessarily identical to, the peak reflection frequency of the grating, but the central propagation for the wave packet is taken to be governed by the periodicity of the Bragg grating.

(viii) The field equations coupling the forward and reverse waves are discussed and methods for finding, from Fourier analysis, the correct coupling coefficients given the waveguide structure are briefly outlined and a MATLAB program has been made available for the reader to explore this aspect further.

Chapters 5 and 6 continue with the physics and design of DFB lasers. The detailed numerical modelling is left to Chapter 7.

4.6 References

1 STATZ, H., and DE MARS, G.: 'Transients and oscillation pulses in masers', *in* TOWNES, C.H. (Ed.): 'Quantum electronics' (Columbia University Press, New York), 1960, pp. 530–537
2 IKEGAMI, T., and SUEMATSU, Y.: 'Direct modulation of semiconductor junction laser', *Electron. Commun. Japan*, 1968, **51**(3), pp. 51–58
3 PAOLI, T., and RIPPER, J.E.: 'Direct modulation of semiconductor lasers', *Proc. IEE*, 1970, **58**, pp. 1457–1465
4 LANG, R., and KOBAYASHI, K.: 'Suppression of the relaxation oscillation in the modulated light output of semiconductor lasers', *IEEE J. Quantum Electron.*, 1976, **12**, pp. 194–199
5 CARROLL, J.E.: 'Rate equations in semiconductor electronics' (Cambridge University Press, 1990), chap. 7
6 OSINSKI, M., and ADAMS, M.J.: 'Transient time-averaged spectra of rapidly modulated semiconductor lasers', *IEE Proc. J*, 1985, **128**, pp. 34–7
7 KAWAGUCHI, H.: 'Bistabilities and non-linearities in laser diodes' (Artech House, Boston) Section 2.1.7 and Chapter 5
8 ADAMS, M.J., and OSINSKI, M.: 'Longitudinal mode competition in semiconductor lasers—rate-equations revisited', *IEE Proc.-I*, 1982, **129**, pp. 271–274
9 YABLONOVITCH, E.: 'Photonic band-gap crystals', *J. Phys. Condens. Matter*, 1993, **5**, pp. 2443–2460
10 DEMARTINI, F., MARROCCO, M., MATALONI, P., MURRA, D., and LOUDON, R.: 'Spontaneous and stimulated-emission in the thresholdless microlaser', *J. Opt. Soc. Am. B-Opt. Phys.*, 1993, **10**, pp. 360–381
11 FURUYA, K., SUEMATSU, Y., and HONG, T.: 'Reduction of resonance like peak in direct modulation due to carrier diffusion in injection laser', *Applied Optics*, 1978, **17**, pp. 1949–1952
12 THOMPSON, G.H.B.: 'Physics of semiconductor laser devices' (Wiley, 1980) Section 7.3.4, 'Diffusion of damping laser oscillations', pp. 424–433

13 CARROLL, J.E., WHITE, I.H., and GALLAGHER, D.F.G.: 'Dependence of chirp in injection-lasers on temporal optical pulse shape', *IEE Proc. J*, 1986, **133**, pp. 279–282
14 HENRY, C.H.: 'Theory of line width of semiconductor lasers', *IEEE J. Quantum Electron.*, 1982, **18**, pp. 259–264
15 GLINSKI, J., and MAKINO, T.: 'Mode selectivity in DFB lasers with a second-order grating', *Electron. Lett.*, 1986, **22**, pp. 679–680
16 KAZARINOV, R.R., and HENRY, C.H.: 'Second order distributed feedback lasers with mode selection provided by first order radiation loss', *IEEE J. Quantum Electron.*, 1985, **21**, pp. 144–150
17 HARDY, A., WELCH, D.F., and STREIFER, W.: 'Analysis of second order gratings' *IEEE J. Quantum Electron.*, 1989, **25**, pp. 2096–2105
18 JEFFREYS, H., and JEFFREYS, B.S.: 'Mathematical physics 3rd edn., 1956' (Cambridge University Press), Green's functions, p. 493
19 STREIFER, W., SCIFRES, D.R., and BURNHAM, R.D.: 'Coupling coefficient for distributed feedback single- and double-heterostructure diode laser', *IEEE J. Quantum Electron.*, 1975, **11**, pp. 867–873
20 GHAFOURI-SHIRAZ, H., and LO, B.: 'Distributed feedback laser diodes' (Wiley, Chichester, 1996), section 2.5

Chapter 5
Basic principles of lasers with distributed feedback

5.1 Introduction

The rationale for inserting the frequency-selective 'Bragg' grating into a semiconductor laser has been met in Sections 1.5 and 1.6 and the effect of a periodic permittivity on the energy exchange has been considered in Section 4.4. The next two chapters provide a relatively self-contained account of the physics and some numerical modelling for uniform and phase-shifted 'Bragg' lasers. There is a slightly arbitrary labelling of 'basic' and 'advanced' features before the detailed numerical time-domain modelling is discussed in Chapter 7. This present chapter starts with a more physical derivation of the coupled-mode equations than the mathematics of Chapter 4 and Appendix 5 and moves on to new features such as the eigenmodes for the analytic solutions, the influence of grating parameters on the dispersion diagram and the 'stopband' of nonpropagating frequencies.

Readers familiar with energy-band–distance diagrams in semiconductor devices will not be surprised to find that stopband–distance diagrams are similarly helpful when discussing Bragg lasers. The deficiencies of the uniform Bragg laser over single-mode selection have already been mentioned and are pursued further here to show the benefits of the insertion of phase shifts at judicious places into an otherwise uniform Bragg grating. The chapter concludes with an outline of the frequency-domain-modelling technique using transfer matrices. This is one of a number of numerical methods for simulating the performance of DFB lasers. The reader will be able to contrast these with time-domain techniques discussed in Chapter 7.

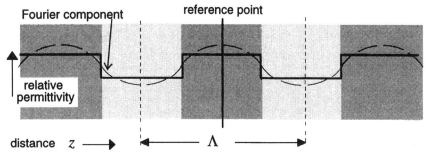

Figure 5.1 Choice of phase position for equal κ_{rf} and κ_{fr}

5.2 Coupled-mode equations for distributed feedback

5.2.1 Physical derivation of the coupling process

The first key paper to propose distributed feedback (DFB) in semiconductor-laser structures was farsighted by investigating theoretically not only the use of real-refractive-index ('index') gratings but also imaginary-refractive-index ('gain') gratings [1]. The derivation here can assume an arbitrary mixture of gain and index coupling though initially it is assumed that the grating is a lossless first-order grating formed from a periodic index change. As seen in Chapters 1 and 4, the reflectivity is distributed to give a mean reflectivity per unit length of κ_{rf} of the reverse wave into a forward wave and, similarly, a reflectivity of κ_{fr} of the forward wave into a reverse wave. When the reference point is taken at an idealised step change in the dielectric as shown in Figure 5.1, then one knows that the field reflections of a forward wave into a reverse wave, and vice versa, are both real such that $\Delta\rho_{rf} = -\Delta\rho_{fr}$ (see Appendix 1 discussing the reflection at the interface between two dielectrics).

When the reflection reference position is moved an arbitrary distance a forward, the reflection of the forward wave into a reverse wave changes so that $\Delta\rho_{rf} \to \Delta\rho_{rf} \exp(2j\beta_b a)$ and a reverse wave into a forward wave changes by $\Delta\rho_{fr} \to \Delta\rho_{fr} \exp(-2j\beta_b a)$. The factor of 2 allows for both the extra distance a travelled in the opposite direction by any reflection as well as the extra distance a travelled by the main wave. However, the total physical distance $2a$ corresponds here to an optical distance $\lambda_m/4$, changing $\Delta\rho_{rf} \to j\Delta\rho_{rf}$ and $\Delta\rho_{fr} \to -j\Delta\rho_{fr}$. It therefore follows from this phase argument that the reflection at the reference centre of symmetry, as shown in Figure 5.1, is purely imaginary and, as also expected from symmetry, this reflection argument leads to the conclusion that $j\kappa_{rf} = j\kappa_{fr}$. The product of the real terms (κ_{rf}, κ_{fr}) at the centre of symmetry for this type of grating is observed always to be positive and remains the same if the reference point moves.

Now consider a forward field F propagating in a waveguide at an angular frequency ω_0 with an axial complex propagation coefficient β_0. The complex amplitude of F at $z=0$ is transformed to amplitude $F\exp(-j\beta_0\Lambda+g\Lambda)$ at $z=\Lambda$ and similarly a backward wave of amplitude R incident at $z=\Lambda$ is transformed to amplitude $R\exp(-j\beta_0\Lambda+g\Lambda)$ at $z=0$. The gain g is the net field gain per unit length making allowance for losses and confinement factor. The value of the propagation coefficient may be written as

$$\beta_0=\beta_b+\delta \tag{5.1}$$

where β_b gives the Bragg wavelength for maximum reflectivity and therefore, as in Section 4.3.3, δ is a measure of the detuning of the optical wave from the central Bragg condition. Allowing for gain, phase shifts and coupling, one may write, for one Bragg period,

$$F_{z=\Lambda}=\{1+(g-j\delta)\Lambda\}F_{z=0}+j\kappa\Lambda R_{z=\Lambda} \tag{5.2}$$

$$R_{z=0}=\{1+(g-j\delta)\Lambda\}R_{z=\Lambda}+j\kappa\Lambda F_{z=0} \tag{5.3}$$

F and R are defined at the reference planes one grating period apart, while $g\Lambda$ and $\delta\Lambda$ are assumed to be sufficiently small compared with unity so that the approximations in eqns. 5.2 and 5.3 hold, which for practical gratings and gains is usually correct. The 'finite-difference' scheme of eqns. 5.2 and 5.3 can then be rearranged with only a slight change of notation into the differential form in the steady state:

$$+\frac{dF}{dz}=(g-j\delta)F+j\kappa R \tag{5.4}$$

$$-\frac{dR}{dz}=(g-j\delta)R+j\kappa F \tag{5.5}$$

Adding in time dependence, by recognising that the fields travel with the group velocity, and adding in terms for the excitation by spontaneous emission (as in eqns. 4.22 and 4.24) one obtains

$$\frac{1}{v_g}\frac{\partial F}{\partial t}+\frac{\partial F}{\partial z}=(g-j\delta)F+j\kappa R+i_{spf} \tag{5.6}$$

$$\frac{1}{v_g}\frac{\partial R}{\partial t}-\frac{\partial R}{\partial z}=(g-j\delta)R+j\kappa F+i_{spr} \tag{5.7}$$

The spontaneous excitation uses random spontaneous 'currents' i_{spf} and i_{spr} (which include the spontaneous coupling β_{sp} along with any appropriate confinement factor Γ_{sp} and are discussed in more detail in Chapter 7 and Appendix 9). The forward random spontaneous excitation is completely independent with respect to the reverse excitation and at any one point and time the excitations are statistically independent relative to the excitation at other points in space or time (to a first approximation). The excitation is similar to white noise except that, to be more accurate, spontaneous excitation will have some correlation with previous values on the same propagating light cone; the weaker the correlation then the broader is the spontaneous-emission bandwidth. These properties can be expressed by the complex (auto) correlations as follows:

$$\langle i_{spf/r}(z_1, t_1) i_{spf/r}(z_2, t_2)^* \rangle = C_{sp}(t_1 - t_2) \quad \text{for} \quad (z_1 - z_2) = \pm(t_1 - t_2)v_g \quad (5.8)$$

$$\langle i_{spf/r}(z_1, t) i_{spf/r}(z_2, t)^* \rangle = r_{sp}^2 \delta(z_1 - z_2) \quad (5.9)$$

where $C_{sp}(t_1 - t_2)$ describes the autocorrelation properties of the spontaneous emission to allow for the limited bandwidth, and $\delta(z_1 - z_2)$ is a Dirac delta function showing a complete lack of correlation at neighbouring points, while r_{sp}^2 determines the strength of the spontaneous emission. As noted in Section 3.3, phase factors of unit magnitude multiplying $i_{spf}(z_1, t_1)$ etc. do not alter the magnitude of the complex autocorrelation or correlation functions. Finding the magnitude of the spontaneous excitation is considered later.

5.2.2 Complex gratings

When calculating κ from a Fourier analysis of the variations of the refractive index as outlined in Section 4.4, the cross-section of the guided field does not adjust its optical width at each tooth [2], so the average modal distribution in the guide was used in that calculation to find the appropriate harmonic components. Such an analysis can give a complex $\kappa = \kappa_{rf} = \kappa_{fr} = \kappa_{index} + j\kappa_{gain}$ through etching the grating down into the gain region, for example. This complex reflection is still referenced from the centre of symmetry as in Figure 5.1 where κ_{rf} and κ_{fr} are equal but their complex values indicate there is a gain grating or a loss grating in addition to the index grating. This means that the variation in the permittivity of the grating material also has to be complex, say proportional to $\varepsilon_{rr} + j\varepsilon_{ri}$ (with both elements positive). Positive ε_{ri} represents a material with gain and so with the reference position as in Figure 5.1, $\kappa = \kappa_{index} + j\kappa_{gain}$ (with both elements positive) represents a grating where the gain maxima (or loss minima) coincide with the maxima in real index (and vice versa if κ_{gain} were negative). With $\kappa_{index}>0$ and $\kappa_{gain}>0$ one may refer to a gain grating 'in phase with' the index grating (or a loss grating 'in antiphase with' the index grating). With $\kappa_{index}>0$ and $\kappa_{gain}<0$ one refers to a gain grating 'in antiphase with' the index grating (or a loss grating 'in phase with' the index grating).

As before, the phase of the reference point can change the phase of the imaginary component so that it is easy to be confused between loss and gain in a grating, and it is safest to stick to writing the equations with a consistent reference point. A pure index grating had a positive $\kappa_{rf}\kappa_{fr}$ product and, using similar arguments, a pure gain grating has a negative $\kappa_{rf}\kappa_{fr}$ product regardless of the point of phase reference.

5.3 Coupled-mode solutions and stopbands

5.3.1 Eigenmodes

An elegant method for solving the coupled-mode equations is to find the two eigenmodes, each consisting of a forward- and backward-wave pair that propagate indefinitely along the grating with a fixed amplitude ratio (s_+ and s_-, respectively) [3]. The angular frequency ω determines the offset parameter δ which describes the detuning from the Bragg condition:

$$\delta = \frac{n(\omega - \omega_b)}{c} \quad (5.10)$$

The two eigenmodes then become

$$A\{\exp(-j\beta_b z) + s_+ \exp(j\beta_b z)\} \exp(-j\beta_e z) \quad (5.11)$$

$$B\{\exp(j\beta_b z) + s_- \exp(-j\beta_b z)\} \exp(j\beta_e z) \quad (5.12)$$

where β_e is a propagation coefficient offset from the Bragg value, *in the presence of coupling*; β_e is to be found for each δ and g. The solution to any problem with a uniform grating can then be obtained by matching the eigenmodes to the boundary conditions. The solutions for $s_{+/-}$ are found by substituting eqns. 5.11 and 5.12, and into eqns. 5.4 and 5.5 to obtain

$$s_+ = \frac{[(\delta+jg) \pm \{(\delta+jg)^2 - \kappa_{fr}\kappa_{rf}\}^{0.5}]}{\kappa_{rf}} \quad (5.13)$$

Basic principles of lasers with distributed feedback 133

$$s_- = \frac{[(\delta+jg) \pm \{(\delta+jg)^2 - \kappa_{rf}\kappa_{fr}\}^{0.5}]}{\kappa_{fr}} \quad (5.14)$$

where the ± sign is chosen so that $|s_+| \leq 1$ and $|s_-| \leq 1$. The dispersion relationship between the frequency parameter δ and the offset propagation coefficient β_e is given from

$$\beta_e = \mp \{(\delta+jg)^2 - \kappa_{fr}\kappa_{rf}\}^{0.5} \quad (5.15)$$

Note that, in the absence of any coupling, the (complex) propagation is just $\delta+jg$. This dispersion relationship, eqn. 5.15, is discussed shortly.

While the eigenmodes form an elegant mathematical technique, analysis in terms of the 'forward'- and 'reverse'-field components F and R as in eqns. 5.4 and 5.5 is usually easier. The eigenmodes of eqns. 5.11 and 5.12 retain the variations $\exp(\pm j\beta_b z)$ while the F and R components have these spatial variations implicit with a uniform β_b over the whole structure and any spatial variations in the Bragg period accounted for by changing the δ parameter. As an example, whenever κ goes to zero, such as in a short section of uniform material or at an antireflection-coated facet which makes air appear like a uniform semiconductor to the outgoing waves, then the continuity of the field components F and R ensure that the amplitude reflectivity ρ that is most straightforward to use is given from $\rho = R/F$. For this reason the eigenmodes, although elegant, are not pursued here.

5.3.2 The dispersion relationship and stopbands

The dispersion relationship given by eqn. 5.15 shows that a pure index grating with zero gain g only has a propagating solution with β_e real for $\delta^2 > \kappa_{fr}\kappa_{rf} > 0$, as shown in Figure 5.2. For $\delta^2 < \kappa_{fr}\kappa_{rf}$, the coupled wave equations yield an evanescent solution with β_e imaginary, and the grating has a strong reflectivity. This region is known as the grating 'stopband'. Within this band, any incident wave is reflected efficiently and the reflections prevent propagation and instead cause a decay in amplitude with distance (so-called *evanescence* with β_e purely imaginary when there is zero gain/loss). Conversely, outside the stopband, when $\delta^2 > \kappa_{fr}\kappa_{rf}$, the phase matching of the grating is too weak to give constructive/destructive reflection and propagation occurs with an oscillatory net amplitude for the (forward+backward) waves and β_e has a real component for all values of gain.

A completely different situation arises for a pure-gain grating for which $\kappa_{fr}\kappa_{rf} < 0$. If the mean gain g is zero then there is a propagating

134 Distributed feedback semiconductor lasers

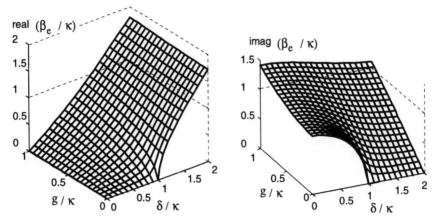

Figure 5.2 Normalised dependence of real and imaginary parts of β_e on δ and the field gain g for a pure index guide

Results are symmetric in δ, and gain has been taken as positive

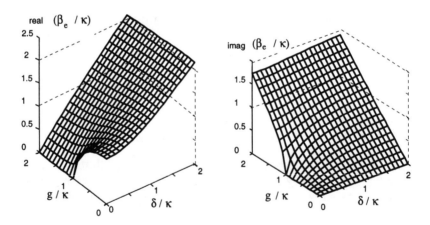

Figure 5.3 Normalised dependence of real and imaginary parts of β_e on δ and the field gain g for a pure-gain grating (real κ)

Results are symmetric in δ, and gain has been taken as positive

solution for all δ (all frequencies), and no stopband exists in the propagation coefficient β_e as shown in Figure 5.3. It is also interesting to see that, when the magnitude of the gain exceeds the magnitude of κ, then a stopband again occurs around the Bragg frequency. Notice how the real and imaginary parts of β_e are effectively interchanged in their roles between Figures 5.2 and 5.3 as are g and δ.

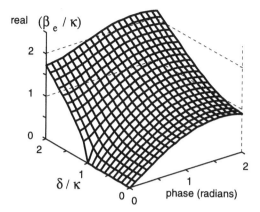

Figure 5.4 Dependence of the real part of β_e on detuning and phase (arg $\kappa_{gain}/\kappa_{index}$) of κ with $\kappa_r > 0$

β can take positive or negative values, and there is symmetry in δ

In Figure 5.4 the real part of the normalised complex propagation coefficient $\beta_e/|\kappa|$ is shown to depend on the normalised detuning $\delta/|\kappa|$ and the phase θ of the complex grating-coupling coefficient $\kappa = |\kappa|\exp(j\theta)$. For zero gain and for pure index-coupling coefficients, a stopband exists in the range for detuning given by $|\delta/\kappa| < 1$ and that for gratings with a complex coupling coefficient there is a range of propagation coefficients which are not allowed.

A program **dispbrag** in directory **grating** can help the reader to investigate the real and imaginary parts of the propagation coefficient normalised to the grating coupling coefficient $\beta_e/|\kappa|$ as a function of the detuning from the Bragg condition normalised to the grating coupling coefficient $\delta/|\kappa|$. The phase of the coupling coefficient is taken so that $\kappa_{rf} = \kappa_{fr}$ but allows for complex $\kappa = |\kappa|\exp(j\theta)$. It will be found that the stopband for a pure-index grating disappears in the presence of any gain or loss coupling. The region of strongest evanescence lies in the stop band for pure-index coupling and this evanescence weakens as the ratio of gain / index increases.

5.4 Matrix solution of coupled-mode equations for uniform grating laser

5.4.1 The field input–output relationships

The optical fields within a finite length of grating, having a uniform coupling coefficient and with perfectly antireflection coated facets at its ends, are now discussed. In Appendix 6 it is shown how the coupled equations may be solved analytically and directly with a matrix formulation where the forward and reverse fields are arranged as a vector with

$$\mathbf{F}(z) = \begin{bmatrix} F(z) \\ R(z) \end{bmatrix} = \mathbf{M}(z) \begin{bmatrix} F(0) \\ R(0) \end{bmatrix} \tag{5.16}$$

The fields at any position z can be put in terms of the fields at $z=0$ through $\mathbf{M}(z)$:

$$\mathbf{M}(z) = \cos(\beta_e z) \begin{bmatrix} 1 & 0 \\ 0 & 1 \end{bmatrix}$$

$$+ \frac{\sin(\beta_e z)}{\beta_e} \left\{ (g - j\delta) \begin{bmatrix} 1 & 0 \\ 0 & -1 \end{bmatrix} - \kappa \begin{bmatrix} 0 & -j \\ j & 0 \end{bmatrix} \right\} \tag{5.17}$$

where $\beta_e^2 L^2 = \{(\delta + jg)^2 - \kappa^2\} L^2$ and δ gives the 'frequency offset'. For an AR-coated device where $F(0) = 0$ and $R(L) = 0$ then the result simplifies to require that

$$F(z) = (j\kappa/\beta_e) \sin(\beta_e z) R(0);$$
$$R(z) = [\cos(\beta_e z) - \{(g - j\delta)/\beta_e\} \sin(\beta_e z)] R(0) \tag{5.18}$$
$$\tan(\beta_e L) = \{\beta_e/(g - j\delta)\} \tag{5.19}$$

More generally, the fields can be also rearranged to give the output fields at each end of a laser of length L in terms of the input fields. This rearrangement requires a little care which is again discussed in Appendix 6. It is shown that, on writing

$$a = \cos(\beta_e z); \quad d = \{(g - j\delta)/\beta_e\} \sin(\beta_e z); \quad b = (\kappa/\beta_e) \sin(\beta_e z) \tag{5.20}$$

one obtains

$$a^2 - d^2 - b^2 = 1 \tag{5.21}$$

This permits the matrix inversion to be performed analytically, and one can write

$$\begin{bmatrix} F(t, L) \\ R(t, 0) \end{bmatrix} = \frac{1}{a - d} \begin{bmatrix} 1 & jb \\ jb & 1 \end{bmatrix} \begin{bmatrix} F(t, 0) \\ R(t, L) \end{bmatrix} \tag{5.22}$$

Note that the idealised mathematical 'oscillation' condition is given when there can be a significant output in the left-hand terms even though there is a negligible (zero) input (right-hand terms), i.e. where $a - d = 0$ or

$$\tan(\beta_e L)/(\beta_e L) = 1/\{(g-j\delta)L\} \qquad (5.23)$$

In this form, the choice of the sign of the square root to determine β_e is manifestly of no consequence and of course eqn. 5.23 is the same as eqn. 5.19. The reader may check that with $\delta=0$ and real κ, β_e is imaginary with $|\beta_e L| > gL$ and it is not possible to solve eqn. 5.23 for any gain g (of course g is real). However, two solutions are possible for $|\delta| > \kappa$ having $\beta_e = \beta_{er} + j\beta_{ei}$ and $\beta_e = \beta_{er} - j\beta_{ei}$ giving two modes, one at the upper- and the other at the lower-wavelength edge of the stopband.

5.4.2 Reflections and the observed stopband

With input fields F_{in} on the left and R_{in} on the right, then with facet coatings giving negligible reflections one matches $F_{in} = F(0)$ and $R_{in} = R(L)$ for a laser of length L. A powerful matrix technique for undertaking this task is given in Appendix 6. The program **refl** in directory **grating** allows the reader to explore $\rho = R/F$ at the input for real κ. Figure 5.5 shows this reflectivity as a function of $\delta/|\kappa|$ for a real κ so that one can see how the reflectivity falls off as $\delta/|\kappa|$ diverges from zero.

The stopband is, strictly speaking, the region where the fields evanesce. With this definition the width is determined solely by the grating-coupling coefficient, as described in Section 5.4.1 and, for real-index gratings, is given by

$$|\delta| \leq |\kappa| \qquad (5.24)$$

In practice the term 'stopband' is used for the spectral separation between the two peaks in reflection which become much larger than unity just before lasing. For a uniform grating, these peaks are situated symmetrically just outside the strict stopband by an amount which depends on the coupling coefficient and length of the grating [4,5]. Figure 5.5 with normalised gain of 0.5 and $\kappa L=2$ shows one of these peaks just beginning to emerge around $\delta/|\kappa| \sim 2$ outside the strict mathematical 'stopband'. The other peak would be with the negative value of $\delta/|\kappa| \sim -2$.

However, there is a further complication in this analysis in that it will later be found desirable, when designing for a single lasing mode, to introduce short additional sections of uniform laser which change the phase of the forward and reverse fields with respect to one another. The classic case is a phase shift of $\lambda_m/4$ situated in the middle of the

138 *Distributed feedback semiconductor lasers*

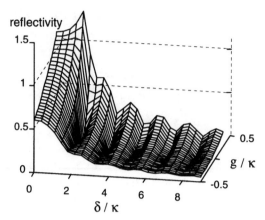

Figure 5.5 *Reflectivity of a uniform grating of length $\kappa L = 2$ as a function of gain and frequency deviation*

The reflection is symmetrical in δ so only one-half of the diagram is drawn to avoid too much detail

laser. Other devices introduce, for example, two $\lambda_m/8$ phase shifts appropriately displaced about the centre of the laser. For such devices, the lasing line lies within a band of amplified spontaneous emission which has significantly lower amplitude where the reflection sculptures the emission spectrum. It is this band of low spontaneous emission which is commonly referred to as the 'stopband' and can be found by numerical calculations as discussed later.

Figure 5.6 shows the dependence of this extended definition of a normalised 'stopband width' (measured as a change in free-space wavelength $2\Delta\lambda_{stop}$) on the κL product of the grating for:

(i) a uniform grating laser,
(ii) a symmetric $\lambda_m/4$-phase-shifted structure, and
(iii) a DFB laser with $2 \times \lambda_m/8$ phase shifts positioned symmetrically in the cavity and spaced by 70% of the cavity length (see Section 5.6.5).

The curves are plotted as dimensionless groups so as to generalise them:

$$\Delta_{normalised} = \left(\frac{2\pi}{\lambda_0^2}\right) n_{eff} \Delta\lambda_{stop} L \qquad (5.25)$$

The functional form of the normalised stopband width $\Delta_{normalised}$ as a

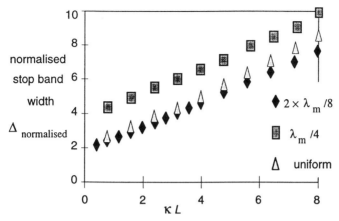

Figure 5.6 Normalised 'stopband width' against κL

The stopband width is defined here as the separation between the two main peaks of spontaneous emission at high gain and so varies for lasers with different structures as discussed in Section 5.5

function of κL means that, for low values of κL, $\Delta\lambda_{stop} \propto 1/L$, and for high κL products, $\Delta\lambda_{stop} \propto \kappa$, i.e. for very short grating lengths the 'stopband width' →infinity because there is no frequency selectivity, while for high κL products the 'stopband width' $\propto \kappa$ as one might expect [6].

5.5 DFB lasers with phase shifts

5.5.1 Phase shifts

This concept of introducing phase shifts into a uniform laser, which has been mentioned briefly before, is now examined further by considering steady-state models of such lasers. The discussion begins by considering the physics of the oscillation condition where, for lasing action to occur, the net complete round-trip complex gain must be unity. The phase change must therefore be equivalent to an integral number of wavelengths. Recalling eqn. 1.4 for a Fabry–Perot laser with reflectivities ρ_{left} and ρ_{right}:

$$1 + j0 = \rho_{left}\rho_{right} \exp\left\{(\Gamma g_m - a_m) - j\frac{2\pi}{\lambda}\right\}2L \qquad (5.26)$$

The confinement factor Γ has now been inserted and modifies the material gain; Γ had not been introduced at the start of Chapter 1. This general principle that the round-trip complex gain is unity applies to a DFB laser as to any laser. The facet reflectivities ρ_{left} and ρ_{right} are replaced by equivalent grating reflectivities with appropriate phases of reflection, and the *actual* cavity length has to be changed to an *equivalent* cavity length. This is the essential physics which is hidden in the initial lasing equation (eqn. 5.23).

One might think that the strong reflection at the Bragg frequency f_{bragg} would ensure oscillation at f_{bragg} but eqn. 5.23 showed no solution at this central frequency for real κ but instead indicated modes towards the edges of the 'stopband'. This paradox arises because, although the reflection and gain are strong enough at f_{bragg}, the feedback is in antiphase which prevents lasing action. This phasing can be understood by considering a microcavity formed by one grating 'tooth' (taken to be $\lambda_m/4$ long with a low index value) exactly at the centre of the device, for example with the index stepping to the higher value at each end of this tooth. All grating reflections at the left-hand end of this tooth are in phase with each other and can be lumped into one real reflection ($\rho_{tooth\ left} \propto \kappa$) at the step to the adjacent higher-index tooth. Likewise, all the reflections on the right-hand end can be collected into a reflection ($\rho_{tooth\ right}$) at the step to the adjacent right-hand higher-index tooth. From symmetry $\rho_{tooth\ right} = \rho_{tooth\ left} = \rho_{net}$ gives the net reflection at the ends of this central tooth of the pure index grating. Each single pass is $\lambda_m/4$ long (a phase shift of $\pi/2$) giving the net round-trip feedback with one pass there and one pass back as

$$\{\rho_{net} \exp(-j\tfrac{1}{2}\pi)\}^2 < 0 \tag{5.27}$$

This is negative and so prevents lasing even if $\rho_{net} > 1$. An additional 180° phase change happens if there is some wave propagation within the grating (β_e real) which can occur at frequencies just outside the stopband. For oscillation, this propagation and gain must be such that ρ_{net} forms an imaginary value of j or $-j$. In the ideal uniform laser, two main lasing frequencies occur where $|\beta_e L| \sim \pi/2$, symmetrically placed about the central Bragg frequency.

An alternative method of obtaining the additional 180° at the Bragg frequency with a real index grating can be found by inserting a $\lambda_m/4$ phase shift at the centre of the cavity, thus adding 180° to the round-trip phase change within the microcavity but now keeping ρ_{net} real. Now at the centre of the stopband $\rho_{net} \exp\{-j(\tfrac{1}{2}\pi+\tfrac{1}{2}\pi)\}^2 > 0$, so there is

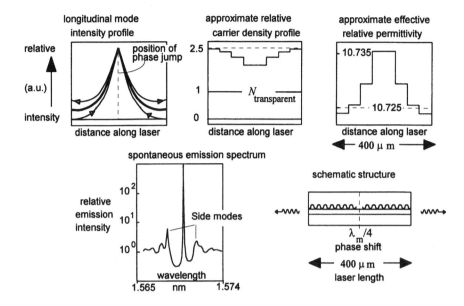

Figure 5.7 Simulated performance of high-κL $\lambda_m/4$ phase-shifted DFB laser operating at low output power

Output power at each facet 0.37 mW, $\kappa L=3$, zero facet reflectivity, 2000 A/cm^2 drive current, effective recombination time 4 ns, transparency density 10^{18} cm^{-3}

positive feedback giving lasing. At this central frequency the waves were seen to be evanescent (β_e imaginary) and for moderate to high values of grating feedback (i.e. κL product >1.25) the summed intensities of these waves then peaks at the $\lambda_m/4$ phase shift and decays 'exponentially' towards each facet, giving a very nonuniform longitudinal-mode-intensity profile, as has been calculated (from numerical analysis) in Figure 5.7.

With the $\lambda_m/4$-phase-shifted laser for $\kappa L>2$, the strong axial variation in the mode's field is found to give several unsatisfactory features in spite of providing a clean single lasing line at low output powers (see Section 5.6). The concept of introducing phase shifts into the structure at points other than the centre has been found to be crucial in providing a good engineering solution to some of these difficulties. It is therefore necessary to examine the analysis of inserting phase shifts into structures and obtaining the oscillation conditions.

5.5.2 Insertion of phase shifts: the transfer-matrix method

Section 5.4.1 shows that by writing

$$\mathbf{F}(z) = \begin{bmatrix} F(z) \\ R(z) \end{bmatrix}^{tr}$$

with $\mathbf{F}(0)$ known then it is possible to write a 'transfer matrix' $\mathbf{M}$ such that

$$\mathbf{F}(z) = \mathbf{M}(z)\mathbf{F}(0) \qquad (5.28)$$

Now the fields F and R are the forward and reverse waves in a very short planar-laser section. It follows that, for a phase change of ϕ, one can replace this section with a phase jump where $F \to F\exp(-j\phi)$; $R \to R\exp(+j\phi)$. The transfer matrix $\mathbf{M}_\phi$ for this phase shift gives

$$\mathbf{F}_{(\text{right})} = \mathbf{M}_\phi \mathbf{F}_{(\text{left})} \qquad (5.29)$$

where

$$\mathbf{M}_\phi = \begin{bmatrix} \exp(-j\phi) & 0 \\ 0 & \exp(j\phi) \end{bmatrix} \qquad (5.30)$$

A phase shift of ϕ in the middle of the laser is therefore incorporated by concatenating transfer matrices as follows:

$$\begin{bmatrix} F_L \\ R_L \end{bmatrix} = [\mathbf{M}_1][\mathbf{M}_\phi][\mathbf{M}_1]\begin{bmatrix} F_0 \\ R_0 \end{bmatrix} \qquad (5.31)$$

where $\mathbf{M}_1$ is as given in eqn. 5.17 but with a length of $L/2$. If both facets have zero reflectivity, the following equation for oscillation may be derived after some algebra:

$$\frac{\kappa_\beta \kappa_{rf}}{\beta_e^2} \sin^2\left(\frac{\beta_e L}{2}\right) = -\left\{\cos\left(\frac{\beta_e L}{2}\right) + j\left(\frac{\delta + jg}{\beta_e}\right)\sin\left(\frac{\beta_e L}{2}\right)\right\}^2 \exp(2j\phi) \quad (5.32)$$

If $\phi = 0$, then the oscillation condition (eqn. 5.23) for the uniform DFB laser may be recovered. A $\lambda_m/4$-phase-shifted DFB laser is modelled with $\phi = (\pi/2)$ or $\exp(j\phi) = j$ which is found to yield

$$\pm \sqrt{(\kappa_{fr}\kappa_{rf})} = \beta_e \cot\left(\frac{\beta_e L}{2}\right) - (g - j\delta) \quad (5.33)$$

The technique may be extended in an obvious manner for a laser with two phase shifts placed symmetrically in the cavity where now one would write

$$\begin{bmatrix} F_L \\ R_L \end{bmatrix} = [M_1][M_\phi][M_3][M_\phi][M_1] \begin{bmatrix} F_0 \\ R_0 \end{bmatrix} \quad (5.34)$$

where M_1 and M_3 are given by eqn. 5.17 for the appropriate section lengths. An analytic solution for eqn. 5.34 is not straightforward and it is preferable to compute the results. However, one can see the utility of transfer matrices for studying complex structures.

Figure 5.8 shows the simulated field and electron-density profiles along the cavity of a $2 \times \lambda_m/8$ laser analysed as a concatenation of uniform grating sections. The significant point about the $2 \times \lambda_m/8$-phase-shift DFB laser is that it provides a more uniform field and electron density along its length over a wider range of current inputs than the single $\lambda_m/4$-phase-shift structure for appropriate κL products ~ 2, provided that the phase shifts are positioned correctly at appropriate values of ΔL from the facets as shown. With a flat field profile, the gain is more effectively utilised, there are narrower line widths because of small variations in the refractive index along the structure and there is less variation in the output wavelength with power level than for many other types of Bragg laser.

While considering transfer matrices, it is worthwhile making a short digression to find the transfer matrix for N pairs of teeth of a Bragg grating, at for example its central Bragg frequency. A transfer-matrix formulation gives

$$M = (M_{\pi/4} M_{+(\frac{1}{2})\Delta\rho} M_{\pi/2} M_{-(\frac{1}{2})\Delta\rho} M_{\pi/4})^N \quad (5.35)$$

where a reflection $+\frac{1}{2}\Delta\rho$ from a step change in refractive index at one position is joined to a reflection of $-\frac{1}{2}\Delta\rho$ by a transfer matrix M_ϕ as in eqn. 5.30 but with a phase shift of $\phi = \pi/2$ corresponding to the optical length of $\lambda_m/4$. This unit is then completed with transfer matrices of phase shift $\pi/4$ on either side to complete the half-wavelength-long unit which concatenates with N other such Bragg elements. The elemental transfer matrix for a step change of refractive index, assuming plane-wave fronts, is $M_{+\frac{1}{2}\Delta\rho}$ given from

144 *Distributed feedback semiconductor lasers*

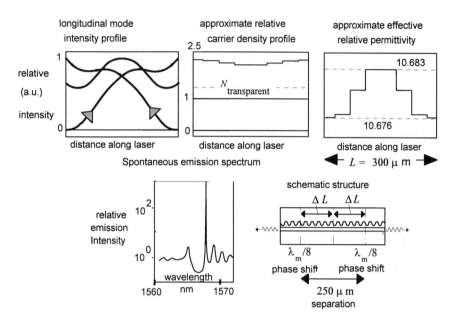

Figure 5.8   *Simulated performance of a moderate κL 2× $\lambda_m/8$-phase-shifted DFB laser emitting 3 mW per facet with a moderate κL (=2.25)*

> Device parameters are such that there is 3.2 mW output at an input current density of 5000 A/cm². The transparency density is 1.5×10^{18} cm^{-3}; κL=2.25 with a 300 μm overall length, bimolecular recombination constant of 1×10^{-10} cm³/s, Auger recombination constant of 3×10^{-29} cm⁶/s; waveguide attenuation of 40 cm^{-1}, Henry's linewidth-enhancement factor of 3, $a=3 \times 10^{16}$ cm². Operation of the device depends on having appropriate values of $\Delta L/L$

$$M_{+(\frac{1}{2})\Delta\rho} = (1/T) \begin{bmatrix} 1 & -\Delta\rho/2 \\ -\Delta\rho/2 & 1 \end{bmatrix} \quad (5.36)$$

where $T=\sqrt{\{1-(\frac{1}{2}\Delta\rho)^2\}}$ and $\frac{1}{2}\Delta\rho=(n_1-n_2)/(n_1+n_2)$. Then looking ahead to a future section on Bragg reflectors for VCSELs (Section 8.5), an estimate for the transfer matrix of 20 pairs of teeth with $n_1=3.5$ and $n_2=3$ (very approximately the values for GaAs and AlAs) is given from

$$M = (M_{\pi/4} M_{+(\frac{1}{2})\Delta\rho} M_{\pi/2} M_{-(\frac{1}{2})\Delta\rho} M_{\pi/4})^{20}$$

$$= \begin{bmatrix} -1.0119 & -j0.1548 \\ +j0.1548 & -1.0119 \end{bmatrix}^{20}$$

$$\begin{bmatrix} F_L \\ R_L \end{bmatrix} = M \begin{bmatrix} F_0 \\ R_0 \end{bmatrix} = \begin{bmatrix} 10.9349 & j10.8891 \\ -j10.8891 & 10.9349 \end{bmatrix} \begin{bmatrix} F_0 \\ R_0 \end{bmatrix} \quad (5.37)$$

so that, with zero input on the right-hand side giving $R_L=0$, $|R_0/F_0|^2 = |M_{21}/M_{22}|^2 = 0.9916$ or 99% power reflectivity. This is an optimistic calculation, neglecting loss and assuming that the optical energy is confined fully to the regions containing the dielectric steps, giving the maximum reflection possible. This type of estimate will be used later in Section 8.5.

5.6 Longitudinal-mode spatial-hole burning

5.6.1 The phenomena

'Spatial-hole burning' is a phrase which is used for the depletion of the injected-charge carriers where the depletion is caused by strong stimulated recombination in regions of high photon density. Spatial-hole burning was first identified in the lateral (horizontal) direction for broad stripe lasers which relied on gain to guide the optical mode and also in structures where the lateral-index guiding was weak. Because of the link between refractive index and charge-carrier density, the depletion of the charge carriers in regions of strong photon density increases the real refractive index and reduces the gain or imaginary refractive index compared with the regions of lower photon density. This can lead to self focusing in laterally guided structures with a built-in real-index waveguide, instability in gain-guided structures, and filamentation (constriction) of the optical beam in broad-area devices. The strength of the mechanism is, in general, significantly reduced by lateral diffusion of the charge carriers because diffusion smoothes out changes in the carrier concentration. Lateral carrier-hole burning can be detected by changes in the near- and far-field patterns as the drive current is increased. Strong lateral spatial-hole burning in general invalidates the approximation that the confinement factor Γ for the active region is constant. However, with sufficiently well designed index guiding, the lateral and transverse spatial-hole burning effect is not too important and is safely neglected.

Stimulated recombination will cause longitudinal carrier-hole burning in both Fabry–Perot and DFB lasers and create nonuniformities in the carrier density $N(z)$ and photon density $S(z)$ along the z direction. The changes are relatively slow, so that diffusion does not significantly moderate the effect and diffusion is ignored. The coupling of electron density and photon density is found from eqn. 4.2 divided by the volume $\mathcal{V}$. The drive-current density replaces the drive current because one now has to allow for nonuniformities in the electron density $N(z)$. The gain $G_m(N)$ is replaced with the differential gain expanding in N about the transparency density. The linear-to-cubic recombination terms of eqn. 4.4 are all considered to give

$$\frac{dN(z)}{dt} = \frac{J}{qd} - AN(z) - BN(z)^2 - CN(z)^3 - G'_m\{N(z) - N_{tr}\}S(z) \quad (5.38)$$

In Fabry–Perot lasers the redistribution of carriers along the cavity with increasing optical output power has only 'second'-order effects and, indeed, if the refractive index and gain vary linearly with carrier density, the emission wavelength is found not to alter. With conventional facet reflectivities there is only a modest nonuniformity in the longitudinal photon-density profile and spatial-hole burning is not usually a problem because the gain and index changes are effectively averaged out by the requirement that the round-trip complex gain remain at unity.

The effects of spatial-hole burning in DFB lasers are in marked contrast to those in FP lasers because local variations in the carrier density, and hence variations in the real refractive index and gain, give rise to changes in the magnitude and phase of the feedback from each section of the grating. All this changes the longitudinal-mode intensity distribution and also alters the gain suppression of side modes relative to the lasing mode. The lasing mode then exhibits a nonlinear light/current characteristic which is accompanied by a frequency shift or 'chirp', as such a shift is commonly called. The output power also takes time to stabilise following a transient in the current as a consequence of the time constant associated with the carrier hole burning. This in turn leads to amplitude-patterning effects under digital modulation. Another way of viewing the problem of spatial-hole burning is that one cannot simply move up and down the *static* light/current characteristic, with rapid changes in current. Spatial-hole burning means that the dynamic characteristics of the laser require time to relax back to the static characteristic on a change of current. The laser designer has to minimise this time by minimising the hole burning.

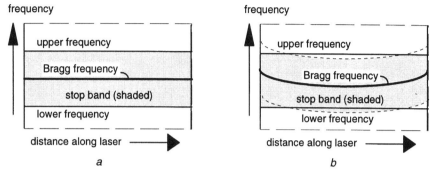

Figure 5.9 Schematic of stopband/distance diagram

a Uniform DFB
b With central spacial hole burning
Changes in the carrier density change the refractive index along the laser, thereby changing the edges of the stopband, for example as in (*b*). Waves inside the stopband experience stronger reflections and less gain than those outside the stopband. Phase adjustments to keep the round-trip phase can more readily occur at the edge of outside the stopband. Oscillation therefore occurs at both upper and lower edges of stopband in (*a*) but the upper frequency is preferred in (*b*)

5.6.2 The stopband diagram

The description and physics of longitudinal spatial-hole burning is considerably enhanced through the concept of a longitudinal stopband profile, as one example will show shortly. If the electron density were uniform then there would be a uniform stopband within the uniform grating, as indicated in Figure 5.9*a*. The stopband/wavelength profile then has a pair of lines symmetrically placed with respect to the Bragg frequency delineating the 'stopband edges' defined as the frequencies of the two ideal lasing modes for the perfect uniform-grating laser with zero facet reflectivities. Moving away from the stopband, the rate of change of phase per unit length increases while the reflectivity reduces. In contrast, moving into the stopband increases the reflectivity and the propagation changes to evanescence so that the number of phase changes per unit length becomes negligible unless a phase shift is introduced. To use the stopband diagram, one must remember that oscillation requires a round-trip gain of unity and multiples of 2π round-trip phase.

For a uniform laser with a high κL when the laser is first turned on, the electron density is uniform and the two modes on either side of the stopband can start to oscillate (Figure 5.9*a*). As the optical power builds up, the optical intensity is highest in the middle of the laser and there is significant spatial-hole burning with the carrier density reducing and therefore increasing the refractive index in the centre of the device. The stopband diagram then shows a marked curvature towards shorter wavelengths (higher frequencies) at the edges of the laser (Figure 5.9b). The short-wavelength (high-frequency) mode then experiences a strong reflectivity near the facets of the laser and maintains propagation with some phase shift and gain in the middle of the laser, so that oscillation can be maintained. The long-wavelength mode, in contrast, now experiences reduced reflectivity at the facets and reduced gain and phase shift in the middle of the laser so that the oscillation conditions cannot be maintained.

Stopband diagrams can, with a little experience, be powerful indicators of the underlying physics in a grating structure, showing rapidly why a DFB laser changes mode and which mode to expect. As well as explaining the performance of uniform lasers with high and low κL [7], stopband diagrams have proved to be most helpful in understanding high-frequency modal oscillations within uniform gratings [8] and also in aiding the understanding of push–pull lasers [9] (see Section 8.3).

5.6.3 Influence of κL product on spatial-hole burning

Uniform-grating DFB laser structures, with perfectly antireflection-coated facets, have longitudinal-mode-intensity profiles which peak strongly at the cavity centre for κL, say around 2 or more, and peak at the facets for values around unity or less. The longitudinal-mode-intensity distribution is plotted in Figure 5.10 for κL products of 2.4, 1.7 and 1 with the middle value giving a nearly uniform distribution.

However, one finds in practice that the mode profile is highly dependent on minor defects in the laser or on the facet reflections, and that in modelling, with a perfectly uniform device with no reflections, the oscillation mode can vary from run to run depending on the random build up of spontaneous emission. Indeed longitudinal spatial hole burning can make the perfectly uniform laser very unstable with respect to which of its two modes will oscillate and there have been a number of ideas to force uniform DFB lasers into lasing on one stable mode such as high–low facets, longitudinal-current variations, complex gratings and phase shifts.

5.6.4 Influence of phase shifts on spatial-hole burning

DFB lasers with a centrally placed $\lambda_m/4$ phase shift in general exhibit strong nonuniformity in their longitudinal photon density. This

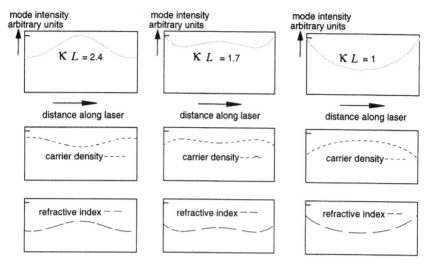

Figure 5.10 Dependence of longitudinal-mode-intensity profile of uniform-grating DFB lasers on the κL product

The photons stimulate carrier recombination so that the change in carrier density very approximately mirrors the change in photon density, as shown schematically here, and this in turn mirrors the changes in refractive index which controls the Bragg frequency and Bragg stopband/distance diagrams

nonuniformity gains in strength as κL increases. When $\kappa L \sim 1.25$, the longitudinal mode profile is at its flattest. For $\kappa L < 1.25$ the highest photon density occurs at the facets. Unfortunately, such a grating strength can pose fabrication problems because imperfections can give reflections comparable with the grating. Although longitudinal hole burning is minimised around $\kappa L \sim 1.25$, the evanescent nature of the travelling-wave solutions at the Bragg frequency still gives rise to significant longitudinal nonuniformity in the mode profile.

Multiple-phase-shift structures, and in particular the $2 \times \lambda_m/8$ DFB structure, have been particularly successful in minimising spatial-hole burning and maintaining a near uniform field profile. This $2 \times \lambda_m/8$ DFB structure normally has the two phase shifts positioned symmetrically about the centre of the cavity, as indicated schematically in Figure 5.8. The fields can be evanescent or propagating depending on whether the phase shifts are near the centre or ends of the laser structure. Detailed trade-offs in the operational parameters have been carried out as the position of the phase shifts and the strength of the grating (κL) are changed [10, 25]. Figure 5.11 shows contours of the nonuniformity, or variance, of the longitudinal-mode intensity dis-

150 *Distributed feedback semiconductor lasers*

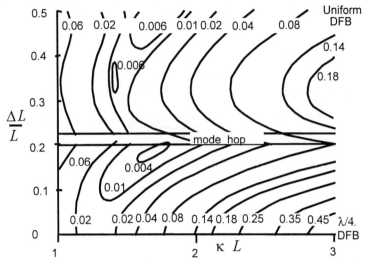

Figure 5.11 Longitudinal-intensity-variance contours for $2\times \lambda_m/8$ DFB lasers

Parameters give variance as measured by $\Delta|E|^2/|E|^2_{max}$ and ordinate $\Delta L/L$ gives position of phase shifts: note that the values of $\Delta L/L$ give two distinct modal points; see Figure 5.13

tribution as a function of κL and the separation of the $2\times \lambda_m/8$ phase shifts determined from $\Delta L/L$ (see Figures 5.8 and 5.13). The plot in Figure 5.11 includes all devices ranging from a $\lambda_m/4$-phase-shifted DFB laser through to a uniform-grating DFB device, in other words all symmetric $2\times \lambda_m/8$ designs.

5.6.5 Spectrum and spatial-hole burning

As indicated earlier, a uniform intensity avoids spatial-hole burning and leads to a more uniform electron density and uniform complex refractive index, giving more effective utilisation of the gain and a narrower linewidth. However, although it is important to reduce the nonuniformity of the intensity distribution, it is also important simultaneously to examine the side-mode-suppression ratio. Figure 5.12 shows the side-mode-suppression ratio mapped for the same variables as in Figure 5.11. The optimum design will then have to consider how a uniform mode distribution can coincide with large side-mode suppression.

Figure 5.13 illustrates, for all devices between a $\lambda_m/4$ phase shifted and a uniform DFB, the trade-off between flatness of the longitudinal mode intensity distribution and spectrum for varying position of the $2\times \lambda_m/8$ phase shifts for a κL product of 1.7.

Basic principles of lasers with distributed feedback 151

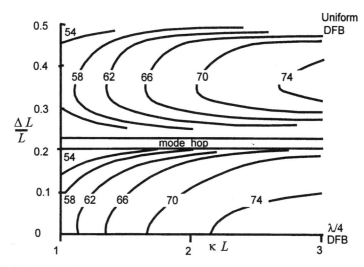

Figure 5.12 Side-mode-suppression ratio (dB) at 1 mW for $2 \times \lambda_m/8$ DFB lasers

Parameters give side-mode-suppression ratio in decibels, ordinate $\Delta L/L$ giving position of the phase shift

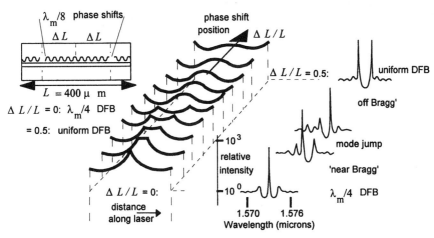

Figure 5.13 Optimisation of longitudinal-mode-intensity distribution and emission spectrum of $2 \times \lambda_m/8$ DFB lasers

The laser is 400 μm long with a drive current of 3000 A/cm² and $\kappa L = 1.7$

Detailed mapping of these parameters as a function of phase-shift position and κL product shows that a good optimisation is achieved for $\Delta L/L=0.35$ and $\kappa L=1.6$. In practice, the advantage of such a structure, over a $\lambda_m/4$-phase-shifted design with $\kappa L=1.25$, is a more uniform longitudinal-mode-intensity distribution. This is obtained for a higher-κL product which is, in fact, better matched to the fabrication technology so that imperfections and errors in fabrication remain small compared with the reflections from the grating and the wider stopband also means that the periodicity does not have to be quite so precise.

5.7 Influence of series resistance

The varying photon density along the cavity of a DFB laser gives rise to a longitudinal variation in the stimulated emission-current density which then changes the carrier density through the electronic/photonic interaction (eqn. 5.38). However, until now, each section of the laser has been driven with a fixed current. In reality, the current will depend on the series resistance of the electronic-drive supply and the series resistance of the semiconductor material between the metal contact and the actual resistance of the p–n junction. The internal series resistances can influence the longitudinal-mode spatial-hole-burning mechanism [11–13]. Even if there were the normal 50 Ω source impedance for the power drive to the laser, but there was a zero contact resistance to the n- or p-side metallic contacts with negligible voltage variation along the contacts, then there would be a flat electron or hole quasi-Fermi level, respectively, and the electron or hole densities would also be constant. There could then be no spatial-hole burning, although in regions of high photon density the local current density drawn from the contacts would be high in order to maintain the uniform carrier density. A high resistance to both contacts leads to a constant injected current density independent of junction voltage and the carrier-density variation along the cavity is then maximised.

In practice, the resistance lies between these extremes, and with present technology it is too large to reduce hole burning effects significantly, and the current-drive model is appropriate. The effect can be assessed straightforwardly. If the voltage at the junction reduces by $\Delta V_J \sim kT/q \sim 25$ mV along the cavity, the carrier density reduces by a factor ~2. It can be seen that to avoid spatial-hole burning it is necessary to keep $\Delta V_j \ll kT/q$. Now the relationship between changes in injected-current density ΔJ and junction voltage ΔV_j is given by

$$\Delta J = \frac{\Delta V_j}{\rho_c t_c} \tag{5.39}$$

where ρ_c is the resistivity and t_c is the thickness of the contact layer. For p-InP-doped material, the resistivity $\rho_c (= 1/q P \mu_{hole}) \sim 8 \times 10^{-4}$ Ωm for a hole density $P \sim 1 \times 10^{18}$ cm^{-3} and mobility $\mu_{hole} \sim 75$ cm^2/V s [14]. Hence for a contact layer of such material with $t_c = 2$ μm and $\Delta J \sim 1.5$ kA/cm^2 we find that $\Delta V_j \sim 25$ meV. The variations of the injected carrier density along the laser are determined by the variations in the stimulated photon rate. To avoid significant spatial-hole burning in such a laser, then, it is necessary for these variations to be much less than 1.5 kA/cm^2. Typical threshold-current densities are of this order of magnitude, so that spatial-hole burning is likely when one is operating at two or more times threshold.

If the device was fabricated on a p-type substrate, it would be appropriate to consider the voltage drop arising from injection through n-type material. A typical value of the resistivity of n-InP doped at 1×10^{18} cm^{-3} is about 3×10^{-5} Ωm [15] which would imply that longitudinal hole burning could be significantly reduced as a result of the low series resistance for such a device.

A more detailed analysis involves relating the junction voltage to the injected carrier density along the laser, taking into account the resistivity and thickness of the intervening layers, and the voltage difference between the contact and junction. The relationship between junction voltage and carrier density could be obtained using the conventional Fermi–Dirac functions for the occupation probability, but this approach requires an integration to obtain the total carrier density. An alternative is to use the approximate but usefully accurate analytic relationship between Fermi-level energy and carrier density for electrons and holes, respectively [16], defining contact parameters A_{con} and B_{con} from

$$\frac{N}{N_c} \sim A_{con}(1 + 0.15 A_{con}) \qquad \frac{P}{N_v} \sim B_{con}(1 + 0.15 B_{con}) \tag{5.40}$$

and

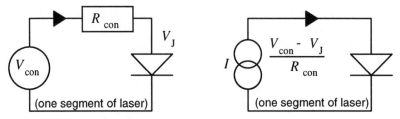

Figure 5.14 Voltage drive against current drive

The current $I = (V_{con} - V_J)/R_{con}$ where V_{con} is the contact voltage and R_{con} is the contact resistance with V_J the junction voltage. If changes in I lead to changes in V_J given by δV_J and these are such that $\delta V_J \ll (V_{con} - V_J)$, then a constant-current input is an adequate and appropriate approximation. Note that variation of contact resistances from segment to segment of the laser will also cause variations in the current

$$A_{con} = \log_e[1 + \exp\{(\mathscr{E}_{fN} - \mathscr{E}_c)/kT\}]$$
$$B_{con} = \log_e[1 + \exp\{(\mathscr{E}_v - \mathscr{E}_{fP})/kT\}] \qquad (5.41)$$

where $(\mathscr{E}_{fN} - \mathscr{E}_c)$ is the quasi-Fermi energy difference for the conduction band and $(\mathscr{E}_v - \mathscr{E}_{fP})$ is the quasi-Fermi energy difference for the valence band. The total junction voltage V_J is given from

$$V_J = \{\mathscr{E}_g + (\mathscr{E}_{fN} - \mathscr{E}_c) + (\mathscr{E}_v - \mathscr{E}_{fP})\}/q \qquad (5.42)$$

where $\mathscr{E}_g$ is the bandgap energy. The carrier-rate (eqn. 5.38), now with $dN/dt = 0$ in the steady state and an external applied contact voltage of V_{con}, is then modified to become

$$\frac{J}{qd} = \frac{(V_{con} - V_J)}{\rho t_c qd} = AN + BN^2 + CN^3 + G'_m(N - N_{tr})S(z) \qquad (5.43)$$

As technology improves to reduce the contact resistances, it will become important to incorporate this type of voltage drive into laser models. One can rapidly assess the relative importance by taking any section of the laser and looking at the Thevenin/Norton transformation from a voltage drive to a current drive, as in Figure 5.14. If, as the current changes, the changes in the active junction voltage $\delta V_J \ll (V_{con} - V_J)$, then the current-drive model remains a valid model because variations in V_J will have little effect on determining the current. This is equivalent to saying that the dynamic contact resistance $R_c \gg R_{junction}$ ($= \delta V_J/\delta I$).

5.8 Simulating the static performance of DFB lasers

5.8.1 Light/current characteristics

The dynamic time-domain modelling to be given in Chapter 7 can be used to simulate the static performance by switching the current from zero to its drive value and letting the laser settle down. The time taken to reach equilibrium is limited by the time taken for the electron densities to equilibrate and can be artificially speeded up multiplying dN/dt by a factor of around 3–5. Too large a factor for this speed-up will reduce the natural damping too much and can also induce numerical instabilities so that caution is required with this approach. Spontaneous emission can be switched off numerically once the photon density has been established and this gives 'ideal' profiles for the field, electron density and spectrum. Small steps in the drive help the laser to resettle more rapidly and so steady-state-light current characteristics can be obtained.

However frequency-domain modelling also has an important function, especially when discussing optical systems over many kilometre lengths where time-domain techniques require many millions of time steps to be stored. In frequency-domain analysis, the essence is to compute the performance over a large number of individual optical frequencies. It is usually only straightforward for conditions where there is negligible mixing of frequency components, allowing orthogonality of the spectral components to be assumed. This section outlines this approach to numerical modelling of lasers which has been especially useful in examining the detailed side-mode spectrum at different output powers from the laser facets at specified drive currents.

The first stage in any numerical analysis is to carefully define the laser's structure, i.e. dimensions, defining the sections of planar guide and grating, along with phase shifts between grating sections, and the coupling coefficient and period of grating sections etc. The two facet reflectivities and their phases relative to any adjacent grating section are also important [17]. In the numerical modelling, all devices are subdivided into subsections which will be assumed to have a uniform carrier density within them, but with adjacent subsections allowed to have differing carrier densities. The lengths of the subsections are chosen to be appropriate to the spatial resolution required. Unlike time-domain modelling, this spatial step size does not control the numerical stability, and large spatial steps can be used with adequate simulation of the device physics with the controlling factor being the

156 *Distributed feedback semiconductor lasers*

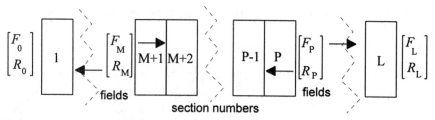

Figure 5.15 Subdivided laser for transfer-matrix methods

Electron density N is uniform in each subsection

nonuniformity of the gain along the laser's length. For example, a 400 μm long DFB laser with $\kappa L < 2$ in time-domain modelling requires several tens of subsections but often less than ten subsections will model the same laser satisfactorily in the frequency domain. As κL increases, the increasing nonuniformity of the gain may require more sections.

The key feature of frequency-domain modelling is the transfer matrix, described earlier. The laser is divided into a number of sections with a uniform electron density, κ, g and δ as already seen in Section 5.4, but allowing these to vary if required from section to section. The fields along the structure may be related through the transfer matrices at a given frequency. If the fields are labelled as in Figure 5.15, the fields in the middle may be related at a number of frequencies through the concatenated matrices:

$$\begin{bmatrix} F_m \\ R_m \end{bmatrix} = [M_{m+1}] [M_{m+2}] \ldots [M_{p-1}] [M_p] \begin{bmatrix} F_p \\ R_p \end{bmatrix} \qquad (5.44)$$

At the facets, the appropriate field reflectivities ρ_{right} and ρ_{left} give conditions $R_L = \rho_{right} F_L$ and $F_0 = \rho_{left} R_0$.

The method is iterative and, if there are large numbers of significant modes, then the iteration takes longer. It is therefore suited to DFB lasers where there are perhaps only three or four significant spectral modes which need to be tracked. The iteration starts by ignoring stimulated emission and calculating the carrier density in each subsection, allowing for the appropriate carrier-recombination terms (as in eqn. 4.2 but for static conditions). The effective refractive index of each subsection is then calculated allowing for the plasma effect and Henry's alpha factor (Chapter 2) [18] which relates the change in the real and imaginary refractive index with carrier density:

$$\alpha_H = \frac{-\left(\dfrac{dn_r}{dN}\right)}{\left(\dfrac{dn_i}{dN}\right)} \tag{5.45}$$

Locally this expression is a function of the properties of the waveguide, in terms of both the material properties and of the waveguiding characteristics [19]. The change in the active-region refractive index has to be diluted by the confinement factor of the mode to the active region in order to obtain the change in the effective guide refractive index:

$$\Delta n_r = \Gamma \frac{dn_r}{dN}\Delta N = \frac{-(\Gamma\alpha_H\lambda_0 \dfrac{dg}{dN}\Delta N)}{2\pi} \tag{5.46}$$

Once the carrier density and refractive index have been found in each subsection, the round-trip gain and phase, taking some reference point near the centre of the laser, are computed as a function of frequency using eqn. 5.44. From these transfer-matrix operations the forward and reverse fields at each frequency can be computed along the laser and from these fields the mean photon density at any frequency in each section may be found.

The peaks in the gain spectrum are the potential lasing modes. All the potential modes then have to be examined carefully to see which of these modes are at or above their own threshold gain for lasing. For each mode that is above its threshold gain, it will subsequently increase in power and therefore its associated forward and backward field amplitudes, at the modal frequency, in each subsection are stored along with the total photon density over all (say, M_L) such modes. The frequencies of these modes around threshold are also stored along with the photon densities in each section. Using this model, as discussed later in Section 6.2, the linewidth is 'Lorentzian' and this fact permits one to fit the solutions from a few frequencies around a potential mode so as to find the net optical power, thereby reducing the calculation time. Finding this photon density is an iterative process in itself, adjusting the carrier density in each subsection to allow for the mean stimulated emission rate for a self-consistent solution. Note that the refractive index is not altered at this stage because this would change the frequency of the mode significantly. The change in gain introduces only a small frequency shift.

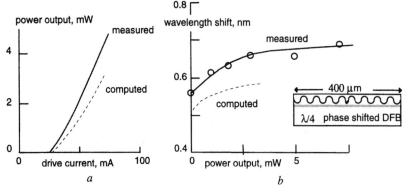

Figure 5.16 Static characteristics

For high κL (=3) $\lambda/4$ phase-shifted DFB showing curvature of the light/current characteristics, along with the proportional shift of the emission wavelength across the stopband

The next step is to find the electron density in each subsection from the steady-state carrier-rate equation, eqn. 4.2 (see also Appendix 4) modified to take into account the different recombination mechanisms and the total sum of all the photon densities S_M in all the M_L modes with different differential gains G'_{mM} and transparency densities $N_{tr\,M}$. The electron rate equation in the steady state now reads as

$$\frac{J}{ed} = AN(z) + BN(z)^2 + CN(z)^3 + \sum_{M=1}^{M_L} G'_{mM}\{N(z) - N_{tr\,M}\}S_M(z) \quad (5.47)$$

The process is iterated until the modal amplitudes and electron densities have stabilised. Now the refractive indices in each section are adjusted to allow for the changed carrier densities, the frequencies then recalculated and stored and the whole cycle reiterated until the longitudinal refractive-index profile has stabilised, giving the solution for the selected drive current.

This approach to solving an arbitrary DFB laser structure works very well where the device gives stable lasing in a single mode (see Figure 5.16). However, when a side mode increases in intensity, for example because of longitudinal-mode spatial-hole burning, the model can sometimes be unable to find a stable distribution of power between the main lasing mode and the side mode. Even with high side-mode-suppression ratios, intrinsic longitudinal instabilities in some grating structures occur [20, 21] and then it is necessary to use, say, a time-domain analysis to understand the dynamics of the instability.

Basic principles of lasers with distributed feedback 159

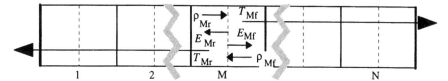

Figure 5.17 Spontaneous fields in Mth section (reflection and transmission)

5.8.2 Simulation of emission spectrum

When one requires detail for the spectra with kilohertz resolutions around the central line, time-domain methods at present give unacceptably long computation times, at least for desktop computing by impatient researchers! An alternative method examines the frequency domain and concentrates the computational resources around the required limited range of frequencies. Simulations are possible of both subthreshold spectra [22] and spectra of devices that are lasing [23]. The laser is first broken down into short subsections where the electron densities can be considered to be sufficiently 'uniform' and the self-consistent steady-state field solution of a DFB laser at a specified current can be found, as in Section 5.8.1.

The approach then is to calculate, over some requisite number of frequencies, the complex-field reflectivities of the whole laser structure, for both the forward- and backward-propagating waves (ρ_{Mf}, ρ_{Mr}), as seen from the centre of the Mth subsection into which the laser has been divided (Figure 5.17). In addition, the complex transmission coefficients (T_{Mf}, T_{Mr}) from the centre of the subsection to the emission from each facet must be calculated. These reflection factors and transmission factors through to the output should include any facet reflectivities (ρ_{right}, ρ_{left}) at the right- and left-hand facets, respectively, which contribute to the overall reflectivities and transmissivities (in some of the referenced work facet reflectivities are put in separately).

The spontaneous emission is considered to be excited by random uncorrelated spontaneous fields (E_{Mf}, E_{Mr}) generated at the centre of each subsection. These excitations are linked to the random uncorrelated spontaneous excitation 'currents' $i_{sp\,f}$ and $i_{sp\,r}$ in eqns. 5.6 and 5.7 by (E_{Mf}, E_{Mr}) $\simeq$ $\mathsf{S}_M(i_{sp\,f}, i_{sp\,r})$ where S_M is the length of the Mth section. The reverse wave E_{Mr} is immediately reflected into a forward wave $\rho_{Mr} E_{Mr}$ which adds to the forward wave E_{Mf} to give a net forward component ($\rho_{Mr}E_{Mr}+E_{Mf}$) which is then transmitted to the right output as $T_{Mf}(\rho_{Mr}E_{Mr}+E_{Mf})$ and also reflected into a reverse-wave component

$\rho_{Mf}(\rho_{Mr}E_{Mr}+E_{Mf})$. This latter field component is again reflected as ρ_{Mr} ρ_{Mf} $(\rho_{Mr}E_{Mr}+E_{Mf})$ and also transmitted to the output as $T_{Mf}\rho_{Mr}\rho_{Mf}$ $(\rho_{Mr}E_{Mr}+E_{Mf})$ etc. This process of reflections bouncing back and forth is sometimes referred to as a 'bounce diagram'. A geometric series of all these reflections and transmissions builds up to give an output, at the right-hand facet, of

$$E_{Ff} = \sum_M [T_{Mf}(\rho_{Mr}E_{Mr}+E_{Mf})\{1+(\rho_{Mr}\rho_{Mf})+(\rho_{Mr}\rho_{Mf})^2+\cdots\}]$$

$$= \sum_M [T_{Mf}(\rho_{Mr}E_{Mr}+E_{Mf})/\{1-(\rho_{Mr}\rho_{Mf})\}] \qquad (5.48)$$

The convergence of the series is assured because physically $|\rho_{Mr}\rho_{Mf}|^2<1$ for practical lasers with a finite spontaneous emission although $|\{1-(\rho_{Mr}\rho_{Mf})\}|^2 \gg 1$. The correlations of eqns. 5.8 and 5.9 are simplified to give time-averaged products of the random spontaneous fields from

$$\overline{|E_{M''r}^*E_{M'r}}=\overline{E_{M''f}^*E_{M'f}}=\overline{E_{M'f}^*E_{M'r}}=0$$

where $\qquad M' \neq M''$ and $|\overline{E_{Mr}^2}|=|\overline{E_{Mf}^2}|=|\overline{E_{Msp}^2}|$ $\qquad (5.49)$

The value of the mean square spontaneous emission $|\overline{E_{Msp}^2}|$ depends on the electron densities in the Mth section and evaluation of such spontaneous emission is discussed in Appendix 9 and Chapter 7. As each section produces uncorrelated fields excited by the spontaneous emission, the total output power from the right-hand facet is the sum of all the spontaneous emission powers from each section (including forward and reverse fields):

$$\sum_M \frac{|\overline{E_{Msp}^2}|\,|T_{Mf}|^2(1+|\rho_{Mr}|^2)}{|1-\rho_{Mr}\rho_{Mf}|^2} \qquad (5.50)$$

For the output from the left-hand facet, the subscripts f and r are interchanged.

It can be seen that, if there is zero spontaneous emission, then a finite output requires $|1-\rho_{Mr}\rho_{Mf}|^2=0$. The fact that $(\rho_{Mr}\rho_{Mf})$ is a regular complex function of frequency means that this is zero only at discrete real frequencies which define the permissible lasing frequencies. With finite spontaneous emission, one discovers, around the lasing frequency ω_0, that $|1-\rho_{Mr}\rho_{Mf}|=1-G_{Mr}-j\,2(\omega-\omega_0)\,Y_{Mr}$ where G_{Mr} and Y_{Mr} are parameters associated with the round-trip gain (see

Basic principles of lasers with distributed feedback 161

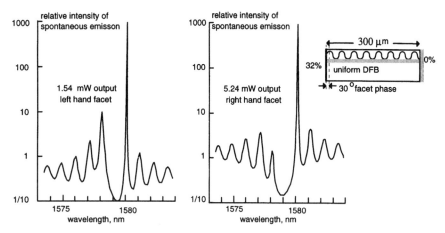

Figure 5.18 Typical emission spectra

Uniform DFB with high-low power reflectivities of 32% and 0%, $\kappa L=2.45$, $I_{thr}=23.8$ mA (2263 A/cm^2), spectra shown for current drive $I=42$ mA

also Section 6.2) as seen from the centre of the Mth section and ω_0 is the same 'lasing' frequency for all values of M. This form of the denominator, as a function of frequency, ensures that the line shape of any semiconductor laser is effectively a Lorentzian line shape within the limitations of the model; and fitting a few frequencies to this shape permits the prediction of the output around the lasing frequency, without the need to compute the precise maximum output frequency. Moreover, it is found that, if $|1-\rho_{Mr}\rho_{Mf}|^2=0$ for the Mth cell, then it is also zero for the M'th cell and the calculation of the lasing frequency is independent of which cell is chosen as the 'central' cell but is determined by the total effect of all the sections. However, the closeness of the denominator to zero in eqn. 5.50 sometimes leads to errors unless the local gain is iterated appropriately. Figure 5.18 shows typical emission spectra calculated by this method, which has also been formalised into a technique referred to as the power-matrix method [24]. Figure 5.16b shows the spectral relative shifts within the stop band for a $\lambda_M/4$ phase shifted DFB.

5.9 Summary

The chapter started with a discussion of the physics of the grating-coupling process, a section which should be read in conjunction with work in Sections 1.6 and 4.4. This section gave alternative insights into this 'coupling' physics and extended the concepts to gratings with periodic gain variations as well as periodic index variations. The eigenmodes of the coupling equations provide an elegant solution but, when concatenating different sections with phase shifts and facets, it is usually more straightforward to use the 'forward' and 'reverse' field components of the coupled-wave equations. Inside a 'stopband', these field components evanesce rather than propagate. The term 'stopband' was extended from this mathematical concept of evanescing fields to a term indicating the range of frequencies between the two spectral modes at the edges of the stopband in a DFB. These two modes can usually be observed in a careful spectral measurement close to threshold looking at the external spontaneous emission which is sculpted by the frequency selective reflections within the laser cavity. Stopband/distance diagrams were introduced giving analogies with the semiconductor counterpart of energy-band/distance diagrams. Matrix methods of solving the fields at arbitrary frequencies were developed and used to explore the stopbands along with the addition of phase shifts within the laser structure. The concept of spatial-hole burning was then discussed along with the use of the phase shifts to reduce the effects of spatial-hole burning and increase the field uniformity and also improve efficiency and linearity.

Practical laser technology is always improving, and in modelling some modern well constructed lasers it is important to note that the very low contact resistance can lead to voltage-driven lasers rather than current-driven lasers and methods that can be used to extend the work in this book (where laser models are primarily current-driven) were discussed.

Finally, a discussion was given on the use of matrix methods of reflection and transmission to construct the light/current characteristics and emission spectrum starting from the random spontaneous emission fields that excite the laser.

5.10 References

1 KOGELNIK, H., and SHANK, C.V.: 'Coupled-wave theory of distributed feedback lasers', *J. Appl. Phys.*, 1972, **43**, pp. 2327–2335
2 THOMPSON, G.H.B.: 'Physics of semiconductor laser devices' (John Wiley & Sons, 1980), pp. 480–481
3 WANG, S.: 'Principles of distributed feedback and distributed Bragg-reflector lasers', *IEEE J. Quantum Electron.*, 1974, **10**, pp. 413–427
4 THOMPSON, G.H.B.: 'Physics of semiconductor laser devices' (John Wiley & Sons, 1980), pp. 501–502

5 WHITEAWAY, J.E.A., THOMPSON, G.H.B., COLLAR, A.J., and ARMISTEAD, C.J.: 'The design and assessment of $\lambda/4$ phase-shifted DFB laser structures', *IEEE J. Quantum Electron.*, 1989, **25**, pp. 1261–1279
6 IGA, K.: 'On the use of effective refractive index in DFB laser mode separation', *Jpn. J. Appl. Phys.*, 1983, **22**, p. 1630
7 ZHANG, L.M., and CARROLL, J.E.: 'Dynamics and hole burning in uniform DFB semiconductor lasers', *Int. J. Optoelectron.*, 1993, **8**, pp. 279–291
8 MARCENAC, D.D., and CARROLL, J.E.: 'Comparison of self-pulsation mechanisms in DFB lasers'. Presented at (Baltimore) *IEEE Lasers and Electro-Optics Society 7th annual meeting*, 1994, Paper SL8.5
9 NOWELL, M.C., CARROLL, J.E., PLUMB, R.G.S., MARCENAC, D.D., ROBERTSON M.J., WICKES, H., and ZHANG L.M.: 'Low chirp and enhanced resonant frequency by direct push pull modulation of DFB lasers', *IEEE J. Sel. Top. Quantum Electron*, 1995, **1**, pp. 433–441
10 WHITEAWAY, J.E.A., GARRETT, B., THOMPSON, G.H.B., COLLAR, A.J., ARMISTEAD, C.J., and FICE, M.J.: 'The static and dynamic characteristics of single and multiple phase-shifted DFB laser structures', *IEEE J. Quantum Electron.*, 1992, **28**, pp. 1227–1293
11 LASSEN, H.E., WENZEL, H., and TROMBORG, B.: 'Influence of series resistance on modulation responses of dfb lasers', *Electron. Lett.*, 1993, **29**, pp. 1124–1126
12 CHAMPAGNE, Y., and MCCARTHY, N.: 'Influence of the axially varying quasi-Fermi-level separation of the active region on spatial hole burning in distributed-feedback semiconductor-lasers', *J. Appl. Phys.*, 1992, **72**, pp. 2110–2118
13 BANDELOW, U., WENZEL, H., and WUNSCHE, H-J.: 'Influence of inhomogeneous injection on sidemode suppression in strongly coupled DFB semiconductor-lasers', *Electron. Lett.*, 1992, **28**, pp. 1324–1326
14 HAYES, J.R., ADAMS, A.R., and GREENE, P.D.: 'Low-field carrier mobility', *in* PEARSALL, T.P. (Ed.): 'GaInAsP alloy semiconductors' (John Wiley & Sons, 1982), chap. 8, p. 204.
15 HAYES, J.R., ADAMS, A.R., and GREENE, P.D.: 'Low-field carrier mobility' *in* PEARSALL, T.P. (Ed.): 'GaInAsP alloy semiconductors' (John Wiley & Sons, 1982), chap. 8, p. 202
16 UNGER, K. 'Spontaneous and stimulated emission in junction lasers', *Z. Phys.*, 1967, **207**, pp. 322–331
17 SODA, H., and IMAI, H.: 'Analysis of spectrum behaviour below the threshold in DFB lasers', *IEEE J. Quantum Electron.*, 1986, **28**, pp. 637–641
18 THOMPSON, G.H.B.: 'Physics of semiconductor laser devices' (John Wiley & Sons, 1980), pp. 535–537
19 OSINKSI, M., and BUUS, J.: 'Linewidth broadening factor in semiconductor lasers - an overview', *IEEE J. Quantum Electron.*, 1987, **23**, pp. 9–28
20 SCHATZ, R.: 'Longitudinal spatial instability in symmetric semiconductor lasers due to spatial hole burning', *IEEE J. Quantum Electron.*, 1992, **28**, pp. 1443–1449

21 GOOBAR, E., RIGOLE, P.J., and SCHATZ, R.: 'Correlation measurements of intensity noise from the two facets of DFB lasers during linewidth rebroadening', *Electron. Lett.*, 1992, **28**, pp. 1542–1543
22 SODA, H., and IMAI, H.: 'Analysis of spectrum behaviour below the threshold in DFB lasers', *IEEE J. Quantum Electron.*, 1986, **22**, pp. 637–641
23 WHITEAWAY, J.E.A., THOMPSON, G.H.B., COLLAR, A.J., and ARMISTEAD, C.J.: 'The design and assessment of $\lambda/4$ phase-shifted DFB laser structures', *IEEE J. Quantum Electron.*, 1989, **25**, pp. 1261–1279
24 ZHANG, L.M., and CARROLL, J.E.: 'Large signal dynamic model of the DFB laser', *IEEE J. Quantum Electron.*, 1992, **28**, pp. 604–611
25 KINOSHITA, J., and MATSUMOTO, K.: 'Yield analysis of SLM DFB lasers with an axially-flattened internal field', *IEEE J. Quantum Electron.*, 1989, **25**, pp. 1324–1332

Chapter 6
More advanced distributed feedback laser design

6.1 Introduction

Chapter 5 considered some basic features of DFB lasers, concentrating on the static performance. However, several key features for high-performance lasers were not discussed such as linewidth, the influence of reflections and especially the phase of weak-facet reflections, the role of complex gratings and what happens on designing for power levels in the hundreds of milliwatt range rather than the milliwatt range of conventional communication lasers. These more advanced problems of the static design of lasers are outlined here and the chapter ends with a discussion on some results of modelling the dynamic performance of DFB lasers, considering problems associated with carrier transport into quantum wells. The dynamic performance of DFB lasers highlights yet further the problems that have already been met with a uniform grating in a uniform DFB laser. Change of mode with time, dynamic instabilities, yield of devices with the right mode etc. prove to be major problems unless the laser has additional features.

As seen in Chapter 5, the use of one or more phase shifts offers considerable improvement in the performance of a laser, and this chapter concentrates on the dynamic performance of a DFB laser with two phase shifts. It is generally agreed that one phase shift [1], although stabilising the mode at low power levels, does not offer the solution to all the dynamic problems. Three phase shifts have been advocated [2], and continuous changes in the grating period/coupling constant along the laser can be of interest [3, 4] but, from a practical manufacturing view, two phase shifts offers a good compromise between keeping the design as straightforward as possible while maintaining excellent dynamic performance over a range of power levels [5]. While the theoretical advantages for gratings with periodic changes in gain as well as periodic changes in refractive index have long been recognised [6], it is only relatively recently that the technology has indicated that such complex gratings are another practical way of achieving the required improvements in dynamic performance of the uniform DFB [7]. These then are some of the topics investigated in this chapter.

6.2 Linewidth

6.2.1 General

The spectral channels in wireless-communication systems are determined by precise radio or microwave carrier frequencies with the spacing between these channels determined by the linewidth of the *modulated* carrier. At radio frequencies, the modulation bandwidth dominates this linewidth so that one can say (for the amplitude modulation format at a maximum frequency of f_M and allowing for the upper and lower sidebands) that the modulated carrier linewidth is close to $2f_M$. The principle remains with other modulation techniques—the modulated linewidth is determined by the modulation and its format. Channel spacing follows with appropriate spectral 'guard bands' to ensure that negligible interference from neighbouring channels can statistically occur. The DFB laser, with its ability to determine the optical frequency, can define channels in an optical communication system so that its linewidth and the accuracy of specifying its frequency become important system parameters in determining minimum optical channel spacing for a WDM system.

Unfortunately, even when the modulation frequencies $f_M \sim 0$ the linewidth of a DFB laser is significant and cannot be ignored. But worse, as the modulation frequency increases then the carrier's linewidth broadens much more than the value of $2f_M$ that would be considered for the classical radio system mentioned above. The laser exhibits frequency shifts called dynamic chirp because of changes in refractive index caused by changes of gain as the laser is modulated, as outlined in Sections 2.5 and 4.2 (Figure 4.4). On switching the laser on and off there are changes in gain, modulated by the photon–electron resonance; hence the laser is frequency-modulated at the photon–electron resonance frequencies by an amount which depends on the

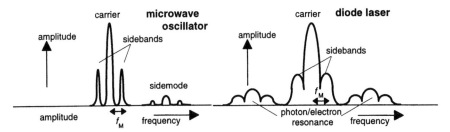

Figure 6.1 Schematic spectrum of modulated microwave oscillator and diode laser

linewidth-broadening factor α_H and the change in gain (eqn. 4.15). The different sidebands to the modulated laser's spectrum are then very different from those of a classical oscillator and play a more dominant role. Figure 6.1 indicates schematically these differences of spectrum of a modulated classical (microwave) oscillator and a modulated semiconductor laser taking the modulation frequency f_M to be a few hundred megahertz in both examples.

Consider initially the unmodulated laser. The complex optical amplitude from a laser can, to a first approximation, be thought of as performing a random walk (see Figure 7.5). Fourier analysis of the laser-output field $E_{out}(t) = E_0(t)\cos\{2\pi ft + \phi(t)\}$ typically gives a Lorentzian *power spectrum*, at least to a useful first approximation, as shown in Figure 6.2 (showing both linear and logarithmic representations) along with a schematic side mode for discussion shortly.

In standard electrical systems, the full width at the half-maximum-power points or FWHM measurement of the spectrum is often used. In optical systems, it is sometimes the -20 dB spectral width which is of concern. In well constructed DFB lasers, the steady-state -20 dB linewidth can be as low as 1–10 MHz or, in terms of the incremental wavelength, $\Delta\lambda \sim 1/1000$ Å (10^{-13} m). As already noted, on modulating the laser drive current, the linewidth-enhancement factor α_H causes the modulated linewidth to increase many times over the ideal value of $2f_M$ discussed above but also in a laser, with a high α_H it can increase the steady-state linewidth by an order of magnitude or more. This section aims to explain some of these effects using fundamental concepts.

Another important feature for any spectrum is its sidemode-suppression ratio (SMSR). Sidemodes always appear in any optical spectrum but, provided that the net power associated with these is more than 30 dB below the power in the main optical mode, they are not usually significant in the performance of most systems. Figure 6.2

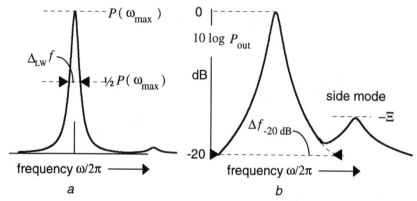

Figure 6.2   Linewidth and SMSR

A Lorentzian line gives a mean power-frequency relationship according to

$$P(\omega) = P(\omega_{max})/\{1 + (\omega - \omega_{max})^2/(\pi\Delta_{LW}f)^2\} \quad (\omega = 2\pi f)$$

so that the separation of the half-power points gives the FWHM of $\Delta_{LW}f$:
(a) Linear ordinate (b) Logarithmic ordinate—frequency scale the same.
The sidemode-suppression ratio (SMSR) is the ratio of the power in the largest sidemode to the power in the fundamental. SMSR = $-\Xi$ decibels in (b)

indicates an unsatisfactory SMSR of around −15 dB. Sidemodes are also considered in this chapter.

6.2.2 Calculation of linewidth under static conditions

A full dynamic time-domain model as that outlined in Chapter 7 (which allows for spatial distribution of noise sources and fields and all significant frequencies) can automatically determine the linewidth under both static and dynamic conditions, and demonstrate effects of the laser's structure on the linewidth along with phenomena such as linewidth broadening and rebroadening. The TLLM method [8] has provided an appropriate dynamic model demonstrating the key effects but, for the present, the aim is to give the reader a good understanding of the physics as a well as an ability to compute laser behaviour.

The detailed explanation of the linewidth for a semiconductor laser requires considerable discussion [9] pioneered by Henry [10–12]. The analysis here gives a simplified physical explanation which is sufficiently general to apply to most semiconductor lasers. By way of introduction it is easier to consider first an idealised model for the linewidth of a uniform laser (represented approximately by a single-moded Fabry–Perot laser). The calculation is done in stages. The first stage is to regard the laser output as amplified spontaneous emission, with feedback centred around a single spectral line at ω_0. Consider then the right hand optical field $\tilde{E}(\omega)$ just inside the laser with its net round-trip complex-field gain given from the expression $G_{round}\exp\{-2j\beta(\omega)L\} \approx \{G_{round} - j(\omega-\omega_0)(2G_{round}\ L/v_g)\}$ for small changes in frequency $(\omega-\omega_0)$ from the lasing frequency ω_0. Here, to obtain the right phasing of the feedback, $\exp\{-2j\beta(\omega_0)L\}=1$, i.e. a whole number of wavelengths for the round trip in the medium. In the absence of any significant spontaneous emission, the ideal lasing condition is that the total round-trip complex gain is exactly unity. However, in the presence of spontaneous emission—represented here as additional fields $\tilde{E}_{sp}(\omega, z)$ which also experience a round-trip gain G_{round} but with a relative phasing which depends on the time and position of emission—the oscillation requirement to match the net field inside the laser on one round-trip pass is given from an expression of the form:

$$\tilde{E}(\omega)\{G_{round} - j2(\omega-\omega_0)Y_r\} + \Sigma_z G_{round}\exp\{-2j\beta(\omega)(L\pm z)\}\tilde{E}_{sp}(\omega, z)\ \delta z$$
$$= G_{thr}\tilde{E}(\omega) \qquad (6.1)$$

where, for the steady-state oscillation, one may write the threshold round-trip gain, $G_{thr}=1$ and $Y_r=(G_{round}\ Ln_g/c)$. The term (Ln_g/c) is the effective value of the group transit time through the laser and, as this increases, the stored electromagnetic energy U increases in the laser. As noted in Sections 3.3 and 5.2, phase factors, multiplying random and spontaneous emissions which are uncorrelated over the length of the laser, do not alter the spatial autocorrelation and correlation functions so that, to a first order, where $G_{round} \approx G_{thr}=1$, one can average the spontaneous emission over the whole laser and write the internal field close to a facet in the general form

$$\tilde{E}(\omega) = \tilde{E}_{sp\ tot}(\omega)/\{G_{thr} - G_{round} + j2(\omega-\omega_0)Y_r\} \qquad (6.2)$$

This result can also be interpreted as a circuit theory for a laser [13] where $(G_{thr} - G_{round})$ is the net conductance and $2(\omega-\omega_0)Y_r$ is a circuit susceptance. Work by Slater [14] shows that many electromagnetic oscillators can be interpreted using such an equation, where $\omega_0 Y_r = Q\ G_{thr}$ with Q giving the quality factor of the circuit in the absence of material gain or loss. Hence Q gives a measure of the stored energy which oscillates back and forth between electric and magnetic forms at twice the oscillation (lasing) frequency. This circuit theory then confirms the link already made between stored energy U and the term Y_r through the simpler argument using the group transit time.

The round-trip complex-gain terms G_{round} and $Y_r = G_{thr} Q/\omega_0$ are properties of the oscillation/lasing mode centred at the optical frequency ω_0. The normalised output field may then be written as

$$\tilde{E}_{out}(\omega) = T\tilde{E}_{sp\,tot}(\omega) / \{1 - G_{round} + j2(\omega - \omega_0) Q/\omega_0\} \quad (6.3)$$

where T is an appropriate transmission term linking the interior fields with the fields emitted from the facet. At this stage it is helpful to normalise $|\tilde{E}_{out}|^2 \delta\omega$ to be proportional to power in the range $\delta\omega$. With spontaneous emission noise approximately uniform and covering a large enough band of frequencies and T varying slowly with frequency, eqn. 6.3 reveals that all such oscillators are expected to have a 'Lorentzian' line shape which is a term given to any laser where the net power output per unit frequency as a function of frequency is given by the general form

$$P_{out}(\omega) = |TE_{sp\,tot}|^2 / [\{1 - G_{round}\}^2 + \{2Q(\omega - \omega_0)/\omega_0\}^2] \quad (6.4)$$

The -3 dB linewidth $\Delta_{LW} f$ may be found directly:

$$2\pi \Delta_{LW} f = \omega_0 (1 - G_{round})/Q \quad (6.5)$$

The net power output $P_{net\,out}$ is found, using a standard integral $\int_{-\infty}^{\infty} d\Omega/(1+\Omega^2) = \pi$, from

$$P_{net\,out} = \int_{-\infty}^{\infty} P_{out}(\omega)\, d\omega = P_{net\,sp}/(1 - G_{round}) \quad (6.6)$$

where $P_{net\,sp} = \pi\omega_0 |TE_{sp\,tot}|^2/2Q$ gives the net spontaneous emission over a spontaneous emission-noise bandwidth $\sim \pi\omega_0/2Q$ in this model. Hence the linewidth may also be written as

$$\Delta_{LW} f/f_0 = P_{net\,sp}/(QP_{net\,out}) \quad (6.7)$$

The result of eqn. 6.7 is essentially the well known Schawlow–Townes formula [15] for the linewidth of a laser. This circuit model shows that the linewidth decreases with increasing output power (provided that the laser remains in a single mode) and decreases as the ability of the laser to store energy increases (e.g. increased length of laser so that the laser's equivalent circuit Q-factor increases). Unfortunately, this straightforward theory that is useful for gas and solid state lasers as well as microwave oscillators usually gives over an order-of-magnitude *too low* an estimate of the static linewidth and a totally inadequate approximation to the dynamic linewidth, and this must be explained.

6.2.3 Linewidth enhancement

In Section 2.5 and Appendix 7, the relationship between gain and phase was discussed and, in particular, it was shown that, because of the Kramers–Kronig relationships, gain changes have to be accompanied by phase changes. In the first instance, the relative changes in gain and phase are associated purely with the properties of the material. An increase in electron density ΔN results in an 'increase' Δn in the complex refractive index, $\Delta n = \Delta n_r + \Delta n_i$, where for small enough changes,

$$-\left(\frac{\Delta n_r}{\Delta N}\right) \bigg/ \left(\frac{\Delta n_i}{\Delta N}\right) = \alpha_H \qquad (6.8)$$

with $\alpha_H (>0)$ known as Henry's linewidth-enhancement factor. The negative sign shows that an increase in electron density decreases the magnitude of the refractive index. Values of α_H vary around 4–8 for bulk material and around 1–3 for quantum-well material and also vary with wavelength. In eqns. 2.42 and 4.15 it was recalled how the value of α_H caused changes in gain to induce frequency chirp, directly broadening the modulated linewidth over and above the conventional Fourier limit. Less obviously, α_H also enhances the linewidth in the steady state and the previous analysis can be modified to help explain this effect.

The first essential feature to understand is that clamping the round-trip gain to unity makes the feedback process in the laser stabilise the intensity of the optical fields. The spontaneous emission events $[\tilde{E}_{sp}(z, \omega) \, \delta z]$ increase the optical intensity and to stabilise the intensity the sum of these emission events creates a small decrement $\delta G_{round \, sp}(\omega)$ in the round-trip field gain. Eqn. 6.3 therefore is rewritten as

$$\tilde{E}_{out}(\omega)\{1 - G_{round} + j2(\omega - \omega_0) Q/\omega_0\} - \delta G_{round \, sp} \tilde{E}_{out}(\omega) = 0 \quad (6.9)$$

with $\delta G_{round \, sp} \tilde{E}_{out}(\omega)$ replacing $T \tilde{E}_{sp \, tot}(\omega)$ in eqn. 6.3 and recognising the slight change in the round-trip gain caused by the spontaneous emission. However, once one recognises such a change in the real part of the round-trip gain one sees from eqn. 6.8 that there has to be an associated phase change modifying eqn. 6.9 further to become

$$\tilde{E}_{out}(\omega)\{1 - G_{round} + j2(\omega - \omega_0) Q/\omega_0\} - \delta G_{round \, sp}(1 + j\alpha_H) \tilde{E}_{out}(\omega) = 0 \ (6.10)$$

This phase change in the spontaneous term is now no longer random as in eqn. 6.1, but is caused by the effect of the combined spontaneous emissions. Replacing $\delta G_{round\ sp}\tilde{E}_{out}(\omega)$ in eqn. 6.10 with $T\tilde{E}_{sp\ tot}(\omega)$ once again then modifies eqn. 6.4 to give

$$P_{out}(\omega) = (1+\alpha_H^2) \mid T\tilde{E}_{sp\ tot}\mid^2 / [\{1 - G_{round}\}^2 + \{2Q(\omega - \omega_0)/\omega_0\}^2] \quad (6.11)$$

and with similar assumptions as before leads to a linewidth

$$\Delta_{LW} f / f_0 = (1+\alpha_H^2) P_{net\ sp} / (Q P_{net\ out}) \quad (6.12)$$

the linewidth enhancement by a factor of $(1+\alpha_H^2)$ being caused by additional phase noise through the term $j\alpha_H$ in eqn. 6.10. In spite of this enhancement factor, practical linewidths for good DFB lasers can be reduced to the 100 kHz range with frequency selective feedback.

Recapitulating, a random increase in the spontaneous emission reduces the carrier density and hence reduces the gain which reduces the mode intensity allowing the carrier density to recover to re-establish the round-trip gain close to unity. Fluctuations in the net gain are limited, and it is the phase fluctuations which are important and add to the linewidth through frequency modulation. Although the round-trip gain G_{round} is clamped close to unity, the net overall gain for the spontaneous emission is determined from $(1 - G_{round})^{-1}$ and this value peaks at values $\sim 10^3 - 10^4$ at frequencies close to the centre of any modal frequency and this large value of net gain will explain why reflections even as low as 0.1% can still have considerable effect on a laser's performance. The reader wishing for more rigour is advised to read Henry's original work [10–12].

6.2.4 Effective linewidth enhancement

To explain the values of α_H, as measured by an observation of linewidth broadening in DFB-laser structures, the structural effects caused by the grating have to be recognised as also changing the effective value of α_H from the intrinsic value for the basic material. The problem is that the gain and phase terms, in the denominator of eqn. 6.11, have to be calculated for a complete pass around the laser cavity. However, in a DFB laser, changes in the carrier density alter the strength and phase of the feedback from the grating. The effective value of α_H in eqn. 6.12 is then a composite value which takes into account the round-trip average over the whole structure and includes the effects from the material, from the waveguide and from spatial-hole burning.

The generality of eqns. 6.10 and 6.11, which go back to the roots of electromagnetic theory, forms a basis to estimate or measure the effective value of α_H for arbitrary laser structures. Consider a change in output intensity which requires a small change δG_{round} in the round-trip gain of the laser; then this has to be accompanied by a phase change to give a total round-trip change of complex gain of $\delta G_{round}(1+j\alpha_H)$. The oscillation/lasing condition (neglecting changes in the spontaneous emission) is therefore effectively

$$\tilde{E}_{out}(\omega)\,[(1 - G_{round} - \delta G_{round}) + j2Q\{(\omega - \omega_0)/\omega_0\} - j\delta G_{round}\alpha_H]$$
$$- (1+j\alpha_H)\,T\tilde{E}_{sp\,tot} = 0 \qquad (6.13)$$

From eqn. 6.13, the change in intensity causes a change in the central frequency of $\delta f_0/f_0 = -\alpha_H(\delta G_{round}/2Q)$. Further, because the *intrinsic* 3 dB linewidth is $(\Delta_{LW}f) = \omega_0(1 - G_{round})/2\pi Q$, this also has to change by $\delta(\Delta_{LW}f) = -\omega_0 \delta G_{round}/2\pi Q$. The effective value of α_H therefore may be determined from δf_0 and $\delta(\Delta_{LW}f)$ to give

$$\alpha_{H\,eff} = \frac{\delta f_0}{\delta(\Delta_{LW}f/2)} \qquad (6.14)$$

This value usually will not be the same as the material's intrinsic linewidth-enhancement factor given in eqn. 6.8 but is modified by the field distributions and waveguiding within the laser.

To use eqn. 6.14 to find numerically the effective linewidth-enhancement factor, first calculate the steady-state solution (for example as in Figure 6.3a for a $\lambda_m/4$ phase-shifted laser with $\kappa L>2$). The mode intensity everywhere is, say, enhanced by ±1% and the new charge carrier densities $N_+(z)$ and $N_-(z)$ are recalculated from, say, eqn. 5.38. These then change the local refractive index $n_+(z)$ and $n_-(z)$ which in turn leads to new central frequencies f_{0+} or f_{0-} where there is unity complex round-trip gain. The two linewidths for these two conditions are calculated so that the net change in the central frequency, given from $\delta f_0 = f_{0+} - f_{0-}$, along with the net change in the full bandwidth $\delta(\Delta_{LW}f) = \Delta_{LW}f_+ - \Delta_{LW}f_-$ then give the effective value of $\alpha_{H\,eff}$ from eqn. 6.14.

This approach has been used successfully to predict the effective linewidth-enhancement factor for $2\times\lambda_m/8$ DFB laser structures with varying κL product and varying phase-shift position $\Delta L/L$. Figure 6.4 plots the effective linewidth-enhancement factor for a uniform ($\Delta L/L=\frac{1}{2}$) and $\lambda_m/4$-phase-shifted ($\Delta L/L=0$) DFB laser as a function of output power and the intrinsic linewidth-enhancement factor of the material. The DFB structure significantly increases the effective α_H at

174 *Distributed feedback semiconductor lasers*

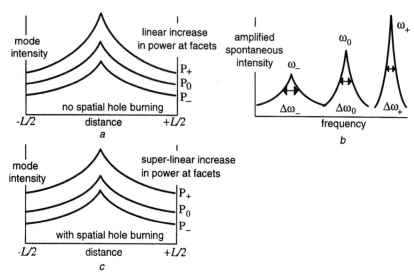

Figure 6.3 Schematic diagrams illustrating general method for calculating effective linewidth-enhancement factor allowing for structural effects

low power levels where a change in the power level creates more change in spatial-hole burning than at high power levels. With the phase-shifted DFB the differences between the effective and intrinsic values of α_H are small at both high and low powers, with larger differences mainly at intermediate powers.

Another striking example of structural effects on the effective value of α_H is given by a DFB laser when there is a complex coupling coefficient $\kappa_{index}+j\kappa_{gain}$. For such a grating with the notation used in this book, the peaks in the gain and the refractive-index periodicity coincide when κ_{index} and κ_{gain} are both positive ('in phase'). With $\alpha_H>1$ an increase in the carrier density increases the strength of the gain grating, but weakens the index grating by a greater amount and reduces the round-trip feedback or round-trip gain. The net differential round-trip gain with respect to carrier density is reduced thereby increasing the effective linewidth-enhancement factor. Alternatively, with $\kappa_{index}>0$ but $\kappa_{gain}<0$ then both the gain and index gratings increase in strength giving a lower cavity loss. The net differential mode gain is enhanced thereby reducing the linewidth-enhancement factor and hence the linewidth.

A more sophisticated analysis of the result given in eqn. 6.14 has been given for a Fabry–Perot laser [17]. Provided that one uses the effective α_H, eqn. 6.14 still holds regardless of the laser being a DFB or FP. However, care is needed if one attempts to estimate the linewidth

More advanced distributed feedback laser design 175

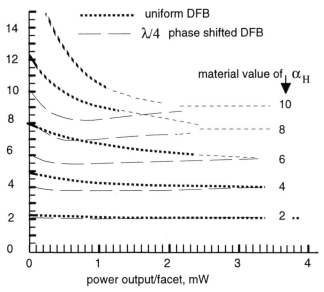

Figure 6.4 Estimated effective linewidth-enhancement values for uniform and λ/4 phase-shifted DFB lasers for different values of material linewidth-enhancement values

Laser is 400 μm long, with $\kappa = 75$ cm^{-1}. Results at 'zero' power agree with Amann [16] who assumed no spatial-hole burning

for a DFB laser on a similar basis to the calculation for a FP laser: the DFB can have half an order of magnitude larger linewidths. As outlined in Section 2.4.2, when using standard rate equations, it was necessary to modify the spontaneous-emission terms by the Petermann factors, K_{tr} to account for enhanced spontaneous emission transversely, and K_z to account for enhanced effects of spontaneous emission because of longitudinal distributions of optical intensity. The net effect is a spontaneous-emission factor $K_{tot} = K_{tr} \times K_z > 1$. The reasons for a significant K_z in certain types of DFB lie in these having end sections acting more as amplifiers rather than as feedback elements. The factor K_{tot} is then very approximately a multiplier of the spontaneous emission that would be obtained in an ideal uniform FP laser of similar material and length. A similar factor, K_{tr}, which applies to gain guiding that can occur horizontally under the injecting contact of a laser structure [18] where also there is an increased spontaneous-emission rate into the lasing mode leading to a larger linewidth–power product.

Now it must be emphasised that, if one uses a dynamic time-domain model, similar to that in Chapter 7 where longitudinal spatial distributions are taken into account and also good index guiding is assumed, then one inserts the material's intrinsic value of α_H into the calculation and the main effects referred to above automatically produce the right overall effective value of α_H with an appropriate K_{tot} hidden in the numerical analysis. There is then no need to refer to an equivalent FP though the comparison may help in understanding the physics.

6.2.5 Effective dynamic linewidth enhancement

The dynamic linewidth is increased well over the value of the static linewidth of eqn. 6.12, and the discussion here merely refers the reader back to Section 4.2. To gain an estimate without detailed numerical modelling, the appropriate value of α_H will again be the effective value averaged over the laser. From eqn. 4.15, the linewidth will then be increased by $\delta(\Delta\lambda_{+LW}) \sim (\lambda_{fs}^2/2\pi)\,(\alpha_H \Delta g/n_r)$. To gain an estimate of this increase, one observes that, as the modulation frequency increases, so the change in gain has to increase to change the waveform sufficiently rapidly within the modulation period. One might require 10 dB power gain within a quarter of a modulation period, so that with a period T one finds $\exp(2\Delta g\, v_g\, T/4) > 10$ giving roughly $\Delta g \sim 5\,(f_M/10^{10})$ cm^{-1} for a modulation frequency f_M. With this estimate, and even with $\alpha_H = 1$, $\delta(\Delta\lambda_{+LW}) \sim 0.5\,(f_M/10^{10})$ nm or 0.5 nm for a 1.55 µm laser modulated at 10 GHz.

6.2.6 Linewidth rebroadening

Although the modifications to the Schawlow–Townes formula increase the linewidth by $(1+\alpha_H^2)$, the linewidth is still predicted to decrease monotonically with increasing output power. In practice, however, the linewidth rebroadens at some critical power (for typical communication lasers this critical power level ~10–30 mW) and this requires a more detailed consideration of the physics which is only overviewed here.

Three main mechanisms have been identified which can give rise to this rebroadening of the laser linewidth. To understand the first mechanism, recall the effects of clamping of the round-trip gain to near unity on stabilising the amplitude fluctuations caused by the spontaneous emission (recall the derivation of eqn. 6.11). However, the subthreshold modes, with round-trip gains significantly less than unity, do not clamp the gain and, in such modes, the amplitude fluctuations induced by the spontaneous emission give them a relatively broad linewidth. At high enough powers, the negatively cross-correlated noise in the main and side modes, combined with a difference in their nonlinear gain coefficients, give rise to rebroadening of the main lasing mode. A detailed analysis [19] then predicts an extra linewidth contribution $\Delta_{LW+}f$ which varies with the cube of the sidemode intensity P_{SM}, and as the square of the difference $\Delta\epsilon$ in the nonlinear gain coefficients ϵ for the main and side modes:

$$\Delta_{LW+}f = \frac{\alpha_H^2 \Delta\epsilon^2 \langle P_{SM}\rangle^3}{2\pi n_{inv} K_{tot} hf\eta_{ext}} \qquad (6.15)$$

where n_{inv} is the inversion factor, K_{tot} is the Petermann spontaneous-emission factor referred to in Section 6.2.4, and η_{ext} is the external quantum efficiency.

A second mechanism requires that the sidemode has a different intensity distribution along the laser (e.g. the '-2 mode' has an asymmetric field distribution and the '-1 mode' has a symmetric field distribution) so that the average clamping of the electron density which occurs for the laser as a whole does not apply locally. Then if the laser frequency changes with the local current density along the cavity [20, 21] any fluctuations in the *distribution* of the longitudinal carriers create dither in the central frequency and hence the linewidth broadens. This process does not apply to Fabry–Perot lasers where each mode has virtually the same longitudinal mode *intensity* distribution and the emission wavelength is not affected by redistributions of carrier density along the cavity.

Finally, in DFB lasers it has been observed, with certain grating designs at high enough power levels, that perturbations in the carrier/photon-density feedback reinforce the original changes leading to intrinsic longitudinal instabilities of the mode which broaden the linewidth [22–25].

6.3 Influence of reflections from facets and external sources

6.3.1 Reflections and stability

The laser manufacturer, having perfected the design of the grating structure for the desired static- and dynamic-performance characteristics, does not wish to find random extraneous reflections interfering with the grating's feedback. This is an important problem because facets can easily reflect 1–5% of the incident power, even with an 'antireflection' coating, but as noted at the end of Section 6.2.3, the effective gain as a result of feedback is $\{1/(1-G_{round})\}$ which can exceed 1000, hence reflected power from the facets can be greatly enhanced. One usually has to improve the AR coating from a single 'blooming' layer to two or more layer coatings to achieve less than 0.1% power reflectivity, at which level facet reflections can have little significant effect in appropriate laser structures.

The key problem is not so much the magnitude of the reflection from the facet but that the phase of the reflection, with respect to the grating period, is difficult to control with present technology and is often simply a random variable. A high-κL laser has a greater internal reflectivity than a low-κL device, and hence a lower field near the facets for a given stored energy, thereby reducing the sensitivity to facet reflections. Unfortunately, the efficiency is also then reduced and a trade-off has to be made, suggesting that designs which maximise the ratio of (stored energy/threshold gain) will have an optimum sensitivity to reflections.

The importance of a low sensitivity to reflections extends beyond consideration of those from the facets. Reflections *external* to the laser, particularly coherent reflections, can induce a wide range of effects and five regimes have been identified [26] which broadly apply to both DFB and Fabry-Perot lasers:

(i) For power reflectivities of -50 dB or less adjacent to the laser (falling to -90 dB or less at 300 cm from the laser) the device remains single mode but the linewidth increases or decreases depending on the phase of the feedback [27, 28].

(ii) Reflections ~ -45 dB at any distance and close to the upper power limit of (i) cause the oscillation line to split for out of phase feedback, giving two low-linewidth modes whose spectral separation depends on the strength of the reflection. The laser will mode hop between the two modes at a rate which decreases with increasing reflection amplitude and stops altogether above -45 dB, leading then to (iii).

(iii) A third regime is now entered for a small range of feedback levels with a stable single mode with a narrow line for all feedback phases.

(iv) Above -39 dB reflection, the stability in (iii) breaks down again and relaxation oscillation occurs. As the reflection is increased, the relaxation oscillation increases until coherence collapse occurs.

(v) Finally, for power reflectivities above −8 dB, the device becomes single-mode again with the external reflection level being typical of that obtained in external-cavity lasers. The reader is referred to a useful analytic discussion of these feedback phenomena [29] which gives a good physical understanding of the detailed mechanisms involved.

6.3.2 Facet reflectivity and spectral measurements

A laser's spectrum is one key property of interest in optical-communication systems, and parameters which control this spectrum must also be measured and controlled by the manufacturer. One such parameter is the value of κL and the measurement of this parameter is in principle established from the 'stopband' width as determined from the modal separation of the two DFB modes (say, just around threshold) just on either side of the mathematical stopband as described in Section 5.4. However, errors in this measurement process arise because facet reflections and their phase with respect to the grating significantly alter the laser's spectrum. Reflectivities as low as 0.05% can be required [30] in order to estimate a value of $\kappa L \sim 1$ to an accuracy of about ±0.25, although this improves at higher values of κL. Again this can be understood in general terms by remembering the high internal optical gain with feedback so that minuscule changes in the round-trip gain G_{round} greatly affect the output through the term $1/(1 - G_{round})$. To understand the process in detail, the static characteristics of DFB lasers found in Chapter 5 need to be extended to account for facet reflectivity.

Consider, for example, the transfer-matrix equation (eqn. 5.16) with no input and no facet reflections. With nonzero facet reflectivities, the boundary conditions at $z=L$ on the right and at $z=0$ on the left become $R_L = \rho_{right} F_L$ with $F_0 = \rho_{left} R_0$. The phase of reflection relative to the maximum index in the Bragg grating can be accounted for by forming complex values of ρ_{left} and ρ_{right}. Remember that the reflection phase is twice the phase shift of the facet relative to the grating, since the propagating wave passes once through the phase-shifting region adjacent to the facet and then back again on reflection. The determinantal equation for the oscillation frequencies for a two-phase-shift device is now

180 *Distributed feedback semiconductor lasers*

$$\begin{bmatrix} F_L \\ \rho_{right}F_L \end{bmatrix} = [M_1][M_\phi][M_3][M_\phi][M_1] \begin{bmatrix} \rho_{left}R_0 \\ R_0 \end{bmatrix} \quad (6.16)$$

Algebraically tedious, but numerically straightforward, the transfer-matrix product can be calculated as a single matrix

$$M = [M_1][M_\phi][M_3][M_\phi][M_1] = \begin{bmatrix} M_{11} & M_{12} \\ M_{21} & M_{22} \end{bmatrix}$$

so that eqn. 6.16 then reduces to solving numerically for those frequencies where

$$\frac{\rho_{left}M_{21} + M_{22}}{\rho_{left}M_{11} + M_{12}} = \rho_{right} \quad (6.17)$$

The difficulties with a perfectly uniform DFB will be discussed shortly and the conclusion is that adding in '$2 \times \lambda_m/8$' phase shifts at appropriate positions, as seen already for example in Figure 5.8, provides a good route to a useful device. The sensitivity of a DFB to reflections is well exhibited by calculations which have been made on such devices, with $\kappa L = 2.1$, over a range of facet reflectivities of 0.1–6% corresponding to practical reflectivities for TiO_2/SiO_2 to single Al_2O_3 coatings, respectively. The detailed parameters for the device structure and operation are given in Table A11.2 of Appendix 11.

Figure 6.5a shows the marked change in 'stopband width' (i.e. the separation of the two band-edge modes) around critical phases for a facet reflectivity of 3.0%, while Figure 6.5b indicates the relative frequencies of obtaining different stopband widths as the facet phase varies randomly. This illustrates well the difficulties of precise control of laser parameters because of small reflections, and the near impossibility of determining κL from the stopband when there are residual reflections.

6.3.3 *Influence of facet reflectivity on SMSR for DFB lasers*

Another parameter of practical manufacturing importance is the yield of devices with a sufficiently high sidemode-suppression ratio (SMSR) for use with communication systems (say < -50 dB), but again facet reflections seriously degrade the initially expected values. The simulations had an uncoated reflecting rear facet; an antireflection-

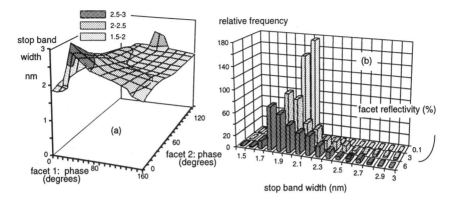

Figure 6.5 Influence of facet phase and reflectivity on stopband width

> Figure 6.5b shows a calculated histogram of stopband width over all possible facet phases against facet reflectivity, showing the expected increase in the spread of measured stopband widths as the facet reflectivity increases. The devices are DFBs with $2 \times \lambda_m/8$ phase shifts

coated front facet with a random phase with respect to the grating (corresponding to most manufacturing conditions); allowed for longitudinal-mode spatial-hole burning and examined the main mode and principal side mode. Values of $\kappa L \sim 0.75$ optimised the single-mode yield for high power [31] and the rear-facet phase had a strong influence on the maximum output power of this single mode. Later work [32] examined the effects of a small reflectivity for the front facet, showing that in general this reduced the single-mode yield for $\kappa L \sim 0.75$, but that this yield was nearly independent of κL provided that the front-facet reflectivity <5%. Threshold current and linewidth–power product are also influenced by facet reflections [33].

Using the previous range of facet reflectivities, with phases stepped at 45° intervals, the performance of a $2 \times \lambda_m/8$ DFB laser was examined when modulated with a 40 mA peak-to-peak current drive at 2.5 Gbit/s and with a bias level selected to give an extinction ratio of about 10:1, values appropriate to real systems. The simulation considered the main mode and principal side mode together with spacial hole burning. Figure 6.6a shows the distribution of threshold currents obtained and Figure 6.6b shows the relative frequencies, for given values of sidemode-suppression ratio, as the facet phase is stepped.

The simulations above took into account submount parasitic impedances to produce realistic current drives into the laser at this frequency. This device was modelled with parameters given in Table A11.2 and had significant speed limitations because of the time taken

182 *Distributed feedback semiconductor lasers*

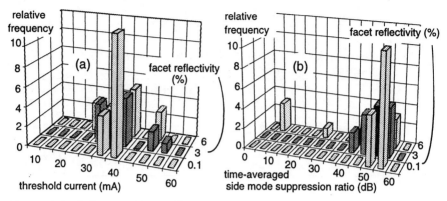

Figure 6.6   Influence of facet-reflectivity phase
   a   On threshold
   b   On side-mode suppression ratio (laser modulated at 2.5 Gbit/s)
   The devices are DFBs with $2 \times \lambda_m/8$ phase shifts

to transport the charge carriers from the contacts to the quantum wells. Excessively high facet reflectivities changed the yield of devices achieving less than 50 dB sidemode-suppression ratio under modulation of the laser current. This is one of the more sensitive tests for the degradation of performance induced by facet reflections. Note that there was a change from -53 dB down to -44 dB and even down to -4 dB if the reflectivity increased from 0.1% to 3% to 6%, respectively, for the 'worst' facet phase. In addition to the features discussed above, adverse facet reflectivities combined with a statistically varying facet phase change the yield of good devices in production because of changes in the following properties:

(i)   the optical extinction ratio—this needs to be greater than 10:1 at the required on/off current drives;
(ii) the curvature of the light-current characteristic, defined here as the ratio of the incremental efficiencies at high and low power—the smaller this curvature the better;
(iii) the spectral linewidth—broadening is caused by excessive facet reflections combined with certain facet phases. A 6% reflectivity can give a statistical spread that can be up to 2.5 times larger than from lasers with a lower reflectivity of 0.1% (this causes serious dispersion penalties for devices operating in standard fibre systems);
(iv) amplitude patterning in the transmitted digital optical waveform—the patterning induced by facet reflections with certain phases is related to the difference between the curvatures in the

More advanced distributed feedback laser design 183

static and dynamic L/I characteristics. A measure of this is the transmitted eye-diagram shape, and a hexagonal eye-mask test pattern is used to determine whether a given eye diagram is sufficiently 'open' and hence is acceptable. (This is important in determining whether a system can distinguish between ones and zeros for digital modulation. Typically the opening of the 'eye' has to be larger than an appropriate hexagonal test pattern but this test is not found to be a sensitive indicator of unsatisfactory facet reflectivity.)

The overall conclusion for the modelled $2 \times \lambda_m/8$ DFB laser is that 0.1% reflectivity has little influence on device performance, which is good, no matter what phase of reflection, but 3% reflectivity gives serious cause for concern while 6% at certain facet phases can wreck the performance of the laser.

6.4 Complex grating-coupling coefficients

6.4.1 General

The concept that deep periodic etching into the active-gain region can give a gain grating has already been met. In terms of Fresnel plane-wave reflections at each step, as outlined in Section 1.7.1, with a complex refractive index change between adjacent teeth of a grating, the reflectivity $\Delta \rho$ per Bragg period (two reflections per period) in Section 1.7.1 becomes:

$$\Delta \rho = 2 \frac{([n_{r1} + jn_{i1}] - [n_{r2} + jn_{i2}])}{([n_{r1} + jn_{i1}] + [n_{r2} + jn_{i2}])} \quad (6.18)$$

However, this gives a significant overestimate per step because the deep etch does not give an abrupt junction penetrating fully into the optical fields but rather a modulation of the thickness of an active region or an absorbing layer, as discussed next. The numerical techniques such as those outlined in the program **slabexec** in directory **slab** have to be used, in general.

6.4.2 Techniques for introducing complex grating-coupling coefficients

Gain gratings have been fabricated by two methods. Early developments exploited the different rates of growth of material by liquid-phase epitaxy over a periodic grating etched into a substrate [34]. Specifically, the local growth rate for the epitaxial material depended on the local curvature of the substrate and on the composition of the layer being grown. Growth of the active-gain region by this method resulted in a periodic variation in its thickness resulting in both a periodic gain and a periodic index variation. Interestingly, the index variation could be approximately cancelled by a simultaneous thickness variation of a passive-waveguiding layer leaving substantially a periodic gain only. The second approach is to etch partially or completely through the active region prior to regrowing on top of the grating, as seen in Figure 4.8. This may result in low reliability for the device unless appropriate technologies are used. Such deep etching into the active region may alter the distribution of carriers in a multiple-quantum-well stack [35]. Significantly, the imaginary or gain-coupling coefficient is now dependent on the carrier density N in the active region which, under dynamic operation, means that the gain coupling varies temporally with N calculated from the usual rate equations (Section 4.1). Assuming a linear relationship for material field-gain/unit-distance as a function of electron density, the coupling coefficient takes the form:

$$\kappa_{gain} = \frac{\Delta \rho}{\Lambda} \sim \frac{\Delta n_{i,\,eff}}{n_r \Lambda} = \frac{\Delta \Gamma \lambda_0}{\Lambda 2\pi n_r} g'(N - N_{tr}) \tag{6.19}$$

where $\Delta \Gamma$ gives the modulation of the effective fraction of the optical field power within the gain grating which has been modelled by step changes in the complex refractive index.

The introduction of a loss grating [36,37] has been realised, for example, by growing quantum wells within the laser waveguide which are wider than those in the active region in such a way that these additional wells provide photon absorption while those in the active region can give gain. However, the loss mechanism can saturate as a result of the incomplete removal of the photogenerated electron–hole pairs and the complex dynamics for the charge carriers in such quantum wells needs careful modelling [38].

6.4.3 Influence of complex grating-coupling coefficient on static performance

A perfectly antireflection-coated DFB laser with a pure-gain grating will lase exactly at the Bragg frequency, with frequency offset $\delta = 0$, and will not suffer from the mode degeneracy of pure index-coupled devices at low power. This result can be explained by extending the argument concerning the phase change for a single pass round the DFB-laser cavity, as discussed in Section 5.5.1, showing the lack of oscillation at the central Bragg frequency for an index grating. With the gain grating, the elemental reflection for each period $\Delta\rho$ is now imaginary and so $\rho_{tooth\ right} = \rho_{tooth\ left} = j\rho_{net}$ signifying a purely imaginary single-pass reflection as seen at the edges of the 'central' tooth of the grating. Modifying eqn. 5.27, the round-trip feedback within the central tooth is then given by $\{(j\rho_{net})\exp(j\tfrac{1}{2}\pi)\}^2$ but this is now positive and consequently lasing (oscillation) can occur given sufficient gain so that $\rho_{net}=1$. Such pure-gain gratings are difficult to realise but the design in [34] approaches this ideal. The more usual situation is a mixture of index and gain coupling. Section 5.3 showed that the presence of gain or loss coupling removes the stopband in a strict mathematical sense in the real-eigenmode propagation coefficient. The observable 'stopband' in the emission spectrum reduces with increasing gain until, for a pure gain-coupled device, it is absent.

The static performance of DFB lasers with complex grating-coupling coefficients can be simulated numerically in the same manner as described in Chapters 5, 6 and 7 for pure index-coupled devices. Provided that there is no saturation in loss, the loss-coupled device merely needs to replace the real κ for an index-coupled device with an imaginary constant $j\kappa$. However gain-coupled devices have a coupling coefficient which depends on the carrier density in the active region and this complicates the analysis with both gain and feedback dependent on electron density which also generally varies with space.

Besides the strength of the gain component relative to the index component, one also has the relative phase. Typically, the two situations discussed most in the literature are $\kappa_{index}+j\kappa_{gain}$ where $\kappa_{index}>0$ and $\kappa_{gain}>0$—the 'in-phase' case, or $\kappa_{index}>0$ and $\kappa_{gain}<0$—the 'antiphase' case. In the literature the term 'gain grating' is often taken to include loss gratings, but the terms 'in-phase' and 'antiphase' usually refer to gain gratings, so it is important to be clear on the terminology. An in-phase gain grating and an antiphase loss grating with the same $|\kappa|$ are 'equivalent' though in detail there may be differences caused by the change of gain or loss with electron density. Similarly an antiphase gain grating and an in-phase loss grating are 'equivalent'. The effect of in-phase and antiphase gain gratings on the linewidth has already been noted and indeed the sign also helps to select one of the two main modes in a perfectly uniform DFB [7,17,39] with the in-phase case helping to select the longer-wavelength mode.

Note that for a pure gain grating it is possible to obtain a perfectly 'flat' longitudinal-mode intensity distribution. Using eqns. 5.18 and 5.19 with $j\kappa = \kappa_{gain}$ and $\delta = 0$ so that the frequency is the Bragg frequency with $\beta_e^2 = (\kappa_{gain}^2 - g^2)$:

$$F(z) = (\kappa_{gain}/\beta_e)\sin(\beta_e z)R(0); \quad R(z) = [\sin\{\beta_e(L-z)\}/\sin(\beta_e L)]R(0) \quad (6.20)$$

It can be seen that, if $(\beta_e L) = \pi/2$ and $(\kappa_{gain}/\beta_e) = 1$, then

$$F(z)*F(z) + R(z)*R(z) = \{\sin^2(\beta_e z) + \cos^2(\beta_e z)\}R(0)*R(0) = |R(0)|^2 \quad (6.21)$$

giving a uniform field intensity along the length of the laser, which will therefore not suffer from longitudinal-mode spatial-hole burning [40]. Alarmingly, the net field gain g is zero but the power has to be given from the gain in the gain coupling giving $(\kappa_{gain} L) = \pi/2$ and this is critical and means a critical adjustment of the transparency density at the same time as getting the right gain coupling!

In any one longitudinal section, conventional analysis of DFB-laser structures relates the gain, coupling and stimulated emission-current density to the photon density averaged over a wavelength and, in general, the use of these mean values is well justified. When, for example, referring to a 'flat' field profile in eqn. 6.21, the standing-wave pattern of the type discussed in eqn. 3.5 is neglected. However, the phases of the forward wave and reverse wave could change with respect to each other, giving a standing-optical wave pattern of the form referred to in eqn. 3.5, but neglecting any transverse-field pattern for simplicity:

$$E*E = \{F*F + R*R + 2|FR|\cos(2\beta_b z + 2\phi)\} \quad (6.22)$$

Now with a periodic-gain grating the gain in the device might be written as

$$g(z) = g_{average} + \kappa_{gain\ periodic}\cos(2\beta_b z) \quad (6.23)$$

The stimulated recombination current, averaged over one Bragg period, is then proportional to

$$\text{average}[g(z)E*E] = [F*F + R*R]g_{average} + |FR|\cos(2\phi)\kappa_{gain\ periodic} \quad (6.24)$$

It follows that, depending on the phase ϕ of the standing-wave pattern in relation to the grating's period, the stimulated recombination current can change by an amount up to $\pm \kappa_{gain\ periodic}/g_{average}$. This has been called the standing-wave effect [41, 42]. In this book, this effect has been ignored for simplicity in the modelling and also because for many practical devices the gain-coupling coefficient is weak. The

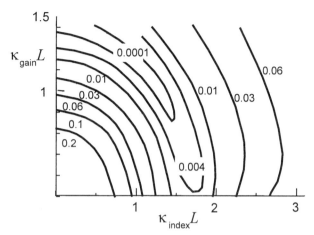

Figure 6.7 Contour map of nonuniformity or variance of longitudinal-mode intensity distribution as a function of real and imaginary κL product

standing-wave effect helps to explain in a clear physical way one of the reasons for the mode selectivity of gain gratings.

In general, the higher the proportion of gain to index coupling the flatter the optical-field profile (once again averaging over the standing-wave patterns) along the cavity can be made by selecting the appropriate κL product. The result is a device which becomes more resistant to longitudinal-mode spatial-hole burning with an increasing proportion of gain coupling, and hence exhibits better stability with respect to the sidemode intensities with increasing output power. In Figure 6.7 the effect is shown on the longitudinal-mode intensity distribution (as measured by the variance of the intensity) with variations in the real and imaginary κL products. This plot includes all uniform grating devices from pure index-coupled to pure gain-coupled structures.

Figure 6.8 shows the change in the longitudinal-mode intensity distribution as a function of the imaginary coupling coefficient for a small real coupling coefficient κ_r of 5 cm^{-1}. In addition, the spectrum is plotted for the two extreme structures showing the reduction in the significance of the stopband with increasing imaginary coupling coefficient.

6.4.4 Influence of complex grating-coupling coefficient on dynamic performance

If the imaginary coupling coefficient has an amplitude comparable with that of the real component then, as already discussed, the

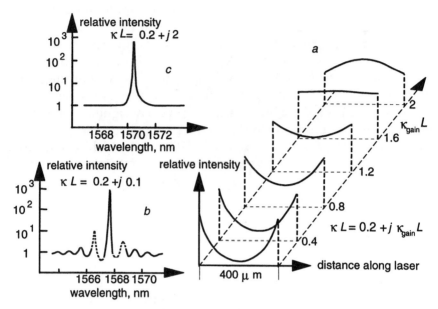

Figure 6.8 Dependence of longitudinal-mode intensity distribution and emission spectrum on magnitude of imaginary coupling coefficient $\kappa_{gain}L$

For a real coupling coefficient of 5 cm^{-1} giving a real part $\kappa L=0.2$. Note the flatness of the intensity profile around $\kappa_{gain}L \sim \pi/2$ in support of the analytic theory. Stopband shown dotted in (*b*) is not present in (*c*)

effective linewidth-enhancement factor is significantly decreased or increased depending on whether the gain grating is, respectively, in antiphase or in-phase with the real grating, assuming that $\alpha_H > 1$ [43–45]. The small-signal AM bandwidth is also changed since the effective differential gain, and hence the slope of the resonance frequency squared against output power, is increased or decreased for antiphase or in-phase gain gratings, respectively. This results in a significant change in the dynamic chirp of the laser under transient conditions. For an antiphase grating, the resulting increase in the effective differential gain can, under certain conditions, lead to instability and self pulsations.

The use of a weak-*gain* grating, which has a strength that is only a few percent of that of the *index* grating, does not significantly affect the dynamic behaviour via any change in the effective differential gain. A more subtle and also useful effect is to increase the yield of devices which lase on either the long- or short-wavelength Bragg mode depending on whether the gain grating is in-phase or in antiphase with the index grating, respectively [36,46,47]. In the absence of gain coupling, the devices as manufactured would be approximately evenly split between the two modes at low power but this symmetry is broken by gain coupling. This might seem insignificant, but in Section 6.6.1 it will be shown that devices with moderate to high κL products exhibit better stability with respect to longitudinal hole burning, and therefore have a reduced wavelength chirp, if they are lasing on the long-wavelength mode. This is of great practical significance for the digital modulation of lasers for long-haul systems.

It is interesting that a second-order DFB laser which suffers from radiation loss (see Section 4.4.3) behaves in principle in similar ways to a gain-coupled DFB laser. The radiation loss leads to a different gain for the two Bragg modes and results in the long-wavelength mode being favoured [48,49]. However, the radiation loss is probably less easy to control in manufacture than making a gain grating.

6.4.5 Influence of facet reflectivity

Sections 6.3.2 and 6.3.3 discussed the spread in several parameters determining the static and dynamic performance for index-coupled DFB lasers, the spread being caused by small randomly phased facet reflectivities caused by fabrication processes. For complex coupled devices, the effect of facet reflections is reduced as the proportion of the total coupling coefficient which is due to gain coupling is increased [50]. Physically, this can be considered to result from the peaks in the mode standing-wave pattern coinciding with the maxima in gain, and thereby stabilising the mode with respect to the perturbing effects of the reflections. It is not at present practical to control the facet phase, but a technique for measuring it in gain-coupled DFB lasers has been reported and correlated with the front-to-back facet-emission ratios [51].

6.5 High-power lasers with distributed feedback

6.5.1 General

Initially, the commercial applications for DFB lasers were almost exclusively for long-haul digital transmission at 1.5 μm wavelength where the optical fibre has a minimum in attenuation. The significant dispersion in the fibre at this wavelength demanded sources with a narrow spectral width. There is now an increasing interest in high-power (>100 mW) DFB-laser structures for applications such as optical pumping of erbium-doped fibre amplifiers. High-power devices for optical carriers over the access network with cable television is another clear application for lasers which give out more power than the conventional 'communication' source. In WDM applications, power combiners for the different wavelengths usually lose significant amounts of power, and again higher power is required.

There is, then, a strong commercial interest in high-speed low-wavelength chirp sources which are composed of narrow linewidth CW DFB-laser sources coupled monolithically, or in a hybrid manner, to external modulators. The modelling of devices with external modulators is left until Chapter 8 and for the present it is noted that this is a tried and trusted system designer's choice for controlled low-chirp modulation, though it adds complexity and cost. The attenuation in such external modulators can be significant, and in hybrid arrangements there is the additional loss associated with the coupling between the laser and modulator, and consequently the source needs perhaps 10 dB more power than previously.

One way of increasing the power output from a laser is to change from using a symmetric device, which usually wastes power by radiating equally from both facets. A high front-to-back emission ratio can concentrate the power where required but this leads to surprising and significant drawbacks.

6.5.2 Techniques for obtaining high front-to-back emission ratios

Four techniques have been utilised for obtaining high front-to-back-emission-ratio DFB laser structures with one mode clearly selected.

The first approach is to apply an antireflection coating to the front facet, and either leave the rear facet as-cleaved (with a power reflectivity of about 32%) or apply a high-reflectivity coating. This straightforward approach has been employed widely by commercial DFB-laser manufacturers. However, as repeatedly stated, the phase of the rear facet with respect to the grating is difficult to control in a manufacturing environment and the wrong phase has been shown to have serious consequences on sidemode suppression, selection of frequency etc. Present devices have random distributions of the rear-facet phase and need careful screening to meet a required specification. Second-order gratings [52] can help in providing some discrimination, but not sufficiently so as to avoid the necessity of screening.

Secondly, the introduction of gain or loss coupling into the grating structure to give a complex κ significantly helps in the modal discrimination but not, of course, on back-to-front facet-emission ratios. The presence of gain modulation in phase with the index grating leads to a preference for oscillation on the long-wavelength side of the stopband. Likewise, antiphase gain coupling favours oscillation on the short-wavelength side of the stopband. (After reading Chapter 7, try out the program dfbgain in directory dfb and change the sign of the imaginary part of κ.) It is necessary to add significant imaginary coupling to the index grating before the device always oscillates on one side of the stopband only. For example, if $\kappa = 100 + j5$ cm^{-1} for a 300 μm long device with a rear-facet power reflectivity of 32% (i.e. as cleaved), then the distribution of the selected mode over the range of all phase shifts changes from 50%:50% between the two modes to approximately 17%:83% in favour of the long-wavelength mode.

A third approach is to introduce an asymmetry into the DFB grating itself. This can be achieved, for example, by driving one half harder so as to 'tilt' the longitudinal-mode intensity distribution so that more radiation is emitted from one facet than the other, even though both facets are antireflection coated to ensure that the reflection phase is not relevant. However, such asymmetry in the optical-field hole-burns spatial asymmetry into the electron density and gain. Low side mode suppression and even multiple-mode emission or instability can occur.

The fourth, and most satisfactory, approach is to integrate a DFB laser monolithically with an amplifier. The rear facet of the laser and the front facet of the amplifier are antireflection coated and the interface between DFB and amplifier is virtually matched because the amplifier is the same material but with the grating omitted. A single contact can be applied to both DFB and amplifier, allowing the device to be operated p-side down without introducing the complexity of patterned contacts.

6.5.3 Laser-amplifier structures with distributed feedback

The first integrated distributed-feedback laser–amplifier used a DBR-laser structure integrated with an amplifier section and emitted at 1.3 μm [53]. The rear facet had a high-reflectivity coating and a passive grating section provided the feedback from the front of the laser, the combination giving a high front-to-back emission ratio. Regrettably, the DBR laser like a Fabry–Perot laser can operate with different numbers of wavelengths between reflectors if the temperature or current changes; consequently the lasing mode can 'hop' to a different wavelength under changes of drive etc. The structure of Figure 6.9 with a phase-shift section is therefore preferred.

Distributed feedback semiconductor lasers

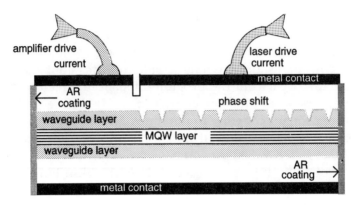

Figure 6.9 Schematic diagram of integrated DFB laser and amplifier

Any residual reflection from the amplifier's facet experiences further gain before returning to the laser, so must be made extremely low. Angling the facet at the output has been investigated for reducing this reflection [54], but modern AR coatings are usually now adequate. The utility of these integrated devices for high power, suitable for pumping erbium-doped fibre amplifiers, has been well illustrated by modules delivering ~300 mW at 1.48 μm [55] and also a DBR laser–amplifier emitting a CW single longitudinal mode with 110 mW at 0.98 μm [56], the two different wavelengths appropriate for pumping an EDFA.

Several variant structures have been made successfully using $\lambda_m/4$-phase-shifted grating DFB lasers with an amplifier [57] to $2 \times \lambda_m/8$-grating DFB lasers with antireflection coatings. These latter have given 45 mW at a minimum linewidth of 2.3 MHz at 35 mW [58]. Although with AR coating, the laser section radiates as much power out of the rear facet as the front, the overall efficiency penalty with the amplifier in place is small on account of the asymmetry introduced by the amplifier.

All monolithically integrated laser–amplifier structures experience some broadening of the linewidth because of spontaneous emission from the amplifier which is fed back into the laser. Applying eqn. 6.12 indicates immediately that a higher net spontaneous emission rate into a laser mode broadens the linewidth. An optical isolator placed between the laser and the amplifier could reduce such broadening, but with present isolator constructions [59,60] that is difficult to integrate.

The magnitude of the linewidth broadening can be significant. By way of example [57], consider an 800 μm-long bulk-active-region $\lambda_m/4$

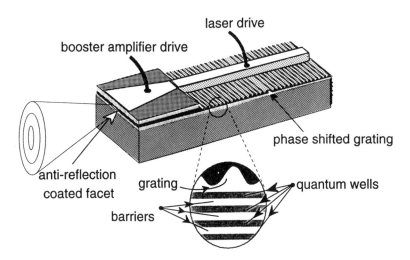

Figure 6.10 Schematic monolithically integrated DFB laser and flared amplifier

Slightly modified from the drawing by M.J. Fice, Nortel [57] with permission

DFB laser with a κL product of 1.2, integrated with a 600 μm-long amplifier. If the laser is injected with 10 kA/cm^2 and the amplifier with 2 kA/cm^2, then modelling predicts a linewidth of about 12 MHz for reasonable values of the materials parameters. However, increasing the amplifier current density to 10 kA/cm^2 results in an increased source linewidth of about 20 MHz while the amplifier gain increases from approximately 1 to 3.5.

A new development for high-power lasers is the use of a flared amplifier integrated with a DBR laser which has resulted in significant increases in the available output power. The advantage of this design is that the mode propagating through the flared amplifier is allowed to spread under an expanding contact which, with a roughly uniform current density, provides an increasing injected current in the regions where the increasing photon flux requires a higher stimulated emission current. Both longitudinal and transverse hole-burning effects are reduced, leading, for example, to 2 W of CW power at 0.97 μm wavelength [62] and 1.3 W CW at 0.86 μm [63]. The use of a phase-shifted DFB laser, using a $2 \times \lambda_m/8$ design, integrated with a flared amplifier, to give stable single-frequency operation combined with high output power, has also been investigated [61] (Figure 6.10). The ability to model such concatenated optical components along with their interaction is the beginning of subsystem and system modelling which is pursued in Section 8.2 and will form an important role for modelling in the future.

6.6 Dynamic modelling of DFB lasers

6.6.1 Uniform-grating DFB laser with reflective rear facet

The importance of the uniform DFB laser is that it is technologically the most straightforward Bragg laser to manufacture so that a considerable range of studies has been carried out to elucidate the role of the device and its two modes. These studies include: longitudinal intensity distribution with κL [64], dependence of stopband width on κL [65], subthreshold spectra and facet effects [66], instabilities caused by longitudinal spatial-hole burning [67,68], FM and AM responses [69,70], high-speed performance in a package [71] and intermodulation distortion [72,73]. The dynamic performance of DFB lasers with a high-reflectivity rear facet and low-reflectivity front facet, has shown experimentally a lower penalty in the received eye diagram after propagation through dispersive fibre when lasing in the long-wavelength mode as compared with those lasing strongly in the short-wavelength mode.* In spite of these studies, the uniform-index-coupled DFB has serious drawbacks which are brought out in this section by reviewing some modelling work for such devices, mounted in realistic packages, and driven at frequencies of 2.5 Gbit/s with current levels adjusted to give a 10:1 extinction ratio. Device parameters are listed in Table A11.1 of Appendix 11. The main difficulty lies in obtaining an adequate yield of single-mode devices operating in the long-wavelength mode where a satisfactory performance for operation in a real communication system can be obtained. It follows that other methods of mode selection have to be found which augment the uniform Bragg grating.

The modelling study here started with an investigation of the static performance of as-cleaved/AR-coated (power reflectivities of 32% and 0%, respectively) uniform-grating lasers with a range of facet phases between 0° and 180° similar to that described earlier. The key features which emerged were:

(i) Emission wavelength near threshold could not be controlled and around 50% would lase at the lower wavelength. With facet phases around 50–100° the short wavelength seemed preferred

---

* LEONG, K.W., Nortel Technology, Ottawa, Canada. Private Communication.

with threshold currents decreasing with increasing facet phase. At facet phases around 0–40° the long wavelength was preferred with lower threshold currents at the lower values of facet phase. Facet phases around 120–150° gave multimoded performance or uncertain behaviour. In general, the side modes increased with increasing current and only at one critical facet phase did the SMSR improve with increasing current.

(ii) Threshold current varied with facet phase by slightly less than 20%.

(iii) Incremental front-facet efficiency was reasonably independent of facet phase, varying by about 5% of its value.

(iv) The maximum single-mode output power predicted by the model before instability occurred was much higher for the devices lasing on the long-wavelength mode but this occurred only over a relatively narrow range of facet phases around 20–40°. At the time of this work, the connection between numerical instabilities of static simulations and dynamic instabilities of DFBs [74] was not well understood but recent evidence suggests a strong connection between the two‡.

The L/I curvature and wavelength chirp are both linked to the spatial-hole burning of the carriers [75], the former being affected by the variation in the efficiency when there is spatial-hole burning, the latter through change of frequency with carrier-density distribution along the cavity. The correlation shown by modelling indicates that it is a high curvature of the *total L/I* output (from both facets) which correlates best with the chirp in the laser.

This work was extended to take into account the dynamic performance [76] and again the short-wavelength devices clearly showed inferior and unacceptable dynamic performance when used over 80 km of normal dispersive fibre link† where, for a 'one', the trailing edge of a pulse gets moved forward in time and pulse compression occurs. Additionally, there is evidence of increased damping of the transient response for devices lasing on the short-wavelength mode which should improve the performance over a dispersive link, but this seems to be less significant than the adiabatic chirp described above. Figure 6.11 compares the received eye diagrams for the different wavelength transmissions indicating a relative closure for the short-wavelength device [76] which although acceptable to some users, makes a clear difference in quality systems.

† LEONG, K.W., Nortel Technology, Ottawa, Canada. Private Communication.
‡ FICE, M.J., Nortel Technology, Harlow, UK. Private Communication, 12 May 1996.

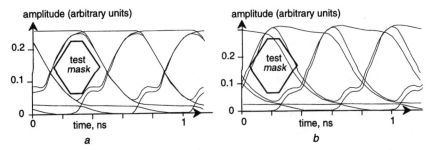

Figure 6.11 Simulated filtered received eye diagrams from uniform-grating DFB lasers emitting on (a) long-wavelength and (b) short-wavelength Bragg modes

The main features discovered in these dynamic studies are briefly:

(a) The time-averaged sidemode-suppression ratio (SMSR) is found to be an inadequate guide to selecting devices with good or poor received eye diagrams. Devices with 50 dB or better SMSR could still have an unacceptable opening of the eye for acceptance in a mask test, indicated in Figure 6.11b.

(b) The penalty at the centre of the received eye diagram after transmission through 80 km of dispersive fibre (17.5 ps/nm km) varied from 0.4 to 1.1 dB dependent again on the facet phase and whether the long or short mode was lasing. This penalty is defined as

$$10.\log_{10}\left[\frac{\{(\text{Lowest 'one' level})-(\text{Highest 'zero' level})\}}{\{(\text{Highest 'one' level})-(\text{Lowest 'zero' level})\}}\right]$$

The work indicated that this measure is not necessarily adequate and a better measure of acceptance is an appropriately designed hexagonal mask which then shows up the inferior eye opening of short-wavelength devices more clearly, and such a test is more reliable as an indication of acceptability in a real system.

(c) Certain values of rear facet phase give totally unacceptable performance. Rear facet phases of 30 and 80° resulted in stable lasing in the long- and short-wavelength modes, respectively, with the long-wavelength mode being acceptable but the short-wavelength mode being unacceptable in its overall performance.

(d) Longitudinal-mode spatial-hole burning is stronger for the lasers emitting on the short than the long-wavelength mode. For the short-wavelength lasers, this hole burning results in a larger chirp to yet shorter wavelengths following an increase in current and gives a larger-dispersion penalty.

(e) If the sign of the linewidth-enhancement factor is reversed (from negative to positive), then the long-wavelength mode becomes less stable and the short-wavelength mode offers better transmission performance. This arises from the inversion in the sign of the longitudinal hole burning in the refractive-index profile which swaps the behaviour of the short- and long-wavelength modes.

The overall conclusions from this extensive modelling is that only the long-wavelength mode of the uniform-DFB lasers offers sufficiently stable operation with good sidemode-suppression ratio and clear eye opening and acceptable systems performance. However, in such uniform lasers, there will be an unacceptably low manufacturing yield of devices operating in this desired mode unless the facet phase can be controlled to within $\pm 10°$ or unless additional features are incorporated to ensure the appropriate mode selection. The use of $2 \times \lambda_m/8$ phase shifts, or of partial gain coupling in phase with the index coupling, are examples with a proven track record in this respect.

6.6.2 Large signal performance of $2 \times \lambda_m/8$ DFB lasers with strong and weak carrier-transport effects

This section gives a case history of large-signal modelling, using a variety of techniques, for one particular type of DFB laser where there are two phase shifts adjusted so as to give a reasonably uniform field and electron density over a wide range of current-drive levels and optical-power outputs.

The example given examines the degradation of the dynamic response that occurs when the charge carriers take too long to be transported into the quantum-well active regions. This work is based on laser structures which have been assessed in some detail [76]. The simulated large-signal dynamic performance of the $2 \times \lambda_m/8$ DFB lasers will be examined *with* and *without* significant carrier transport effects at 2.5 Gbit/s. Table A11.2 in Appendix 11 lists the material and device parameters used for this exercise. The simulation method was essentially that presented in Chapter 7, but with greater use made of frequency-domain techniques to assist in a more detailed separation of the main and sidemode characteristics.

Figure 6.12 shows the simulated performance for the case where carrier-transport delays are around 70 ps and are found to be

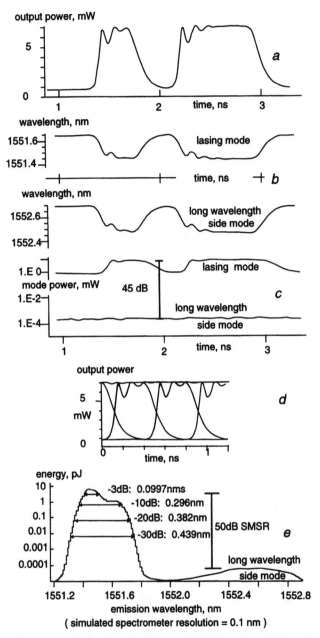

Figure 6.12 Simulated performance of $2 \times \lambda_m/8$ DFB laser with significant delay (~70 ps) in injecting carriers into active regions [79]

(a) Total intensity against time (b) Wavelength of each mode against time (c) Power in each mode against time (d) Unfiltered transmitted eye diagram (e) Time-averaged spectrum including simulated spectrometer resolution significant. In Figure 6.12a the damped turn-on transient is shown while Figure 6.12b shows the shorter lower-emission wavelength in the 'ones' than the 'zeros', both effects arising from the presence of the carriers in the waveguide region transporting into the active material. The excellent sidemode-suppression ratio is seen in Figure 6.12c where values of ~45 dB occur even during the turn-on transient when the round-trip gain exceeds unity. The open-eye diagram in Figure 6.12d shows the weakness of any longitudinal hole burning, but residual effects can be seen in the shortening of the emission wavelength with time in the 'ones' in Figure 6.12b.

This wavelength shift is created by an increasing carrier density and decreasing refractive index near the ends of the cavity, and to a lesser extent a decreasing carrier density at the cavity centre. The simulated eye diagram in Figure 6.12d agrees well with experimentally observed eye diagrams. Figure 6.12e shows a time-averaged spectrum obtained by summing the instantaneous energy over all modes at different wavelengths at each time interval, including Gaussian filtering to simulate a real spectrometer. This technique of taking the instantaneous energy with time allows the intrinsic laser chirp to be evaluated but it does not include Fourier broadening caused by modulation. The important result here is the SMSR of 50 dB.

The longitudinal-mode intensity profile peaks at the centre of the cavity since the κL product is higher than optimum for minimum longitudinal hole burning. The result is dynamic longitudinal hole burning in the longitudinal active-region carrier-density profile as shown in Figure 6.13, indicating the time scale of half a nanosecond to stabilise even though the current may be switched instantaneously.

Some hole burning is not necessarily detrimental to the performance of a device because the hole burning dampens the turn-on transient and hence reduces the transient chirp. It does, of course, extend the duration of the chirp associated with the hole burning, and the two effects must be balanced for any application [80]. The carrier-density profile within the waveguide region also hole burns, but is delayed relative to that in the active region as the latter acts as the driving mechanism for the former. The delay between these two contributions to longitudinal hole burning leads to a complex total effect that can only be accurately analysed with a detailed model. An important point is that, although the carrier density in the waveguide region may be about a factor of ten lower than that in the active region, the thickness of the waveguide region can be sufficiently large so as to approximately compensate. The contribution to chirp from

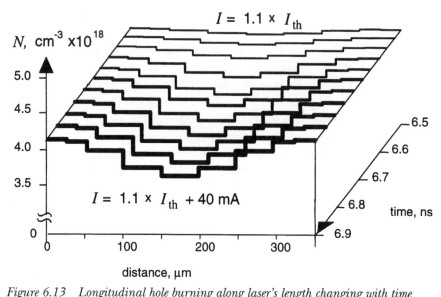

Figure 6.13 Longitudinal hole burning along laser's length changing with time

As illustrated by plot of active-region carrier-density N profile against time. Device is switched from a 'zero' at 1.1 of threshold current to a 'one' level with 40 mA more current

the waveguide region can therefore be very significant depending on the delay associated with the carrier transport in the laser structure.

An increase in the AM bandwidth can be obtained by reducing the delay associated with the transport of carriers into the active region by, for example, thinning the waveguide region or increasing the waveguide-doping level [81]. To clarify the influence of carrier transport, Figure 6.14 shows the modelled behaviour of the same device as in Figure 6.12, but now with the delays caused by carrier transport set to zero.

It can be clearly seen that the intensity transient in Figure 6.14a is less damped than in Figure 6.12a, but also that there is a much smaller difference in the wavelengths of the ones and zeros of the digital modulation as shown in Figures 6.12b and 6.14b. The reduced-wavelength chirp is also very evident in Figure 6.14e which shows a time-averaged −20 dB spectral width of 0.243 nm compared with the 0.382 nm in Figure 6.12e where there is carrier transport delay. Finally, there is less longitudinal hole burning for the structure with no carrier transport. This can be seen from the stable average emission wavelength of the lasing mode during the turn-on transients in Figure 6.14b, whereas in Figure 6.12b this is not the case as evidenced by the

More advanced distributed feedback laser design   201

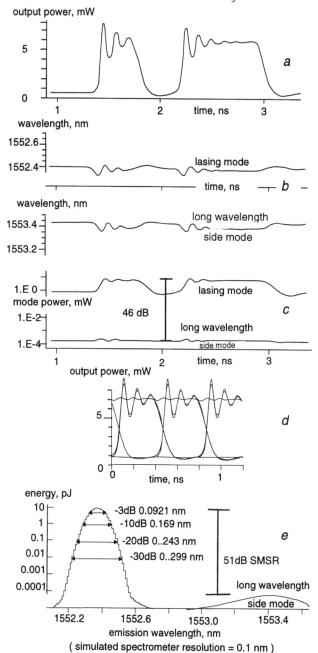

Figure 6.14 *Simulated performance of $2\times \lambda_m/8$ DFB laser with negligible delay in injecting carriers into active regions*

a Total intensity against time *b* Wavelength of each mode against time
c Power in each mode against time *d* Unfiltered transmitted eye diagram
e Time-averaged spectrum including simulated spectrometer resolution

shortening wavelength with time. The average intensity during the turn-on transients in Figure 6.14a and d also shows clear evidence that longitudinal hole burning is negligible as there is no evidence of an underlying delayed increase in intensity over about 200 ps as is seen in Figure 6.12a and d.

Following the lessons learnt from the structure, discussed above, a high-speed low-chirp $2\times\lambda_m/8$ DFB-laser structure with a 22 GHz small-signal AM bandwidth has been reported [79]. This was achieved by using six 1% compressively strained 7.5 nm-thick InGaAsP quantum wells, with 0.28% tensile-strained 15 nm-thick barriers to reduce the overall strain. The barriers were p-doped to 5×10^{17} cm^{-3}. The upper and lower waveguide layers were p-doped at 5×10^{17} cm^{-3} and n-doped at 2×10^{17} cm^{-3} and with thicknesses of 0.1 μm and 0.18 μm, respectively. The upper p-side waveguide layer was kept as thin as possible to minimise hole-transport effects while maintaining a sufficient thickness to support a grating. The differential gain was estimated as being 25×10^{-16} cm^2 from measurements of resonance frequency against output power on Fabry–Perot devices. Additionally, the laser shunt capacitance (C_{laser}) was reduced from about 8 pF to 0.75 pF by patterning the p-side metallisation, and the series resistance reduced to 3 Ω, thereby increasing the RC roll-off frequency from 4 to 70 GHz. The submount used was also substantially improved so that the circuit parasitics were no longer significant up to at least 20 GHz.

6.7 Summary

This chapter began by looking at how to estimate the linewidth starting from a theory which applied to most electromagnetic oscillators but was then found to be inadequate for semiconductor lasers because of the plasma and Kramers-Krönig effects where changes in the gain (changes in electron density) give changes in the refractive index through α_H, Henry's linewidth-broadening factor. This increases both the static linewidth and the dynamic linewidth and in the latter case has detrimental consequences in increasing the dispersion penalty for high-bit-rate communication over long-haul optical fibres. The effective value of the linewidth enhancement as measured by the output spectra from a device was noted often to be larger than the intrinsic linewidth-enhancement factor of the material and this was explained by considering the averaging effects over the laser. The dynamic effects of linewidth enhancement have been noted in Chapter 4.

Perfectly uniform DFB lasers need modifications to be acceptable to ensure that the DFB operates only in the longer-wavelength mode of its two main modes. It is only this mode which is found to have an acceptable performance within an optical-communication system, taking into account SMSR, eye-opening etc. One of these mode-selection methods was seen in Chapter 5 to be the introduction of $2 \times \lambda_m/8$ phase shifts into the grating, but even so the studies discussed in this chapter show that operation is critically affected by the strength and phase of the facet reflections, and reflections should be reduced to below 0.1% for a reliable commercial product; and by the same arguments one has to protect the laser carefully from reflections coming from the system if the best performance is to be obtained.

An alternative change to the uniform DFB laser discussed in this chapter is to introduce some gain coupling into the grating. This latter can be accomplished, for example, by etching the grating through to the gain region thereby giving some periodic gain as well as periodic changes in refractive index. Provided that the gain coupling was 'in phase' with the index coupling, then the desired longer-wavelength mode could be selected. There is evidence that such complex grating lasers may be more tolerant to reflections if the imaginary coupling coefficient is large enough.

There is a need for more optical output power, especially for pumping EDFAs and for use in WDM systems. An increase of at least 10 dB can be obtained by using integrated amplifiers and, when combined with an integrated modulator, this combination is often a 'preferred' source for designers requiring high-bit-rate modulation of DFB lasers with the capability of being used in a WDM system.

If perfectly uniform lasers were to operate consistently in the desired long-wavelength mode, then studies discussed in this chapter show that the phase of the facet relative to the periodic structure of the grating has to be controlled to a fraction of a radian and, with normal manufacturing tolerances, this would give unacceptable yields. The $2 \times \lambda_m/8$-phase-shifted DFB laser is much better in selecting the preferred longer-wavelength mode when appropriately designed. Part of this design when the lasers are to be directly modulated at gigabit-per-second rates comes in making thin enough separate confinement heterojunction layers. The final section on modelling showed the importance of reducing the time taken to transport the carriers across these layers to well below the period of the modulation.

6.8 References

1 WHITEAWAY, J.E.A., THOMPSON, G.H.B., COLLAR, A.J., and ARMISTEAD, C.J.: 'The design and assessment of $\lambda/4$ phase shifted DFB laser structures', *IEEE J. Quantum Electron.*, 1989, **25**, pp. 1261–1279
2 GHAFOURI SHIRAZ, H., and LO, B.S.K.: 'Structural dependence of three phase shift distributed feedback semiconductor laser diodes at threshold using the transfer matrix method (TMM)', *Semicond. Sci. Technol.*, 1994, **9**, pp. 1126–1132
3 ZHOU, P., and LEE, G.S.: 'Mode selection and spatial hole burning suppression of a chirped grating distributed feedback laser', *Appl. Phys. Lett.*, 1990, **56**, pp. 1400–1402
4 OKAI, M., TSUCHIYA, T., UOMI, K., CHINONE, N., and HARADA, T.: 'Corrugation-pitch-modulated MQW DFB laser with narrow linewidth (170 kHz)', *IEEE Photonics Technol. Lett.*, 1990, **2**, pp. 529–530
5 WHITEAWAY, J.E.A., GARRETT, B., THOMPSON, G.H.B., COLLAR, A.J., ARMISTEAD, C.J., and FICE, M.J.: 'The static and dynamic characteristics of single and multiple phase-shifted DFB laser structures', *IEEE J. Quantum Electron.*, 1992, **28**, pp. 1277–1293
6 KOGELNIK, H., and SHANK, C.V.: 'Coupled-wave theory of distributed feedback lasers', *J. Appl. Phys.*, 1972, **43**, pp. 2327–2335
7 FLANIGAN, B.J., and CARROLL, J.E.: 'Mode selection in complex-coupled DFB lasers', *Electron. Lett.*, 1995, **31**, pp. 977–979
8 CHAN, Y.C., PREMARATNE, M., and LOWERY, A.J.: 'Semiconductor laser linewidth from the transmission-line laser model', *IEE Proc.—Optoelectronics*, 1997, **144**,(4), pp. 246–252
9 OHTSU, M.: 'Highly coherent semiconductor lasers' (Artech House, Boston, 1992)
10 HENRY, C.H.: 'Theory of the linewidth of semiconductor lasers', *IEEE J. Quantum Electron.*, 1982, **18**, pp. 259–264
11 HENRY, C.H.: 'Theory of phase noise and power spectrum in a single mode injection laser', *IEEE J. Quantum Electron.*, 1983, **19**, pp. 1391–1397
12 HENRY, C.H.: 'Phase noise in semiconductor lasers', *J. Lightwave Technol.*, 1986, **4**, pp. 289–311
13 PROBERT, P.J., and CARROLL, J.E.: 'Lumped circuit model prediction of linewidth of Fabry–Perot and DFB lasers, including external cavity devices', *Proc. IEE.–J.*, 1989, **134**, pp. 22–32
14 SLATER, J.C.: 'Microwave electronics' (Van Nostrand, 1950; Dover, New York, 1969)
15 SCHAWLOW, A.L., and TOWNES, C.H.: 'Infra-red and optical masers', *Phys. Rev.*, 1959, **112**, pp. 940–949
16 AMANN, M.C.: 'Linewidth enhancement in distributed-feedback semiconductor lasers', *Electron. Lett.*, 1990, **26**, pp. 569–571
17 LU, H., BLAAUW, C., BENYON, B., LI, G.P., and MAKINO, T.: 'High power and high speed performance of 1.3 μm strained MQW gain coupled DFB lasers', *IEEE J. Quantum Electron.*, 1995, **31**, pp. 375–381
18 PETERMANN, K.: 'Calculated spontaneous emission factor for double-heterostructure injection lasers with gain-induced waveguiding', *IEEE J. Quantum Electron.*, 1979, **15**, pp. 566–570

19 KRUGER, U., and PETERMANN, K.: 'The semiconductor laser linewidth due to the presence of side modes', *IEEE J. Quantum Electron.*, 1988, **24**, pp. 2355–2358
20 PAN, X., TROMBORG, B., and OLESEN, H.: 'Linewidth rebroadening in DFB lasers due to weak side modes', *IEEE Photon. Technol. Lett.*, 1991, **3**, pp. 112–114
21 TROMBORG, B., OLESEN, H., and PAN, X.: 'Theory of linewidth for multi-electrode laser diode with spatially distributed noise sources', *IEEE J. Quantum Electron.*, 1991, **27**, pp. 178–192
22 OLESEN, H., TROMBORG, B., LASSEN, H.E., and PAN, X.: 'Mode instability and linewidth rebroadening in DFB lasers', *Electron. Lett.*, 1992, **28**, pp. 444–445
23 SCHATZ, R.: 'Longitudinal spatial instability in symmetric semiconductor lasers due to spatial hole burning', *IEEE J. Quantum Electron.*, 1992, **28**, pp. 1443–1449
24 LOWERY, A.J.: 'Dynamics of SHB-induced mode instabilities in uniform DFB semiconductor lasers', *Electron. Lett.*, 1993, **29**, pp. 1852–1854
25 MARCENAC, D.D., and CARROLL, J.E.: 'Distinction between multi-moded and singlemoded self-pulsations in DFB lasers', *Electron. Lett.*, 1994, **30**, pp. 1137–1138
26 TKACH, R.W., and CHRAPLYVY, A.R.: 'Regimes of feedback effects in 1.5 μm distributed feedback lasers', *J. Lightwave Technol.*, 1986, **4**, pp. 1655–61
27 BUUS, J., and NILSSON, O.: 'A fundamental limit for the feedback sensitivity of semiconductor lasers'. *12th IEEE International Semiconductor Laser Conference*, Davos, Switzerland, 1990, Paper L-22, pp. 216–217
28 NILSSON, O., and BUUS, J.: 'Linewidth and feedback sensitivity of semiconductor lasers', *IEEE J. Quantum Electron.*, 1990, **26**, pp. 2039–2042
29 PETERMANN, K.: *in* 'Semiconductor lasers with optical feedback' 'Laser diode modulation and noise', (Kluwer Academic Publishers, 1991), chap. 9
30 KINOSHITA, J.: 'Validity of κL evaluation by stopband method for $\lambda/4$ DFB lasers with low reflecting facets', *Electron. Lett.*, 1987, **23**, pp. 499–501
31 SODA, H., ISHIKAWA, H., and IMAI, H.: 'Design of DFB lasers for high-power single-mode operation', *Electron. Lett.*, 1986, **22**, pp. 1047–1049
32 SODA, H., ISHIKAWA, H., and IMAI, H.: 'Longitudinal mode selectivity above threshold for asymmetric mirror structure DFB lasers with low front facet reflectivity', *Electron. Lett.*, 1988, **24**, pp. 431–432
33 WHITEAWAY, J.E.A., THOMPSON, G.H.B., COLLAR, A.J., and ARMISTEAD, C.J.: 'The design and assessment of $\lambda/4$ phase-shifted DFB laser structures', *IEEE J. Quantum Electron.*, 1989, **25**, pp. 1261–1279
34 LUO, Y., NAKANO, Y., TADA, K., INOUE, T., HOSOMATSU, H., and IWAOKA, H.: 'Purely gain coupled distributed feedback semiconductor lasers', *Appl. Phys. Lett.*, 1990, **56**, pp. 1620–1622
35 CHAMPAGNE, A., MACIEJKO, R., and MAKINO, T.: 'Enhanced carrier injection efficiency from lateral current injection in multiple-quantum-well DFB lasers', *IEEE Photonics Technol. Lett.*, 1986, **8** pp. 749–751
36 LUO, Y., CAO, H.L., DOBASHI, M., HOSOMATSU, H., NAKANO, Y., and TADA, K.: 'Gain coupled distributed feedback semiconductor lasers with an absorptive conduction-type inverted grating', *IEEE Photonics Technol. Lett.*, 1992, **4**, pp. 692–695

37 TSANG, W.T., CHOA, F.S., WU, M.C., CHEN, Y.K., LOGAN, R.A., TANBUN-EK, T., CHU, S.N.G., SERGENT, A.M., MAGILL, P., REICHMANN, K., and BURRUS, C.A.: 'Gain-coupled long wavelength InGaAsP/InP distributed feedback lasers with quantum well gratings grown by chemical beam epitaxy'. Proceedings of 13th *IEEE International Semiconductor Laser Conference*, 1992, paper B-1, pp. 12–13

38 ZOZ, J., and BORCHERT, B.: 'Dynamic behaviour of complex-coupled DFB lasers with in-phase absorptive grating', *Electron. Lett.*, 1994, **30**, pp. 39–40

39 LU, H., MAKINO T., and LI, G.P.: 'Dynamic properties of partly gain-coupled 1.55 μm strained DFB lasers', *IEEE J. Quantum Electron.*, 1995, **31**, pp. 1443–1450

40 MORTHIER, G., VANKWIKELBERGE, P., DAVID, K., and BAETS, R.: 'Improved performance of AR-coated DFB lasers by the introduction of gain coupling', *IEEE Photonics Technol. Lett.*, 1990, **2**, pp. 170–172

41 DAVID, K., BUUS, J., and BAETS, R.G.: 'Basic analysis of AR-coated, partly gain-coupled DFB lasers: the standing wave effect', *IEEE J. Quantum Electron.*, 1992, **28**, pp. 427–433

42 MORTHIER, G., and VANKWIKELBERGE, P.: 'Handbook of distributed feedback laser diodes' (Artech House Boston, 1997), chap. 3

43 LOWERY, A.J.: 'Large-signal effective α factor of complex-coupled DFB semiconductor lasers', *Electron. Lett.*, 1992, **28**, pp. 2295–2297

44 LOWERY, A.J., and NOVAK, D.: 'Enhanced maximum intrinsic modulation bandwidth of complex-coupled DFB semiconductor lasers', *Electron. Lett.*, 1993, **29**, pp. 461–463

45 PAN, X., TROMBORG, B., OLESEN, H., and LASSEN, H.E.: 'Effective linewidth enhancement factor and spontaneous emission rate of DFB lasers with gain coupling', *IEEE Photonics Technol. Lett.*, 1992, **4**, pp. 1213–1215

46 LI, G.P., MAKINO, T., MOORE, R., and PUETZ, N.: '1.55 μm index/gain coupled DFB lasers with strained layer multiquantum-well active grating', *Electron. Lett.*, 1992, **28**, pp. 1726–1727

47 LI, G.P., MAKINO, T., MOORE, R., PUETZ, N., LEONG, K-W., and LU, H.: 'Partly gain-coupled 1.55 μm strained-layer multiquantum-well DFB lasers', *IEEE J. Quantum Electron.*, 1993, **29**, pp. 1736–1742

48 MAKINO, T., and GLINSKI, J.: 'Effects of radiation loss on the performance of second-order DFB semiconductor lasers', *IEEE J. Quantum Electron.*, 1988, **24**, pp. 73–82

49 BAETS, R.G., DAVID, K., and MORTHIER, G.: 'On the distinctive features of gain coupled DFB lasers and DFB lasers with second order grating', *IEEE J. Quantum Electron.*, 1993, **29**, pp. 1792–1798

50 DAVID, K., MORTHIER, G., VANKWIKELBERGE, P., and BAETS, R.: 'Yield analysis of non-AR-coated DFB lasers with combined index and gain coupling', *Electron. Lett.*, 1990, **26**, pp. 238–239

51 ADAMS, D.M., CASSIDY, D.T., and BRUCE, D.M.: 'Scanning photoluminescence technique to determine the phase of the grating at the facets of gain-coupled DFB's', *IEEE J. Quantum Electron.*, 1996, **32**, pp. 1237–1242

52 KAZARINOV, R.F., and HENRY, C.H.: 'Second-order distributed feedback lasers with mode selection provided by first-order radiation losses', *IEEE J. Quantum Electron.*, 1985, **21**, pp. 144–150
53 KOREN, U., MILLER, B.I., RAYBON, G., ORON, M., YOUNG, M.G., KOCH, T.L, DEMIGUEL, J.L., CHIEN, M., TELL, B., BROWN-GOEBELER, K., and BURRUS, C.A.: 'Integration of 1.3 μm wavelength lasers and optical amplifiers', *Appl. Phys. Lett.*, 1990, **57**, pp. 1375–1377
54 NAKANO, Y., HAYASHI, Y., CHEN, N., SAKAGUCHI, Y., and TADA, K.: 'Fabrication and characteristics of an integrated DFB laser/amplifier having reactive-ion-etched tilted end facets', *Jpn. J. Appl. Phys.*, 1990, **29**, pp. L2430–2433
55 KOREN, U., JOPSON, R.M., MILLER, B.I., CHIEN, M., YOUNG, M.G., BURRUS, C.A., GILES, C.R., PRESBY, H.M., RAYBON, G., EVANKOW, J.D., TELL, T., and BROWN-GOEBELER, K.: 'High power laser-amplifier photonic integrated circuit for 1.48 μm wavelength operation', *Appl. Phys. Lett.*, 1991, **59**, pp. 2351–2353
56 O'BRIEN, S., PARKE, R., WELCH, D.F., MEHUYS, D., and SCIFRES, D.: 'High power singlemode edge-emitting master oscillator power amplifier', *Electron. Lett.*, 1992, **28**, pp. 1429–1431
57 WHITEAWAY, J.E.A., THOMPSON, G.H.B., GOODWIN, A.R., and FICE, M.J.: 'Design of monolithically integrated single frequency laser and booster amplifier', *Electron. Lett.*, 1991, **27**, pp. 2250–2251
58 FICE, M.J., GOODWIN, A.R., THOMPSON, G.H.B., and WHITEAWAY, J.E.A.: 'Realisation of monolithically integrated single frequency MQW laser and booster amplifier at 1.5 μm wavelength', *Electron. Lett.*, 1991, **27**, pp. 2305–2306
59 GREEN, P.E.: 'Fibre optic networks' (Prentice Hall, Englewood Cliffs, 1993), p. 87
60 GOWAR, J.: 'Optical communication systems' (Prentice Hall, 1993), 2nd ed., p. 438
61 FICE, M.J.: 'Development of an optimised high power coherent 1.3 μm–1.7 μm laser diode source'. ESA contract 8189/89/NL/PB(SC)
62 O'BRIEN, S., WELCH, D.F., PARKE, R.A., MEHUYS, D., DZURKO, K., LANG, R.J., WAARTS, R., and SCIFRES, D.: 'Operating characteristics of a high-power monolithically integrated flared amplifier master oscillator power amplifier', *IEEE J. Quantum Electron.*, 1993, **29**, pp. 2052–2057
63 O'BRIEN, S., MEHUYS, D., MAJOR, J., LANG, R., PARKE, R., WELCH, D.F., and SCIFRES, D.: '1.3 W cw, diffraction-limited monolithically integrated master oscillator flared amplifier at 863 nm', *Electron. Lett.*, 1993, **29**, pp. 2109–2110
64 KOGELNIK, H.K., and SHANK, C.V.: 'Coupled-wave theory of distributed feedback lasers', *J. Appl. Phys.*, 1972, **43**, pp. 2327–2335
65 IGA, K.: 'On the use of effective refractive index in DFB laser mode separation', *Jpn. J. Appl. Phys.*, 1983, **22**, p. 1630
66 SODA, H., and IMAI, H.: 'Analysis of the spectrum behaviour below the threshold in DFB lasers', *IEEE J. Quantum Electron.*, 1986, **22**, pp. 637–641
67 LOWERY, A.J.: 'Dynamics of SHB-induced mode instabilities in uniform DFB semiconductor lasers', *Electron. Lett.*, 1993, **29**, pp. 1892–1894

68 ZHANG, L.M., and CARROLL, J.E.: 'Dynamics and hole burning in uniform DFB semiconductor lasers', *Int. J. Optoelectron.*, 1993, **8**, pp. 279–291
69 VANKWIKELBERGE, P., BUYTAERT, F., FRANSHOIS, A., BAETS, R., KUINDERSMA, P.I., and FREDRIKSZ, C.W.: 'Analysis of the carrier induced FM response of DFB lasers, theoretical and case studies', *IEEE J. Quantum Electron.*, 1989, **25**, pp. 2239–2254
70 CHRISTENSEN, B., OLESEN, H., JONSSON, B., LAGE, H., HANBERG, J., ALBREKTSEN, O., and MOLLER-LARSEN, A.: 'Detailed mapping of the local IM and FM responses of DFB lasers', *Electron. Lett.*, 1995, **31**, pp. 799–800
71 MORTON, P.A., TANBUK-EK, T., LOGAN, R.A. CHAND, N., WECHT, K.W., SERGENT, A.M., and SCIORTINO P.F.: 'Packaged 1.55 µm DFB laser with 25 GHz modulation bandwidth', *Electron. Lett.*, 1994, **30**, pp. 2044–2046
72 KITO, M., ISHINO, M., OTSUKA, N., HOSHINO, N., FUJIHARA K., FUJITO F., and MATSUI, Y.: 'Low distortion up to 2 GHz in 1.55 µm multiquantum well distributed feedback laser', *Electron. Lett.*, 1992, **28**, pp. 891–893
73 YAMADA, H., OKUDA, T., SHIBUTANI, M., TOMIDA, S., TORIAI, T., and UJI, T.: 'High modulation frequency low distortion 1.3 µm MQW-DFB-LDs for subcarrier multiplexed fibre-optic feeder systems', *Electron. Lett.*, 1993, **29**, pp. 1994–1995
74 SCHATZ, R.: 'Longitudinal spatial instability in symmetric semiconductor lasers due to spatial hole burning', *IEEE J. Quantum Electron.*, 1992, **28**, pp. 1443–1449
75 WHITEAWAY, J.E.A., GARRETT, B., THOMPSON, G.H.B., COLLAR, A.J., ARMISTEAD, C.J., and FICE, M.J.: 'The static and dynamic characteristics of single and multiple phase-shifted DFB laser structures', *IEEE J. Quantum Electron.*, 1992, **28**, pp. 277–1293
76 WHITEAWAY, J.E.A., WRIGHT, A.P., GARRETT, B., THOMPSON, G.H.B., CARROLL, J.E., ZHANG, L.M., TSANG, C.F., WHITE, I.H., and WILLIAMS, K.A.: 'Detailed large-signal dynamic modelling of DFB laser structures and comparison with experiment', *Opt. Quantum Electron.*, 1994, **26**, pp. S817–S842
77 WRIGHT, A.P., GARRETT, B., THOMPSON, G.H.B., and WHITEAWAY, J.E.A.: 'Influence of carrier transport on wavelength chirp of InGaAs/InGaAsP MQW lasers', *Electron. Lett.*, 1992, **28**, pp. 1911–1912
78 YAMAGUCHI, M., HENMI, N., YAMAZAKI, H., and MITO, I.: 'Analysis of wavelength chirping for $\lambda/4$ shifted DFB LD considering spatial hole-burning along cavity'. 12th IEEE *International Semiconductor Laser Conference*, Davos, Switzerland, 1990, Paper E4, pp. 66–67
79 WRIGHT, A.P., BRIGGS, A.T.R., SMITH, A.D., BAULCOMB, R.S., and WARBRICK, K.J.: '22 GHz bandwidth 1.5 µm compressively strained InGaAsP MQW ridge waveguide DFB lasers', *Electron. Lett.*, 1993, **29**, pp. 1848–1849

Chapter 7
Numerical modelling for DFB lasers

7.1 Introduction

A commonly used starting point for modelling lasers is a small-signal analysis of the rate equations of the form discussed in Section 4.1 where perturbations from the steady state are considered [1–3]. These solutions are often explicitly analytic and/or they can be rapidly computed. Appendix 4 gives an outline of such an analysis including carrier transport [4], from contacts to the radiative recombination region, which is not included in this chapter. Small-signal methods help to elucidate the physics of modulation and noise [5–8], especially around steady-state values which can be computed more readily than large-signal dynamic states. For Fabry–Perot lasers, coupling of electron equations and photon equations has been done in a variety of ways well reviewed by Buus [9] but the power of computers has moved far in the last decade. DFB lasers have more complex structures and have led to new methods specifically to aid in this understanding. *Transfer-matrix* techniques, mentioned in Chapter 5, are also known as transmission matrices [10,11] and are used for the analysis and design of multisection and nonuniform lasers by tracking their performance around specific frequencies. However, random spontaneous inputs to a laser give randomly varying outputs which then need averaging. The *power-matrix* method [12] is a transfer-matrix method which was specially developed to compute the *mean*-square values of optical power even though the laser is excited by a stochastic 'spontaneous' input. As has already been seen from Chapter 5, transfer-matrix techniques operate well for 'single-mode' lasers with only one or two clear weak sidemodes which can be all tracked as the laser changes frequency, but particular care is needed with many modes or where spectral mixing occurs. The techniques can be specially helpful for long or complicated structures where time-domain analysis requires excessive numerical storage.

'Transmission-line laser modelling' (TLLM) [13–15] is a time-domain technique which copes with many-moded operation and with spectral mixing, all at arbitrary power levels. The concept is that electromagnetic fields can be represented by pulses propagating in time along 'transmission lines' [16]. Gain, loss, scattering, dispersion and so on are introduced by using filter theory with transmission lines. The spectral behaviour is recovered from this time-domain modelling by using *fast Fourier transforms* (FFT). Spontaneous emission is modelled by random excitation which then gives randomly changing outputs which need appropriate averaging. Lowery was among the first to recognise the value of computing the travelling-wave properties of lasers [17,18] and a strong debt is owed to TLLM for many aspects of time-domain modelling, but there can be clear philosophical differences in approaches from other time-domain work. In this book, low-order finite-difference strategies, with attention paid to spectral filtering, are used to compute Maxwell's electromagnetic-field equations, using complex fields, along with the rate equations which balance up the energy flow of electrons and photons. In TLLM, the physics [19,20] has to be translated into the terms of 'transmission lines' but the practical differences may be only a matter of taste in numerical modelling.

This chapter presents the basics of large-signal time-domain modelling using the travelling-wave time/distance nonlinear partial differential equations of the laser. The field patterns and electron densities in the laser are computed permitting electron–photon interactions to be visualised. The lasers are excited by random 'spontaneous noise' which leads to outputs which are never precisely the same from run to run. However, the random output is not normally a major drawback but the time-domain modelling of low-frequency noise can require excessively long times of computation. The FFT readily permits changing from a time-domain to a frequency-domain analysis. With N_{fft} points for the FFT, the lowest spectral resolution is determined by a frequency ($v_g/N_{fft}s$) where v_g is the group velocity in the material and s is the space step. With time steps typically in the subpicosecond range, effects around 1 kHz may require a problematic 10^9–10^{10} steps which are not considered here. The highest frequency ($v_g/2s$) which can be modelled is limited by the time step s/v_g which then must be small enough to enable all the important physical frequencies to fall well below this highest frequency.

This chapter aims to provide tutorial material for the reader with limited experience in numerical modelling. The philosophy is to use straightforward low-order techniques where the synergy between the physics and numerical methods is reasonably good. Speed of computation is gained from modern high clock rates and cheap random-access memory rather than the sophistication of high-order computational techniques with large step sizes. The reader is referred to texts and handbooks [21–24] rather than the research literature. Sections on ordinary differential equations and hyperbolic partial differential equations can then be selected. To help build up the reader's confidence, the work starts with first-order ordinary differential equations before moving in graded steps to the coupled travelling-wave equations of the DFB. A series of tutorial MATLAB programs, able to run on the student version of MATLAB [25], is provided via the 'net'. The reader with a full copy of MATLAB 4.0 or better can enhance these programs, interfacing with C++ to gain computational speed if required. In laser diodes, short enough pulses of optical energy can disperse and change their shape through nonlinear physical interactions, and it is important that the programs do not confuse numerically induced gain or distortion with similar physical effects. The good numerical analyst would design algorithms with step lengths determined by error control [26] built into the algorithm, but the ready availability of high-speed computation has seduced the authors into more limited programs, checking that there are no singularities at small step lengths, and then comparing results with different step lengths to determine that a satisfactory value was used.

7.2 Ordinary differential equations

7.2.1 A first-order equation

By way of illustration, start with the straightforward ordinary differential equation

$$dF/dz = gF \qquad (7.1)$$

where F might represent a field with the gain per unit distance given by g, initially taken as constant. The continuous function $F(z)$ is approximated by $F(\mathsf{Z}\,\mathsf{s})$, where Z is an integer, and s is the step length. The '*forward Euler*' method of solving this equation numerically gives

$$F(\mathsf{Zs+s}) \simeq F(\mathsf{Zs}) + \mathsf{s}(dF/dz)_{\mathsf{Z}} \quad \text{or} \quad F(\mathsf{Zs+s}) = F(\mathsf{Zs})(1+\mathsf{s}g) \qquad (7.2)$$

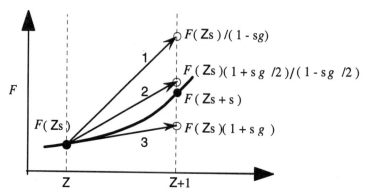

Figure 7.1 Forward, central, and reverse differences (3, 2, 1, respectively)

This is shown schematically in Figure 7.1, and has the merit of simplicity. Another simple technique, also indicated in Figure 7.1, is the '*backward Euler*' method, which is written as

$$F(Zs+s) - s(dF/dz)_{z+1} \approx F(Zs) \quad \text{or} \quad F(Zs+s) = F(Zs)/(1-sg) \quad (7.3)$$

Both methods work satisfactorily given small enough step lengths, and the discussion of their relative merits is limited to considering a purely imaginary $g=jg_i$. Then with the 'forward Euler' $|F(Zs+s)|^2 = \{1+(sg_i)^2\} | F(Zs)|^2$ always leads to some induced numerical growth while with the 'backward Euler' $|F(Zs+s)|^2 = |F(Zs)|^2/\{1+(sg_i)^2\}$ always leads to some induced numerical decay no matter how small the step length, whereas the genuine physical solution for a purely imaginary growth rate gives no growth or decay with $|F(Zs+s)|^2 = |F(Zs)|^2$.

With the previous methods, the amount of growth or decay can be limited by sufficiently small steps but a substantial improvement in accuracy and stability is to use the '*central-difference*' method, which can be regarded as using both of the above methods but for a half step each and leads to a midway prediction between the forward- and backward-difference methods as indicated in Figure 7.1 with

$$F(Zs+s) = F(Zs)\{(1+\tfrac{1}{2}sg)/(1-\tfrac{1}{2}sg)\} \quad (7.4)$$

It can be seen immediately that, when $g=jg_i$, then $|F(Zs+s)|^2 = |F(Zs)|^2$ as required. Variants on this central-difference method when g_i varies with Z do not all precisely preserve the 'energy' as in eqn. 7.4 but are substantially better than the forward- or reverse-difference methods. This central-difference technique is most useful in solution of laser-modelling equations, and is used many times from here onwards. It is also known as the *trapezoidal rule*, and is extended to partial differential equations by use of *Lax averaging* [27] so that when g becomes variable with Z, i.e. $g(Zs)$:

$$dF/dz \to \{F(Zs+\tfrac{1}{2}s) - F(Zs-\tfrac{1}{2}s)\}/s$$

$$= \tfrac{1}{2}\{g(Zs+\tfrac{1}{2}s)F(Zs+\tfrac{1}{2}s) + g(Zs-\tfrac{1}{2}s)F(Zs-\tfrac{1}{2}s)\} \qquad (7.5)$$

The half step may suggest that one is introducing an additional layer of points but this is simply an initial convenient notation to ensure symmetry between the $+\tfrac{1}{2}s$ displacement and the $-\tfrac{1}{2}s$ displacement to give a net step of s. The symmetry with Lax averaging proves to be important in the (spatial or temporal) frequency domain because a finite-difference solution 'samples' the electromagnetic fields leading to the so-called 'Nyquist' limitations on frequency [28] which occur in any discrete signal. This will be discussed later. The half-space intervals may be 'removed' through shifting the origin by writing $Z+\tfrac{1}{2} \to Z'+1$ and $Z-\tfrac{1}{2} \to Z'$ and then dropping the prime to give

$$f(Zs+s) = [\{1+\tfrac{1}{2}sg(Zs)\}/\{1-\tfrac{1}{2}sg(Zs+s)\}] f(Zs) \qquad (7.6)$$

7.2.2 Accuracy

The accuracy of these finite-difference schemes in the end depends on the mathematical result [29] that for all values of the complex 'gain' $\gamma = (g+j\delta)L$

$$\mathrm{Lim}_{N \to \infty} [1+\gamma/N]^N = \exp \gamma \qquad (7.7)$$

while, through halving γ, an even faster convergence is obtained from

$$\mathrm{Lim}_{N \to \infty} [(1+\tfrac{1}{2}\gamma/N)^N / (1-\tfrac{1}{2}\gamma/N)^N] = \exp \gamma \qquad (7.8)$$

Programs **stepr** and **stepj** in directory **diff** are available to help the reader gain a 'feel' for the problem of convergence as the number of steps N is changed. The programs integrate $dF/dz=zF$ (Figure 7.2) and $dF/dz=jzF$, respectively, from $z=0$–3 starting with $F=1$. The test examples of either exponential growth or pure imaginary rate of change in $F(z)$ have been chosen because these (or mixtures) are common in laser physics. It will be found in **stepj** that eqn. 7.5 is slightly modified so as to ensure $d(F^*F)/dz=0$ demonstrating the considerable benefits of appropriate central-difference schemes which can avoid numerically induced growth when there is no physical growth.

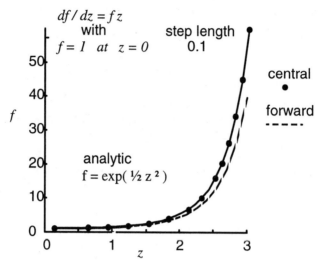

Figure 7.2 Comparison of difference schemes

7.3 First-order wave equations

7.3.1 Introduction

The two equations

$$\frac{1}{v_g}\frac{\partial F}{\partial t}+\frac{\partial F}{\partial z}=0 \qquad \frac{1}{v_g}\frac{\partial R}{\partial t}-\frac{\partial R}{\partial z}=0 \qquad (7.9)$$

are the first-order wave equations for waves propagating in either the forward or reverse directions with group velocities $\pm v_g$. In the literature of partial differential equations, these are often referred to as the *advection* equations. Their physical solutions are straightforwardly given from

$$F=f(t-z/v_g) \quad \text{and} \quad R=r(t+z/v_g) \qquad (7.10)$$

where $f(t)$ and $r(t)$ are arbitrary functions of time. It is possible to consider immediately the equations with gain and phase changes varying as

$$\frac{1}{v_g}\frac{\partial F}{\partial t}+\frac{\partial F}{\partial z}=(g-j\delta)F \qquad \frac{1}{v_g}\frac{\partial R}{\partial t}-\frac{\partial R}{\partial z}=(g-j\delta)R \qquad (7.11)$$

because, on putting

$$F = \exp\{(g-j\delta)v_g t\}\bar{F} \qquad R = \exp\{(g-j\delta)v_g t\}\bar{R} \qquad (7.12)$$

one recovers the straightforward advection equations as in eqn. 7.9 but now as

$$\frac{1}{v_g}\frac{\partial \bar{F}}{\partial t}+\frac{\partial \bar{F}}{\partial z}=0 \qquad \frac{1}{v_g}\frac{\partial \bar{R}}{\partial t}-\frac{\partial \bar{R}}{\partial z}=0 \qquad (7.13)$$

7.3.2 Step lengths in space and time—central-difference method

In computing eqns. 7.13 where the group velocity is constant, one can take the step lengths δt in time and δz in space to be related by $\delta t = \delta z/v_g$. While this simplification is used in all the applications presented in this book, it is useful, for future work, to understand both how more general solutions can be derived with varying group velocities and that potential instabilities occur when attempting to compute information before it can physically have time to arrive. To this end, consider a spatial step $\delta z = s$ with a temporal step $\delta t = s_t/v_g$ with $s_t/s = a$.

The forward function F may be written now as $F(T, Z)$, where T and Z are integers. Taking a cue from Section 7.2, which shows that central differences with Lax averaging offer distinct advantages for complex equations, one particular central-difference system is examined for the advection equation:

$$\tfrac{1}{2}\{F(T+\tfrac{1}{2}, Z+\tfrac{1}{2})+F(T+\tfrac{1}{2}, Z-\tfrac{1}{2}) - F(T-\tfrac{1}{2}, Z+\tfrac{1}{2}) - F(T-\tfrac{1}{2}, Z-\tfrac{1}{2})\}$$
$$+\tfrac{1}{2}a\{F(T+\tfrac{1}{2}, Z+\tfrac{1}{2})+F(T-\tfrac{1}{2}, Z+\tfrac{1}{2}) - F(T+\tfrac{1}{2}, Z-\tfrac{1}{2}) - F(T-\tfrac{1}{2}, Z-\tfrac{1}{2})\}=0$$

$$(7.14)$$

Low values of **a** imply that the pulse travels more slowly while larger values of **a** make the pulse travel faster. Notice the use of 'Lax averaging' in space where

$$F(T+\tfrac{1}{2}, Z) \rightarrow \tfrac{1}{2}\{F(T+\tfrac{1}{2}, Z+\tfrac{1}{2})+F(T+\tfrac{1}{2}, Z-\tfrac{1}{2})\} \qquad (7.15)$$

with similar Lax averaging in time. Different arrangements of Lax averaging can give related finite-difference schemes that can then be explored for stability and accuracy. Eqn. 7.14 can be rearranged with future time values of F given in terms of previous values:

$$(1+a)F(T+\tfrac{1}{2}, Z+\tfrac{1}{2})+(1-a)F(T+\tfrac{1}{2}, Z-\tfrac{1}{2})$$
$$=(1-a)F(T-\tfrac{1}{2}, Z+\tfrac{1}{2})+(1+a)F(T-\tfrac{1}{2}, Z-\tfrac{1}{2}) \qquad (7.16)$$

The half-space values may be 'removed' by a shift of the time and space origin as in the discussion between eqns. 7.5 and 7.6. If a=1 one immediately finds

$$F(T+1, Z+1) = F(T, Z) \quad (7.17)$$

Note that in both this finite-difference solution and the correct physical solution the wave amplitude is preserved as it propagates along at the group velocity.

7.3.3 Numerical stability

There are many ways of arranging the finite difference approximations to the differential equations so that one needs to have another guide to check whether a proposed finite-difference scheme is likely to be satisfactory. Provided that one is cautious about interpreting the results (for reasons discussed by Iserles [30]), it can be helpful to consider a solution $F(T, Z) + f_{error} \exp(j\bar{\omega}T - j\bar{\beta}Z)$ where $F(T, Z)$ is the ideal solution but there is a distribution of small errors where $\bar{\beta}$ is some (arbitrary) normalised propagation coefficient and $\bar{\omega}$ is some resulting frequency. The question then is whether there is some spatial distribution of the errors that leads to unphysical growth with time (i.e. a false complex $\bar{\omega}$) because of the numerical algorithm. In the physical equation (eqn. 7.9), direct substitution shows that

$$\bar{\beta} s = \bar{\omega} s_t \quad (7.18)$$

The errors do not grow physically or distort the signal and one can compare the physical result with the numerical result. If $a \neq 1$, then, from eqn. 7.14,

$$\exp(j\bar{\omega}) = \{\cos(\bar{\beta}/2) + aj\sin(\bar{\beta}/2)\} / \{\cos(\bar{\beta}/2) - aj\sin(\bar{\beta}/2)\} \quad (7.19)$$

The modulus of each side of eqn. 7.19 remains unity for all real $\bar{\beta}$ so that, no matter what Fourier distribution one has of the errors, they do not grow in time and to this extent the computation can be said to be stable and therefore should be a satisfactory algorithm.

However, determining the phase shift per step on both sides of eqn. 7.19 gives

$$\bar{\omega} = 2 \arctan[a \tan(\bar{\beta}/2)] \quad (7.20)$$

It is now only at small enough $\bar{\beta}$ (slowly varying changes with step length) that $\bar{\omega} \to a\bar{\beta}$ as in eqn. 7.18. Numerically induced dispersion/distortion of the pulse shape always occurs to some extent (Figure 7.3). Reducing the step size helps to limit the numerical dispersion. Changing algorithms can also change this numerical distortion.

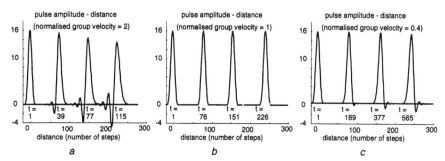

Figure 7.3   Plots of pulse propagation

Figure 7.3 shows how pulses change their shape in space as they propagate. The time of arrival at each point in space depends on the group velocity, and, as the group velocity changes, the numerical methods introduce some distortion. With differences of group velocity of a few per cent less than the normalised value of 1, the distortion is usually acceptable. Here by having a 50% variation in the normalised group velocity one can see significant distortion induced at the pulse edges; this could be reduced by having shorter steps giving more steps but less net distortion over the same total distance

The acceptability of numerically induced distortion is most readily determined in relation to specific applications by testing. Figure 7.3*a* shows the distortion at the trailing edge which occurs for 'short' pulses with the normalised group velocity giving a=0.2. If the pulse moves with a normalised group velocity of 0.4, the numerical distortion is induced at the leading edge (Figure 7.3*c*). If a=1 then there is no distortion (Figure 7.3*b*).

The program **advec** in directory **diff** provides a basic test program where the forward field is stored at all space points but for only two time steps. Unless a=1, the computing scheme requires *two* future points in space to be addressed by the past values of the same two space points. This works for closed systems. There are other methods of solving propagation along open lines [31] with one future point determined by two past points. The normalised group velocity is then required to be less than (or equal to) unity because, in such an open system, any attempt at the prediction of information before it arrives leads to instability (numerically induced growth). For the remainder of the chapter the essential algorithm will use equal normalised space and time steps (a=1, i.e. s=s_t) providing the basis of stable algorithms in either open or closed systems.

The reverse advection equation with the equivalent algorithm is found by changing the sign of **a** to give

$$(1-a)R(T+\tfrac{1}{2}, Z+\tfrac{1}{2}) + (1+a)R(T+\tfrac{1}{2}, Z-\tfrac{1}{2})$$
$$= (1+a)R(T-\tfrac{1}{2}, Z+\tfrac{1}{2}) + (1-a)R(T-\tfrac{1}{2}, Z-\tfrac{1}{2}) \tag{7.21}$$

where again when $a=1$ and the half spaces are 'removed', as explained after eqn. 7.16, to give

$$R(T+1, Z) = R(T, Z+1) \tag{7.22}$$

The requirement of a constant group velocity is not too serious a limitation. It can be relaxed easily if the group velocity does not change with time or frequency but changes only with space. Uniform time steps δt are then maintained but the spatial steps may be adjusted to correspond to $v_g(z)\,\delta t$ at each local space point. Only if $v_g(t, z)$ varies with time as well as space is it necessary to consider a more complicated system, and this is not done in this book.

7.3.4 Gain and phase

Reverting to eqns. 7.11 and 7.12 with the velocity parameter $a=1$ allows the insertion of gain and phase changes, and central differences with Lax averaging may be used to give

$$F(T+1, Z) - F(T, Z-1) = \tfrac{1}{2}(g - j\delta)s\{F(T+1, Z) + F(T, Z-1)\} \tag{7.23}$$

or

$$F(T+1, Z) = \{(1 + \tfrac{1}{2}sg - j\tfrac{1}{2}s\delta) / (1 - \tfrac{1}{2}sg + j\tfrac{1}{2}s\delta)\}F(T, Z-1) \tag{7.24}$$

The simplicity of the algorithm is enhanced by keeping g and δ evaluated at $Z-1$ on the right-hand side of eqn. 7.24, even though this requires smaller steps. The result in eqn. 7.24 is in effect saying that $(1+\gamma s)/(1-\gamma s)$ is a good approximation to $\exp(2\gamma s)$ for complex $\gamma = \tfrac{1}{2}(g - j\delta)$ and indeed, to the level of approximation, one may use either $\{(1+\gamma s)/(1-\gamma s)\}$ or $\exp(2\gamma s)$. However the form of eqn. 7.24 is more useful for later on when shaping the gain with frequency using a numerical filter in the time domain.

The reverse advection equation with gain and phase changes is similarly given from

$$R(T+1, Z) = \{(1 + \tfrac{1}{2}sg - j\tfrac{1}{2}s\delta) / (1 - \tfrac{1}{2}sg + j\tfrac{1}{2}s\delta)\}R(T, Z+1) \tag{7.25}$$

Central differences with Lax averaging improve the accuracy and stability for partial differential equations just as they did for complex ordinary differential equations.

7.4 Coupled reflections

7.4.1 Kappa coupling but no gain or phase changes

With operation at the Bragg frequency ($\delta=0$) and no loss or gain in the waveguide, then the coupled equations of interest are

$$\frac{1}{v_g}\frac{\partial F}{\partial t}+\frac{\partial F}{\partial z}=j\kappa R \qquad \frac{1}{v_g}\frac{\partial R}{\partial t}-\frac{\partial R}{\partial z}=j\kappa F \qquad (7.26)$$

With a constant κ it is straightforward to eliminate R to obtain

$$\frac{1}{v_g^2}\frac{\partial^2 F}{\partial t^2}-\frac{\partial^2 F}{\partial z^2}=-\kappa^2 F \qquad (7.27)$$

One may write an analytic solution as $F=\exp\{\bar{\omega}(v_g t/\mathsf{s}) - j\bar{\beta}(z/\mathsf{s})\}$ where a spatial-step length s is used to normalise the spatial propagation coefficient and temporal frequencies, $\bar{\beta}$ and $\bar{\omega}$, respectively, related by

$$\bar{\beta}^2 = \bar{\omega}^2 - (\kappa\mathsf{s})^2 \qquad (7.28)$$

The physical frequency corresponding to the phase change per time step $\bar{\omega}$ is a modulation frequency ω_M—a measure of the frequency deviation from the central frequency ω_o which has been removed in the formulation of eqns. 7.26. When the phase shift per step is such that $|\bar{\omega}| < \kappa\mathsf{s}$ there are 'evanescent waves' giving an imaginary spatial phase shift per step $\bar{\beta}$ (corresponding to the DFB's *stopband*).

For a straightforward first-order finite-difference scheme, one treats the forward and reverse waves on a symmetrical footing. There are a number of ways of using Lax averaging on the right-hand side of the coupled equations and a useful way forward as judged from later work on energy conservation is given from

$$F(\mathsf{T}+\tfrac{1}{2},\mathsf{Z}+\tfrac{1}{2}) - F(\mathsf{T}-\tfrac{1}{2},\mathsf{Z}-\tfrac{1}{2}) = j\tfrac{1}{2}\kappa\mathsf{s}\{R(\mathsf{T}+\tfrac{1}{2},\mathsf{Z}-\tfrac{1}{2}) + R(\mathsf{T}-\tfrac{1}{2},\mathsf{Z}+\tfrac{1}{2})\} \qquad (7.29)$$

$$R(\mathsf{T}+\tfrac{1}{2},\mathsf{Z}-\tfrac{1}{2}) - R(\mathsf{T}-\tfrac{1}{2},\mathsf{Z}+\tfrac{1}{2}) = j\tfrac{1}{2}\kappa\mathsf{s}\{F(\mathsf{T}+\tfrac{1}{2},\mathsf{Z}+\tfrac{1}{2}) + F(\mathsf{T}-\tfrac{1}{2},\mathsf{Z}-\tfrac{1}{2})\} \qquad (7.30)$$

7.4.2 Matrix formulation

One now wishes to re-arrange eqns. 7.29 and 7.30 into a form which demonstrates that the 'energy' that is scattered by a lossless index grating is conserved. Start with the equations written in matrix terms:

$$\begin{bmatrix} 1 & -j\tfrac{1}{2}\kappa s \\ -j\tfrac{1}{2}\kappa s & 1 \end{bmatrix} \begin{bmatrix} F\{(T+\tfrac{1}{2}),\,(Z+\tfrac{1}{2})\} \\ R\{(T+\tfrac{1}{2}),\,(Z-\tfrac{1}{2})\} \end{bmatrix}$$

$$= \begin{bmatrix} 1 & j\tfrac{1}{2}\kappa s \\ j\tfrac{1}{2}\kappa s & 1 \end{bmatrix} \begin{bmatrix} F\{(T-\tfrac{1}{2}),\,(Z-\tfrac{1}{2})\} \\ R\{(T-\tfrac{1}{2}),\,(Z+\tfrac{1}{2})\} \end{bmatrix} \quad (7.31)$$

Write

$$\frac{1}{\sqrt{(1+\tfrac{1}{4}\kappa^2 s^2)}} \begin{bmatrix} 1 & -j\tfrac{1}{2}\kappa s \\ -j\tfrac{1}{2}\kappa s & 1 \end{bmatrix} = \begin{bmatrix} \cos\tfrac{1}{2}\theta & -j\sin\tfrac{1}{2}\theta \\ -j\sin\tfrac{1}{2}\theta & \cos\tfrac{1}{2}\theta \end{bmatrix} = \mathbf{U}(\theta)$$

(7.32)

with $\tan\tfrac{1}{2}\theta = \tfrac{1}{2}\kappa s$ and eqn. 7.32 then gives a matrix $\mathbf{U}(\theta)$ such that $\mathbf{U}^\dagger(\theta)\,\mathbf{U}(\theta) = 1$ where $\dagger$ implies conjugation and transposing. $\mathbf{U}$ is said to be a *unitary* matrix. Changing the origin of time and space as before to 'remove' the half steps in eqn. 7.31 one can get back almost to the conventional advection relationships linking $F\{(T+1),\,(Z+1)\}$ and $R\{(T+1),\,Z\}$ with $F\{T,\,Z\}$ and $R\{T,\,(Z+1)\}$:

$$\begin{bmatrix} F\{(T+1),\,(Z+1)\} \\ R\{(T+1),\,Z\} \end{bmatrix} = \begin{bmatrix} \cos\theta & j\sin\theta \\ j\sin\theta & \cos\theta \end{bmatrix} \begin{bmatrix} F\{T,\,Z\} \\ R\{T,\,(Z+1)\} \end{bmatrix}$$

(7.33)

where $\sin\theta = \kappa s/(1+\tfrac{1}{4}\kappa^2 s^2)$. Because $\cos^2\theta + \sin^2\theta = 1$, the scattered fields are linked by a unitary matrix in eqn. 7.33 which ensures that the energy, determined by $F^*F_z + R^*R_{z+1}$, is conserved to $F^*F_{z+1} + R^*R_z$ as time steps on one step and so ensures a stable numerical process which reflects the physics of the continuous system. (Eqn. 7.33 is *not* a transfer matrix because the field F on, say, the right-hand side of the matrix is at the opposite ends of a spatial step to the field R on the same side of the matrix.)

There is an interesting corollary which occurs with *gain-coupled* DFB lasers where, in the extreme limit, $\kappa = j\kappa_{gain}$ is purely imaginary so that eqn. 7.33 changes to become

Numerical modelling for DFB lasers 221

$$\begin{bmatrix} F\{(T+1),(Z+1)\} \\ R\{(T+1),Z\} \end{bmatrix} = \begin{bmatrix} \cosh\theta' & \sinh\theta' \\ \sinh\theta' & \cosh\theta' \end{bmatrix} \begin{bmatrix} F(T,Z) \\ R\{T,(Z+1)\} \end{bmatrix} \quad (7.34)$$

where $\sinh\theta' = \kappa_{gain}\,s/(1 - \tfrac{1}{4}\kappa_{gain}^2 s^2)$ the gain-coupling coefficient. Now one finds that $F^*F_{Z+1} - R^*R_Z$ is correctly conserved, again supporting the strength of the central-difference scheme combined with Lax averaging in obtaining a useful approximation to the physics.

7.4.3 Phase jumps replacing scattering

In the design of DFB lasers, it has become common (Chapters 5 and 6) to introduce one or two sections at appropriate points into the Bragg grating adding an extra phase shift $\exp(-j\phi)$ but with no extra scattering of the forward and reverse waves within that section. The phase jump introduces an additional phase delay on top of any normal propagation delay so that a first approximation replaces the scattering matrix (eqn. 7.34) with the phase-jump matrix:

$$\begin{bmatrix} F\{(T+1),(Z+1)\} \\ R\{(T+1),Z\} \end{bmatrix} = \begin{bmatrix} \exp(-j\phi) & 0 \\ 0 & \exp(-j\phi) \end{bmatrix} \begin{bmatrix} F(T,Z) \\ R\{T,(Z+1)\} \end{bmatrix} \quad (7.35)$$

In the simplified 'universal' DFB program which is demonstrated with this tutorial text, phase jumps are included by such substitution of a phase-jump matrix for a scattering matrix at appropriate points within the structure.

7.4.4 Fourier checks

The Fourier techniques used previously to check stability can again be used on eqn. 7.33 by allowing an arbitrary variation $\exp(j\bar{\beta}Z)$ in space (Z integer) and looking at the normalised frequency $\bar{\omega}$ which is required for the variation $\exp(j\bar{\omega}T)$ in time step number T. After some algebra, the dispersion relationship is obtained:

$$\cos\bar{\omega} = \cos\theta\cos\bar{\beta} \quad (7.36)$$

For all real $\bar{\beta}$, the corresponding value of $\bar{\omega}$ is real because $\cos\theta < 1$. For small enough values of s (where $\bar{\omega} \to 0$ and $\bar{\beta} \to 0$ with $\cos\theta \approx 1 - \tfrac{1}{2}\kappa^2 s^2$) then $\bar{\omega}^2 \approx (\kappa s)^2 + \bar{\beta}^2$ so that the dispersion relationship is correct for sufficiently small step lengths. Because $\bar{\omega}$ is real for all distributions of error (all values of $\bar{\beta}$), the method is normally computationally stable having no numerically induced growth with time. These Fourier techniques will need revisiting when there is physical growth of the laser fields. Sufficient tools are now assembled to consider the complete finite-difference scheme for a uniform Bragg laser.

7.5 A uniform Bragg laser: finite difference in time and space

7.5.1 Full coupled-wave equations

The full 'DFB equations' can be recalled from Chapter 4:

$$\frac{1}{v_g}\frac{\partial F}{\partial t}+\frac{\partial F}{\partial z}=j\kappa R+(g-j\delta)F \quad (7.37)$$

$$\frac{1}{v_g}\frac{\partial R}{\partial t}-\frac{\partial R}{\partial z}=j\kappa F+(g-j\delta)R \quad (7.38)$$

As explained in Chapter 4, these equations have removed the rapid optical frequencies at some chosen central frequency of the wave packet close to the Bragg frequency. The parameter δ is a measure of this detuning. The amplitudes F and R then vary at 'microwave' frequencies with the wave packet propagating with the group velocity v_g which is taken here to be approximately constant at its value around the central or Bragg frequency. The net gain, with a confinement factor and loss, has been simplified here to just a single term g independent of deviations from the central frequency of the laser. The value κ represents the reflection per unit length, while $g-j\delta$ represents the field gain and phase change per unit length so that δ can allow for phase-velocity variations caused by changes of refractive index or also changes in the chosen central frequency.

Using the substitution of eqn. 7.12 defines modified fields $\bar{F}$ and $\bar{R}$ apparently without gain or loss as

$$\frac{1}{v_g}\frac{\partial \bar{F}}{\partial t}+\frac{\partial \bar{F}}{\partial z}=j\kappa\bar{R} \qquad \frac{1}{v_g}\frac{\partial \bar{R}}{\partial t}-\frac{\partial \bar{R}}{\partial z}=j\kappa\bar{F} \quad (7.39)$$

The finite-difference scheme then should give

$$\begin{bmatrix} F\{(T+1), (Z+1)\} \\ R\{(T+1), Z\} \end{bmatrix} = \exp\{(g-j\delta)s\} \begin{bmatrix} \cos\theta & j\sin\theta \\ j\sin\theta & \cos\theta \end{bmatrix} \begin{bmatrix} F\{(T, Z)\} \\ R\{T, (Z+1)\} \end{bmatrix}$$

(7.40)

where $\sin\theta = \kappa s/(1+\frac{1}{4}\kappa^2 s^2)$. Frequently in a DFB, one finds that there are long sections with a uniform κ so that a series of space steps can all have the same values of θ.

The stability criterion should be revisited because now, at each space step, the forward field changes by $\exp(j\bar{\beta}Z)$ and with $\bar{\beta} \sim 0$ an increase at each time step is given from $\exp\{(g-j\delta)s\}$. This growth at each step is the correct physics and does not indicate numerical instability. Numerical instability would arise if the rate of increase in time became unbounded or unphysical as the value of $\bar{\beta}$ either increased or took critical values. Here, following the work leading to eqn. 7.36, the dispersion relationship is

$$\cos(\bar{\omega} - s\delta - jsg) = \cos\theta \cos\bar{\beta} \qquad (7.41)$$

The gain remains bounded for the high spatial frequencies and one may expect convergence and adequately low distortion for sufficiently small steps of length s.

7.5.2 MATLAB code

A follow-up to eqn. 7.40 must show how to concatenate the spatial steps of length s. Initially, the field numbers are considered from 1 to N+1 where there are N sections making the laser's length $L = Ns$. For the moment, the laser is considered to be uniform. The 'present' fields (forward and reverse), labelled as ff(n) and fr(n) where n refers to the space-step numbering, are stored as vectors ff and fr. The 'new' fields, created at the next time step, are stored as ffn and frn. The 'complex magnification' per step is

$$mL = \exp\{(gL - j^*dL)/N\} \simeq (2^*N + gL - j^*dL)/(2^*N - gL + j^*dL) \quad (7.42)$$

Here $gL - j^*dL$ is the code for $(g - j\delta)L$ with L the laser length. The last term is the central-difference approximation to the exponential. This magnification is stored as an array mL(n) defining gain and phase changes at each spatial step and is updated at each time step as required. From eqn. 7.33, the Bragg grating introduces a reflection term $\sin\theta$ $\{= \kappa s/(1+\frac{1}{4}\kappa^2 s^2) \sim \kappa s\}$ and a transmission term $\cos\theta \sim \sqrt{(1-\kappa^2 s^2)}$. Numerical 'energy' conservation requires $\cos^2\theta + \sin^2\theta = 1$ so, although rounding $\sin\theta$ to κs only gives a small error (typically <1%), rounding $\cos\theta$ to unity is not recommended. An array of reflections/ transmissions are now formed at each step:

$$\text{sint(n)} = \text{ks(n)} \quad \text{and} \quad \text{cost(n)} = \sqrt{[1 - \{\text{ks(n)}\}^{\wedge}2]} \qquad (7.43)$$

where, in MATLAB terminology, ^p indicates raising to the power of p while for future reference * indicates multiplication. Here ks is the value of kappa×s product for each section and is permitted to vary from section to section. The 'guts' of the difference equations for the fields in the middle of the DFB then may be written as

for n = 1: N; (7.44)

ffn(n + 1) = mL(n)*cost(n)*ff(n) + j*mL(n)*sint(n)*fr(n + 1); (7.45)

nr = N + 1 − n; (7.46)

frn(nr) = mL(nr)*cost(nr)*fr(nr + 1) + j*mL(nr)*sint(nr)*ff(nr); (7.47)

end (7.48)

The value of dL in eqn. 7.42 is determined by the offset frequency of this input field from the Bragg frequency. A program called **dfbamp** in the directory **dfb** 'tops and tails' eqns. 7.44–7.48 with input routines giving zero input on the right and a unit amplitude field entering on the left so that the device acts as a coherent forward-wave 'amplifier'. Comparison with the analytic solutions, discussed in Section 7.5.3, shows that the program works robustly with as few as 16–30 or so sections in a device where $\kappa L=2$, $gL=0.5$ with $\delta L=4$ as an example taking around 10 s on a 166 MHz 486 processor to compute sufficient steps to allow the device to settle down near to its steady state. MATLAB programs often operate more efficiently when 'vectorised' but this can be done later.

7.5.3 Analytic against numeric solutions

Pauli matrices (Appendix 6) provide compact analytic steady-state solutions for eqns. 7.37 and 7.38 to compare with a finite-difference program. From eqn. A6.16 one may write

$$\begin{bmatrix} F(z) \\ R(z) \end{bmatrix} = \begin{bmatrix} a+d & jb \\ -jb & a-d \end{bmatrix} \begin{bmatrix} F(0) \\ R(0) \end{bmatrix} \qquad (7.49)$$

where $a = \cos(\beta_e z)$; $d = (\mu/\beta_e) \sin(\beta_e z)$; $b = (\kappa/\beta_e) \sin(\beta_e z)$; $\mu = (g - j\delta)$; $\beta_e^2 L^2 = -(\mu^2 + \kappa^2) L^2$ and δ gives the 'frequency'-offset parameter. With only a unit input at $z=0$; $F(0)=1$, $R(L)=0$ leading to

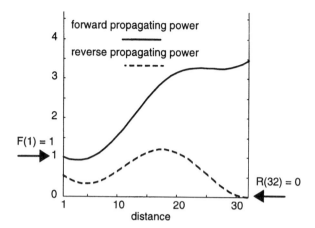

Figure 7.4 Numerical solution for a uniform-DFB-structure amplifier

$\kappa L=2$; $gL=0.5$; $\delta L=4$ with a 31-section laser (32 field points) calculation. On the scale of this drawing one cannot distinguish the difference between the analytic and the numerical solutions from **dfbamp**

$R(0) = jb(L)/\{a(L) - d(L)\}$ so that, at a general value of z, one obtains an analytic result:

$$\begin{bmatrix} F(z) \\ R(z) \end{bmatrix} = \begin{bmatrix} a(z)+d(z) & jb(z) \\ -jb(z) & a(z)-d(z) \end{bmatrix} \begin{bmatrix} 1 \\ jb(L)/\{a(L)-b(L)\} \end{bmatrix} \quad (7.50)$$

This can be calculated and compared with the finite-difference solution from **dfbamp** as in Figure 7.4.

The amplifier's fields take longer to settle to a steady state as one gets closer to the lasing condition (i.e. the effective photon lifetime increases) so that, to avoid apparent errors in steady-state values, a longer time of computation is then required to achieve a steady state. The reader can gain familiarity with this response time by changing the program's parameters. It is suggested that the results of Figure 7.4 be contrasted with results obtained by putting kL=2, gL=0.7; dL=0 where the fields evanesce inside the laser and take longer to settle down.

Different coherent inputs create different interference patterns inside the laser and completely different profiles are obtained for the steady state fields when a DFB is excited coherently in different ways. The reader is encouraged to save **dfbamp** in new files, say dfbamp1 (etc.), and to modify these to allow for inputs at both ends of the laser

226 Distributed feedback semiconductor lasers

to check this phenomenon, remembering that a student edition of MATLAB may limit the number of field points that can be used in a program.

7.6 Spontaneous emission and random fields

7.6.1 Spontaneous noise and travelling fields

As discussed in Chapter 4, a laser is excited through random spontaneous emission. The quasiclassical way of modelling the resulting fields is to start with a single-moded optical field $a \exp j\phi \exp(j\omega_0 t) = (a_x + ja_y) \exp(j\omega_0 t)$ with the fixed single-mode optical frequency ω_0, with amplitude a and phase ϕ giving a complex amplitude $(a_x + ja_y)$. Added to this 'coherent' complex-field amplitude are random Gaussian distributions of (real and imaginary) fields. A program **spont** in directory **spontan** gives the reader an indication of this modelling process. The sequential snapshots in time indicate movement of the complex phase relative to the steady phasor $\exp(j\omega_0 t)$ which is treated as a constant (see Figure 7.5a). The frequency-domain representation is found using the fast Fourier transform and shows white noise (incoherent fields) superimposed on the narrow-line coherent signal (Figure 7.5b). This net random signal is then filtered to give narrow spectral-band noise (Figure 7.5d). On filtering, the (coherent+noise) output phasor performs a random walk as sketched in Figure 7.5c.

As the 'coherent' amplitude gets large relative to the noise, any uncertainty or randomness in either the net amplitude or the phase of the coherent fields becomes negligible. To demonstrate further features of the model, consider a laser amplifier without feedback where there are only forward fields advancing with gain, according to the advection equation. The complex field varies with some constant central frequency which for the purposes of calculation may be 'removed' as in Figure 7.5. The coherent complex-field amplitude is amplified but also has spontaneous emission added as the fields travel:

$$\frac{\partial F}{\partial z} + \frac{1}{v_g}\frac{\partial F}{\partial t} = gF + i_{spf}(t, z) \qquad (7.51)$$

Excitation of the fields is through the random spontaneous emission, represented by the term $i_{sp}(t, z)$ whose magnitude is studied in

Numerical modelling for DFB lasers 227

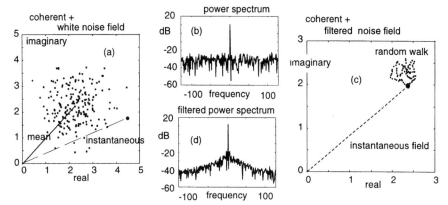

Figure 7.5 Complex coherent field with added noise

> A program spontd provides numerical demonstrations of adding spontaneous emission (e.g. white noise) to a coherent signal (i.e. a constant complex field) at the central frequency. The demonstration shows
> (*a*) A display of the sampled instantaneous fields with time
> (*b*) The white-frequency-domain spectrum apart from the coherent spike
> (*c*) The effect in the time domain of filtering in the spectral domain (discussed in Section 7.9) to give
> (*d*) A coherent signal plus narrowband noise where the coherent vector performs a random walk in the phase plane as in (*c*)

Appendix 9. Initially omitting the spontaneous emission in eqn. 7.51, a difference scheme with space and time steps s and s/v_g, respectively, may be written as

$$F\{(T+1)s/v_g, (Z+1)s\} - F(Ts/v_g, Zs)$$
$$= \tfrac{1}{2}gs[F\{(T+1)s/v_g, (Z+1)s\} + F(Ts/v_g, Zs)] \quad (7.52)$$

which rearranges to

$$F\{(T+1)s/v_g, (Z+1)s\} = \{(1+\tfrac{1}{2}gs)/(1-\tfrac{1}{2}gs)\}F(Ts/v_g, Zs) \quad (7.53)$$

An essential feature is that, at each (T, Z) step, spontaneous fields are added through the spontaneous excitation:

$$s i_{spf}(Ts/v_g, Zs) = \mathcal{R}\, r_{sp} \quad (7.54)$$

where $\mathcal{R} = \mathcal{X} + j\mathcal{Y}$ has $\mathcal{X}$ and $\mathcal{Y}$ each normally distributed random real numbers. The randomness of both $\mathcal{X}$ and $\mathcal{Y}$ ensures there is no correlation between the spontaneous emission at any two points of space and time. The strength of the spontaneous emission (Appendix 9) is related to the local electron density N so that with this scheme it makes sense to link the spontaneous excitation with the mean electron

228 Distributed feedback semiconductor lasers

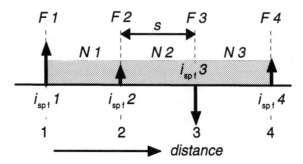

Figure 7.6 Field and spontaneous excitations
At different space points, for one time

density on either side of the field point at which the spontaneous emission is inserted (Figure 7.6) leading to

$$F\{(T+1)s/v_g, (Z+1)s\} = \{(1+\tfrac{1}{2}gs)/(1-\tfrac{1}{2}gs)\}\{F(Ts/v_g, Zs) + si_{spf}(Ts/v_g, Zs)\} \quad (7.55)$$

where, writing $Ne = N/N_{tr}$ for a normalised electron density to avoid confusion with any integer N,

$$i_{spf}(Ts/v_g, Zs) \propto \tfrac{1}{2}[Ne(Ts/v_g, Zs) + Ne\{Ts/v_g, (Z-1)s\}] \quad (7.56)$$

7.6.2 Null correlation for different times, positions and directions

Once a diode laser is lasing, it is found that the electron density typically changes by only a few per cent so that the precise variation of the spontaneous emission with the changes of electron density proves to be of less concern than might initially be thought. Of more concern is the possibility of spurious numerically induced correlations of the randomly excited signals within the laser. It is again emphasised that, to avoid such correlation, both $\mathcal{X}$ and $\mathcal{Y}$ above are determined separately and randomly for both forward- and reverse-travelling spontaneous emission so as to ensure all these excitations are uncorrelated with each other.

Uncorrelated inputs at each time step ensure a white-spontaneous-noise power spectrum over the finite bandwidth of v_g/s which is the Nyquist limit caused by the sampling action. Appendix 9 shows how to tailor this spectrum and, as indicated earlier, such tailoring may be required if v_g/s extends for an excessive range of physical frequencies. For the basic models outlined in this chapter, a 'white' spontaneous spectrum is mostly used because the laser's gain is filtered and the shape of the spontaneous spectrum at outlying frequencies becomes unimportant. The mean-square value of the spontaneous random generator $\mathcal{R}$ is chosen to be $\overline{\mathcal{R}}^2 = 1$ (ensemble or temporal average) so that the spontaneous noise is then determined by r_{sp} in eqn. 7.54.

7.6.3 Spontaneous magnitude

The precise formulation of r_{sp} depends on the normalisation of the fields F. Here the normalisation is such that $F^*F = S_f \mathcal{A} v_g$ gives the numbers of photons per second with S_f, the density of photons associated with the forward flow along the guide of area $\mathcal{A}$. Equation A9.47 shows:

$$s i_{sp}(\mathsf{T} s/v_g, \mathsf{Z} s) = \{(K_{tr}\lambda^2 v_g/8\pi\varepsilon_{rr\,eff}\Delta_{sp}f)\Gamma_{sp}BPN\}^{1/2}\mathcal{R} \quad (7.57)$$

where $(K_{tr}\lambda^2 v_g/8\pi\varepsilon_{rr\,eff}\Delta_{sp}f)$ is an effective volume with an order of magnitude of 10 (μm)^3 for 1.55 μm-wavelength lasers and K_{tr} is the transverse Petermann factor also discussed in Appendix 9 and in Chapter 2 but taken as 1 for good-index guiding lasers. The term BPN is the spontaneous (bimolecular) recombination rate per unit volume given charge-carrier densities P and N with B a parameter that has specific measured values for different materials. The parameter Γ_{sp} is the confinement factor appropriate for spontaneous emission equivalent to (but because of different distributions of photons and charge carriers not necessarily identical to) the confinement factor Γ for gain. Typically, in laser modelling $P \sim N$, reflecting the charge neutrality. Provided that the laser remains single-moded, Appendix 9 shows a slightly surprising result that the increase in spontaneous emission, caused by an increased guide volume per unit length where the area $\mathcal{A}$ is increased, is effectively cancelled because the increase in guide aperture also leads to increased 'aerial gain' for the spontaneous dipoles within the guide, and hence a reduced solid angle for the effective coupling of the spontaneous emission into the guide.

7.6.4 Tutorial programs

In directory **spontan** the programs **amp1**, **amp2** and **amp3** provide three demonstrations based on an 'ideal' amplifier of eqn. 7.55. The stimulated gain is taken as proportional to PN (the charge-carrier product) as well as being proportional to the photon density. The spontaneous emission is similarly proportional to PN and this random emission is inserted between the field points as in Figure 7.6. The first demonstration (**amp1**) shows how the (white) spontaneous-noise

230 *Distributed feedback semiconductor lasers*

power increases linearly with the amplifier length when there is no gain (amp1f filters the noise to give a coloured spectrum but essentially is the same demonstration as in amp1). The second demonstration (amp2) shows how, with a fixed gain per unit length and increasing length of amplifier, the ratio of noise power to signal power reaches a limit in this ideal 'quantum' amplifier. The third demonstration (amp3) shows that the link between gain and spontaneous emission means that there is no benefit in increasing the gain per unit length indefinitely because the (signal/gain)/(noise-power/gain) reaches a limit—the quantum limit. The fields are 'normalised' merely to demonstrate these links between gain, length of amplifier and signal-to-noise power and do not at present relate to physically significant magnitudes.

7.7 Physical effects of discretisation in the frequency domain

7.7.1 Discretisation process—integrals to sums

At first sight, the process of improving the approximation when using finite differences is simply a question of making the step length sufficiently short. However, no matter how short the step length, one is effectively replacing a continuous function, say of time $f(t)$, by its discretised or sampled version $f\{T(s_t/v_g)\}$ where (s_t/v_g) is the temporal step length and T is an integer. Because of the wide use of and teaching about digital signal processing, it is believed that the physics of discrete signals will be relatively well understood by most readers so that only a brief resumé is needed of some points to watch. Figure 7.7a indicates a continuous signal which might represent the modulated amplitude of, say, an optical pulse along with its baseband spectrum (Figure 7.7b), and then indicates the effect of discretisation or sampling on that signal (Figure 7.7c) and spectrum (Figure 7.7d). The computation can only take a finite number of samples, say N (this integer N has no necessary relationship with the number of segments of the laser) leading to a total time of observation $(s_t/v_g)N$. The corresponding smallest spectral interval in angular frequency is then $2\pi(v_g/s_tN)$ and the spectrum ranges from $-\pi(s_t/v_g)$ to $+\pi(s_t/v_g)$, again in angular frequency. The limits of this spectral range are referred to as the Nyquist range or the Nyquist limits.

It may not be apparent immediately that discretising the signal over a limited time and taking the discrete Fourier transform (DFT) forces

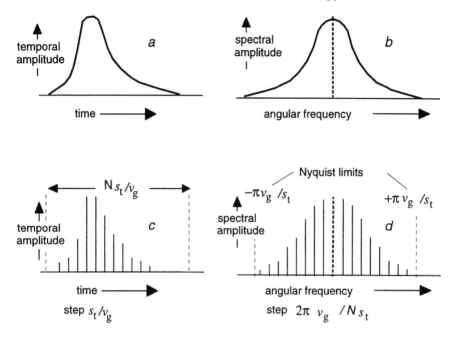

Figure 7.7 Discretisation: schematic relation between time step and spectral limits

the signal to be effectively periodic but this is the effect, as shown schematically in Figure 7.8a [32] (see also [28], for example). There is no physical significance in the signals outside the main interval of either time or frequency. The only reliable information lies within the N time steps of the temporal observation and the N spectral steps (placed symmetrically about the origin). For an adequate approximation to the physics, both the length of observation and the spectral width have to be adequate to cover the required range of physical phenomena. If one attempts to model physical processes with a spectrum wider than the Nyquist limits or tries to obtain information over time scales longer than the observation interval, then one obtains errors. Figure 7.8c shows the effects of modelling high-speed phenomena with an inadequate computational bandwidth so that the physical spectrum extends beyond the Nyquist limits, forcing the spectrum from one spectral period to overlap with the spectrum in the next period (so called aliasing errors). The frequency lines in Figure 7.8c have been slightly displaced to make it clear that there is an overlap of the spectral amplitudes associated with the neighbouring periods of the spectrum. The results for the discretisation of space follow those found with time but with spatial (angular) frequencies or propagation

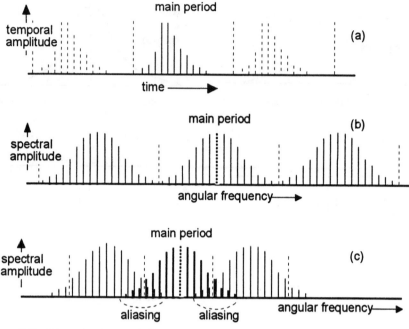

Figure 7.8 *Discretisation: periodicity and aliasing*

coefficients β replacing the temporal (angular) frequencies ω and similar Nyquist limits on propagation.

7.7.2 *Fast Fourier transform (FFT)*

A major advance in DFT analysis came with the fast Fourier transform (FFT) [33] (see also [28,32], for example) where, by choosing the number of steps $N = 2^M$ with M integer, one is enabled to calculate the DFT in exceptionally efficient ways with the calculation time increasing for large N as $\frac{1}{2}MN$ rather than as N^2. FFT algorithms are standard tools and MATLAB has its own version. Programs **spect1** through to **spect7** in the directory **fftexamp** may be helpful in refreshing the reader's mind about some key points (listed below) to bear in mind when using a fast Fourier transform:

(i) There is an equivalent of Parseval's theorem showing that 'energy' in the time domain equals the 'energy' in the frequency domain.
(ii) With a maximum observation time T, the minimum frequency that can be accurately measured is 1/T and this is also the minimum change of frequency that can be measured. If one

wishes to discriminate two frequencies separated by Δf, then $T > 1/\Delta f$ is required.

(iii) With the student MATLAB 4 which has been tested by the authors, one cannot find the FFT for more than $8192 = 2^{13}$ points but there is nothing in the algorithms which prevents a larger number (other than time of computation).

(iv) The maximum frequency that can usefully be measured is $N/2T = v_g/(2 \, s_t)$ where s_t/v_g is the temporal step in the finite-difference scheme. Attempts to measure higher frequencies may appear, erroneously, to give a result, but at a lower frequency (aliasing). In the modelling work here, the potential bad effects of aliasing are reduced by choosing finite-difference algorithms so that the signal is effectively filtered to give a finite spectral bandwidth with optical gain falling to zero at the Nyquist-band limits. However, nothing can invent information which is not present so that the step length still has to be short enough to ensure that the numerical bandwidth encompasses the required physical bandwidth.

7.8 Finite-element strategies for a spectral filter

7.8.1 Lorentzian filter

It is worth opening the discussion with the consideration of representing $g(t)$ by a sampled approximation $\frac{1}{2}[g\{(T+1)(s_t/v_g)\} + g\{T(s_t/v_g)\}]$ where T is an integer. This representation has been referred to as a Lax average. By considering a phase factor of $\exp\{j\omega_M(s_t/v_g)\}$ per temporal step, one can see that, in the spectral domain, this process gives a filter with the characteristic amplitude response $\cos\{\omega_M(s_t/2v_g)\}$. At the Nyquist-band edges, its magnitude falls to zero, so that Lax averaging has a positive beneficial role in limiting the magnitude of any quantity to the central spectral region determined by the sampling time. One could naively then simply change the step length to obtain the best approximate filter shape for the gain. However, a more sophisticated strategy is worth considering where the central frequency and the central gain curvature can be altered even with a fixed step length. This leads to the Lorentzian filter.

There are at least two particular instances in laser-diode modelling where a Lorentzian filter can be helpful:

(i) filtering the spectral shape of the gain spectrum acting on the optical fields, and

(ii) shaping the spectrum of the spontaneous noise.

The gain may be $g(\omega_0)$ at some 'central' frequency ω_0 but this changes as the frequency is changed and reduces to negligible amounts at the edges of the laser-frequency band. One useful filter replaces the single value $g(\omega_0)$ by $g(\omega_0)/\{1+\tau^2(\omega-\omega_0)^2\}^{1/2}$, the so-called 'Lorentzian filter'. Of course the measured gain/frequency relationships, as for example seen in Figures 2.8, 2.9 and 2.11, are not as simple as this, but fortunately only the peak gain value, peak gain frequency and gain curvature are really significant in determining the operation of the main modes in the laser [34] and these features can be matched by an appropriate Lorentzian filter.

In taking this filter shape, the reader is reminded that the central lasing frequency has been removed from the laser fields by writing $E(t)\exp(j\omega_0 t)$ and removing the phase term $\exp(j\omega_0 t)$. The field amplitudes $E(t)$ vary at 'base-band' or modulation frequencies $\omega_M = \omega - \omega_0$ and have to be filtered by $1/(1+\tau^2\omega_M^2)^{1/2}$ taking, for initial simplicity, $g(\omega_0)=1$.

The Lorentzian spectral filter is based on a temporal first-order relaxation equation, linking an output b_{Out} with an input a_{In} through

$$(1+\tau\, d/dt)\, b_{Out} = a_{In} \qquad (7.58)$$

In the frequency domain, the spectral amplitudes at a frequency $\pm\omega_M$ from the central frequency ω_0 are related by

$$|B_{Out}|^2 = |A_{In}|^2/\{1+(\omega_M \tau)^2\} \qquad (7.59)$$

[In the *real* frequency domain in eqn. 7.59, it may appear not to matter if, in place of eqn. 7.58, one used $(1-\tau\, d/dt)\, b_{Out} = a_{In}$, but this leads to any small disturbance growing exponentially in the time domain and therefore is not a physical filter which satisfies the Kramers–Kronig relationships.] To implement eqn. 7.58 digitally, remember yet again that sampling the signal with a normalised time step $s(=s_t)$ automatically limits the useful physical range to the Nyquist limit, and computational errors can occur in attempts to use frequencies outside this limit.

Using central differences with Lax averaging and initially taking half steps:

$$b_{Out}(+\tfrac{1}{2}) + b_{Out}(-\tfrac{1}{2}) + (2\tau v_g/s)\{b_{Out}(+\tfrac{1}{2}) - b_{Out}(-\tfrac{1}{2})\}$$
$$= a_{In}(+\tfrac{1}{2}) + a_{In}(-\tfrac{1}{2}) \qquad (7.60)$$

Writing $K=(2\tau v_g/s)$, and changing to unit time shifts, this filter rearranges to give

$$b_{Out}(1) = \{(K-1)/(K+1)\}b_{Out}(0) + \{1/(K+1)\}\{a_{In}(1) + a_{In}(0)\} \quad (7.61)$$

This program is given in filt1 under the directory filter. The theory can be tested out analytically by noting that, in the normalised frequency domain, there is a phase shift of $\exp(j\bar{\omega}_M)$ per whole time step (i.e. $\bar{\omega}_M = \omega_M s/v_g$) and then use of eqn. 7.60 with half steps gives the spectral form

$$B_{Out} = A_{In} \cos(\tfrac{1}{2}\bar{\omega}_M) / \{\cos(\tfrac{1}{2}\bar{\omega}_M) + jK\sin(\tfrac{1}{2}\bar{\omega}_M)\} \quad (7.62)$$

As already noted, Lax averaging replaces a_{In} with $\{a_{In}(+\tfrac{1}{2}) + a_{In}(-\tfrac{1}{2})\}/2$ as in eqn. 7.60 and filters the signal as $\cos(\tfrac{1}{2}\bar{\omega}_M)$ with the Nyquist limitations $-\pi < \bar{\omega}_M < \pi$. The parameter K allows the curvature of the gain with frequency in the central part of the filter to be changed so that, if $K=1$, there is a spectral power filter varying as $|\cos(\tfrac{1}{2}\bar{\omega}_M)|^2$. If $0 < K < 1$ the spectrum is flattened with respect to the situation with $K=1$ while $K>1$ narrows the gain bandwidth, increasing the gain curvature. The filter is stable provided that $K>0$.

It is frequently more useful to have the spectral filter offset from the central laser frequency by an amount ω_{offset} so that $\omega_M \to \omega_M - \omega_{offset}$:

$$|B_{Out}|^2 = |A_{In}|^2 / [1 + \{(\omega_M - \omega_{offset})\tau\}^2] \quad (7.63)$$

The implementation of this offset frequency in the conventional frequency domain replaces $\exp(j\omega_M t)$ with $\exp[j(\omega_M - \omega_{offset})t]$. In the normalised frequency domain with one positive space step, there is a phase shift of $\exp(-j\bar{\omega}_{offset})$. This phase shift can be incorporated into the digital filter by writing $\Theta = \exp(j\bar{\omega}_{offset})$ and then putting

$$B_{Out}(1) = \{(K-1)/(K+1)\}\Theta B_{Out}(0) + \{1/(K+1)\}\{A_{In}(1) + \Theta A_{In}(0)\} \quad (7.64)$$

7.8.2 Numerical implementation

The checking of this filter theory is done in the program filt1 within the directory filter using either or both analytic theory and a white-noise stochastic excitation, which is another interesting variant for modelling and helps to build confidence that the filter really works on a stochastic input. Figure 7.9a shows a typical result where $K=1.5$ and no phase shift is included while Figure 7.9b shows the same result with a phase shift. In these two figures, the '−3 dB'-gain points are not conventional half-power points but are the normalised frequencies where the *real* (output/input) falls to $1/\sqrt{2}$ of its peak value. The reasons for this are discussed in Section 7.9 on gain filtering.

The same filter can be used if required to filter spontaneous noise. For spontaneous noise, the 3 dB points refer to the frequencies where

236 *Distributed feedback semiconductor lasers*

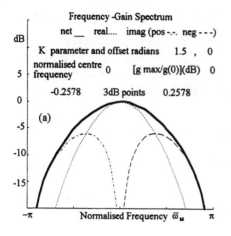

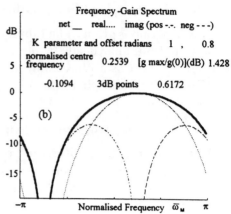

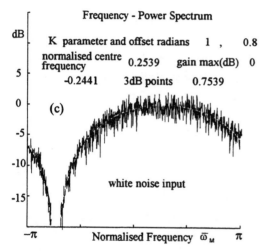

Figure 7.9 Spectral filter for finite-difference scheme

(a) No offset frequency
(b) Offset frequency
(c) White-noise input

the noise power falls to half of its central value. Figure 7.9c shows the output from the same filter as given in Figure 7.9b but now implemented with a white-noise input (using the program filt1n).

On examining the result of Figure 7.9c and considering this applied to gain, the reader may be disturbed at seeing the 'output' rise again at the Nyquist-frequency edges. This problem, if it is a problem, is removed by an alternative filtering demonstrated in the programs filt2. In this alternative program, the gain always falls to zero at the Nyquist-band edges but the gain profile is not symmetrical about the gain peak.

A combination of both techniques in filt1 and filt2, in principle, provides a versatile armoury for tailoring the gain with frequency or shaping the spontaneous spectrum with frequency. Further ideas for filtering the gain, using the principles of TLLM combined with digital filtering, are provided by Nguyen *et al.* [35]. It turns out that, in DFB modelling, it is probably more important to use a filter to tailor the gain with frequency than filter the spontaneous emission. The spontaneous spectrum can be prefiltered in a lookup table without greatly slowing down the computation, but this risks introducing unwanted periodicities if the table is too short.

7.9 Application of the filter theory to gain filtering

7.9.1 General

Care is needed in implementing the gain filter of Section 7.8 because the field gain over a distance z is given from $G_{field} = \exp(\int g\,dz)$ while the power gain over the same distance is given from $G_{power} = \exp(2 \int g\,dz)$. The integrals are not straightforward when the gain varies with time. Now the physics requires the gain g per unit length to be some function of frequency so that the temporal filter must affect g, as in Section 7.8, rather than act on G_{field} or G_{power}, where there would be a more complicated spectral action. Note that, even if $g(\omega) = g(\omega_0)$, a real value at $\omega = \omega_0$, then on moving away from this central frequency, the Kramers–Kronig relationships (Appendix 7) force the gain to take complex values $g(\omega) \rightarrow g_{real}(\omega) + jg_{imag}(\omega)$, where the magnitude of $g_{imag}(\omega)$ depends on the local curvature of $g_{real}(\omega)$. However, it is only g_{real} that changes the power gain per step; the imaginary component merely changes the phase change per step. It was for this reason that, in Figure 7.9*a* and *b*, the bandwidth was related only to the real part of the gain.

A further complication comes from the fact that, as the gain increases with an increase in the electron density, so, at the same time, there is an associated reduction in the permittivity of the material leading to a complex differential gain $g(1+j\alpha_H)$ which changes with frequency *and* electron density. Here α_H is known as Henry's linewidth-broadening factor [36, 37], taking values typically between 1 and 7 depending on the material and structure of the lasing waveguide [38]. However, it has to be recognised that the net complex refractive index is made up of several contributions with each contribution varying in distinctive ways. Appendix 7 addresses this issue with three components: gain, loss and the plasma effect. At the central gain peak $g(\omega_0)$ at angular frequency ω_0 it is changes in the plasma effect (or oscillation of a plasma of nearly free electrons) to which is attributed changes in the phase term $jg(\omega_0)\alpha_H$, with the value at the transparency density taken as the reference level. Over the frequencies where the gain is strong, this phase contribution is assumed to remain approximately constant. The Lorentzian filter is then applied only to the net-gain/loss term and not to the term in α_H:

$$g(\omega) = g(\omega_0)/\{1+j\tau(\omega-\omega_0)\} \qquad (7.65)$$

with

$$g_{real}(\omega) = g(\omega_0)/\{1+\tau^2(\omega-\omega_0)^2\} \qquad (7.66)$$

and

$$g_{imag}(\omega) = -j\tau(\omega-\omega_0)g(\omega_0)/\{1+\tau^2(\omega-\omega_0)^2\} \qquad (7.67)$$

For each spatial step of length s, the complex magnification mL of eqn. 7.42 is given by $\exp\{g(\omega)(1+j\alpha_H)s\}$ and is now modelled as

$$mL = ((1+\tfrac{1}{2}gs)/(1-\tfrac{1}{2}gs)) * \exp(j*ahs) \qquad (7.68)$$

where the MATLAB vector ahs is the array of values of $\{g(\omega_0)\alpha_H s\}$ that are allowed to vary at each step dependent on the central gain associated with the electron density at that step, but are taken to be independent of the frequency of lasing. The real part of this magnification *at the central frequency* is given by $\exp\{g(\omega) s\}$ per step represented in eqn. 7.68 by its central difference approximation of $(1+\tfrac{1}{2}gs)/(1-\tfrac{1}{2}gs)$ as in eqn. 7.42 (with the substitution that L/N=s) but now as the frequency changes, $g(\omega) s$, or equivalently the term gs, will be modelled in accordance with the filter principles just outlined. At the time of writing, the definitive paper in this area of complex gain modelling with frequency has perhaps not yet been written, and more needs to be done numerically, experimentally and physically, but [35] provides important ideas for transmission-line laser modelling.

7.9.2 Filtering the gain in the travelling-wave equations

The application of the filter theory to the gain needs care. Consider eqns. 7.44–7.48 along with the definitions of mL, sint and cost in eqns. 7.42 and 7.43. Initially, consider just a real gain dependent on the electron density which will change much more slowly with time than the temporal changes in the optical fields. Consequently, from eqn. 7.68 one can rewrite the central-difference approximation hidden in eqns. 7.44–7.48 for the effect of gain in the nth section as

$$\{1 - \tfrac{1}{2}g(n)s\}\text{ffn}(n+1) = \{1 + \tfrac{1}{2}g(n)s\}C(n) \qquad (7.69)$$

where

$$C(n) = \exp\{j\ \text{ahs}(n)\}\{\cos t(n)\text{ff}(n) + j\ \sin t(n)\text{fr}(n+1)\}.$$

The programs for the distributed lasers store sets of the current values of the discretised spatial forward/reverse fields {ff(n) fr(n)} along the laser. Then there is also the set of spatial fields {ffo(n) fro(n)} for the old (or previous) time step, and from the current fields and the old fields are generated the updated or new set of fields all along the laser at the next time step {ffn(n) frn(n)}. The assumption is made that the changes in electron density occur sufficiently slowly over two time steps so that, between these three temporal values of 'old', 'present' and 'new', all quantities involving electron densities can be taken to be values at the present time step. The phase changes caused by Henry's α factor are assumed here to vary at a similar rate to that of the electron density. Consequently, in this approximation, all the gains, phase factors from α_H and transmissions/reflections from the grating (i.e. g, ahs, cost and sint) are evaluated at their values for the current time step in eqn. 7.68. For time steps of 0.1 ps, this is usually an entirely reasonable approximation with electron densities varying on the 10 ps scale for 100 GHz modulation frequencies.

Now rearrange eqn. 7.69 so that the gain g(n) is entirely on one side and write

$$B(n+1) = \tfrac{1}{2}g(n)sA(n) \qquad (7.70)$$

where

$$B(n+1) = \{\text{ffn}(n+1) - C(n)\} \text{ with } A(n) = \{\text{ffn}(n+1) + C(n)\}.$$

The indexing of A(n) and relating it to B(n+1) is a matter of arbitrary choice but, having made the choice, eqn. 7.70 determines the definitions. Remember that n denotes the index of the spatial position, and then consider that each 'element' B of the spatial field is the 'output' from an 'input' A with a gain multiplier $\tfrac{1}{2}g(n)s$ rather than unity as previously considered in Section 7.8. Use the notation Bo and Ao to indicate the stored versions of the old (or previous) time step. Applying Section 7.8 to update the output B as in eqn. 7.64, using the same definition of $K(=K)$ and writing $\Theta = \exp(j\ dF) = Ep$ with dF determining the normalised frequency offset as in filt1. All this leads to

$$(K+1)B(n+1) = (K-1)EpBo(n+1) + \tfrac{1}{2}g(n)s\{A(n) + EpAo(n-1)\} \qquad (7.71)$$

where

$$B_o(n+1) = \{ff(n+1) - C_o(n)\} \quad \text{with} \quad A_o(n) = \{ff(n+1) + C_o(n)\}$$

with

$$C_o(n) = \exp\{j\,ahs(n)\}\{cost(n)ffo(n) + j\,sint(n)fro(n+1)\}$$

Eqn. 7.71 is then unravelled, retaining the assumption that the gain and phase changes caused by the electron density vary sufficiently slowly, to give the result

$$ffn(n+1) = ma(n)C(n) - mb(n)C_o(n) + mc(n)ff(n+1) \quad (7.72)$$

where

$$ma = [\{K+1+\tfrac{1}{2}g(n)s\}/\{K+1-\tfrac{1}{2}g(n)s\}] \quad (7.73)$$

$$mb = Ep[\{K-1-\tfrac{1}{2}g(n)s\}/\{K+1-\tfrac{1}{2}g(n)s\}] \quad (7.74)$$

$$mc = Ep[\{K-1+\tfrac{1}{2}g(n)s\}/\{K+1-\tfrac{1}{2}g(n)s\}] \quad (7.75)$$

The updated reverse field $frn(n)$ is found in a similar fashion and gives

$$frn(nr) = ma(nr)D(nr) - mb(nr)D_o(nr) + mc(nr)fr(nr) \quad (7.76)$$

where

$$D(nr) = \exp\{j\,ahs(nr)\}\{cost(nr)fr(nr+1) + j\,sint(nr)ff(nr)\}$$

and

$$D_o(nr) = \exp\{j\,ahs(nr)\}\{cost(nr)fro(nr+1) + j\,sint(nr)ffo(nr)\}$$

Equations 7.72–7.76 achieve the objectives of giving the new fields $\{ffn(n)\ frn(n)\}$ at each time point in terms of the present fields $\{ff(n)\ fr(n)\}$ and old fields $\{ffo(n)\ fro(n)\}$. The gain filter now is a function of modulation frequency and has been designed to give zero gain at the Nyquist-band edges, adjusted for any frequency offset. This significantly reduces the risks of aliasing errors or unphysical oscillations.

7.9.3 Numerical implementation

A program **gain** in the directory **filter** implements the above theory and shows how the noise can be amplified over a controllable narrow band (Figure 7.10). Note that if loss is included then the gain is negative (attenuation) at the Nyquist limits, even though the peak gain has been offset by a considerable fraction of the Nyquist-frequency range. This reinforces the point made towards the end of Section 7.8 that it is the real part of g which determines the net gain. This real net

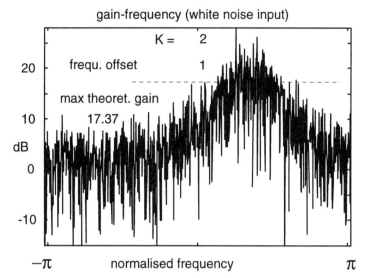

Figure 7.10   Gain/frequency with gain filter
Amplification of spontaneous emission

gain gives strong attenuation at the Nyquist limits unless there are exceptional frequency shifts. Further shaping of the digital filter could be accomplished by making *K* complex as, for example, when using the program filt2 in the directory filter.

7.10 Basic DFB laser excited by spontaneous emission

7.10.1 Introduction and normalisation

This section brings together the previous work and provides numerical models for a range of DFB lasers. Several programs in directory dfb are common to each master program determining the type of DFB (e.g. dfbuni for the uniform laser). Not all MATLAB terms (which are in sans-serif type face here) correspond 1:1 with the mathematical symbols so that the reader must make an appropriate translation with the help of the interface provided by this section and 1readme.txt files available on the web.

Normalisation can aid in evaluating the relative magnitudes of parameters. For example, normalising the electron density N to the value N_{tr} required for gain transparency suggests that with a high-gain and low-loss material, N/N_{tr} (→Ne in programs) will perhaps vary around 1–2 and, once lasing starts, will vary little and slowly compared with the optical fields. At Ne ~ 1, it is also possible to set a recombination time τ_r for the electrons and holes so that the rate of feed of electrons into a given volume $\Gamma \mathcal{A} L$ will be around $N_{tr}\Gamma \mathcal{A} L$ electrons per unit time which can also give a normalising value for the photons emitted per unit time. When a laser lases strongly with negligible losses and only radiative recombination ($\tau_r \sim \tau_{sp}$), any increase $[\Delta(N_{tr}\Gamma \mathcal{A} L/\tau_r)]$ in the input electrons per unit time must give the same change of photons out per time—this is a useful numerical check.

Parameters measured in inverse length are normalised by multiplying by the length of the laser, e.g. (κL) giving the coupling, $(g_m L)$ the material field gain; and (αL) the material optical-field-loss coefficient. In MATLAB code in the programs provided these are written respectively as kL, gL and aL.

This computation is now for a laser and not an amplifier, so that there is no longer a well prescribed central operating frequency, at least not until the computation has been made. The Bragg offset-frequency parameter δ giving a phase change $j\delta$ per unit length in eqns. 7.37 and 7.38 becomes now a measure of how the central Bragg-grating frequency departs from the central gain peak. A prime cause of this change is given by the fact that, as the electron density increases the gain increases by Δg_m and there is a reduction in the refractive index which changes the optical length of the grating and changes the Bragg frequency. All this can be accounted for by a phase change $j\alpha_H \Delta g_m$ per unit length associated with the gain where α_H is Henry's linewidth-broadening factor. A reference level is required, and here this is taken by choosing δL as zero whenever the electron density is at the transparency value N_{tr}. This is not a necessary assumption, but provides a simplification in this book which usefully limits the computed frequency ranges. All this leads then to a complex material gain given by $g_m L - j\delta L$ (written as gL+j*dL in the MATLAB code) with $\delta = \alpha_H \Delta g_m$ where Δg_m measures deviation of gain from the transparency value. The fields F and R are coded as arrays ff and fr with N values giving the fields at each of the N spatial steps at the current time. Typically, as orders of magnitude, (κL) and $(\alpha L) \sim 1$ with $\Gamma(g_m L) \sim 10$ where Γ is the confinement factor ~0.3 for bulk material lasers and $\alpha_H \sim 4$, again dependent on the material. For quantum-well materials, the confinement factor is smaller because the gain is confined to the narrow quantum wells, but the gain is higher, leading to similar or larger values of $\Gamma(g_m L)$. The electromagnetic fields F and R are normalised so that F^*F and R^*R give the photons per second flowing from left to right and from right to left, respectively, at a local position z so that the photon density per unit length is $(F^*F+R^*R)/v_g$ taken over the effective guide area $\mathcal{A}$.

7.10.2 Field equations

The field-advection equations at the position z may then be written as

$$L\left(\frac{1}{v_g}\frac{\partial F}{\partial t}+\frac{\partial F}{\partial z}\right)=j(\kappa L)R+[\Gamma\{(g_m L)-j(\delta L)\}-(\alpha L)]F+(Li_{sp\,f}) \quad (7.77)$$

$$L\left(\frac{1}{v_g}\frac{\partial R}{\partial t}-\frac{\partial R}{\partial z}\right)=j(\kappa L)F+[\Gamma\{(g_m L)-j(\delta L)\}-(\alpha L)]R+(Li_{sp\,r}) \quad (7.78)$$

These equations are computed numerically, including the gain filter, using the codes that have been developed. The gain peak can shift with electron density and temperature but this is not implemented in any of the sample programs. The value of κL could change at each space step to model graded gratings but is taken here as constant. The terms Li_{sp} represent the spontaneous excitation, evaluated from the bipolar radiative recombination, taken as locally proportional to N^2 and include the spontaneous confinement factor which is similar to the gain confinement factor Γ. The latter gives an effective gain below the material gain and reduces the effective phase shift associated with the complex gain.

7.10.3 Charge-carrier rate equation

The assumption of electrical neutrality means only a single rate equation is required for the carrier density N (electrons) $\sim P$ (holes) in order to determine the local recombination and stimulated emission. This density N changes with distance along the laser and so is a function of the section number. The electron number per unit length of the interaction is then $N\Gamma\mathcal{A}$ where $\mathcal{A}$ is the effective optical cross-section and $\Gamma\mathcal{A}$ is the effective electronic cross-section, with Γ the confinement factor assumed to be independent of optical power level. The electronic rate equation per unit length is then given from

$$\Gamma \mathcal{A}\left(\frac{dN}{dt}+\frac{N}{\tau_r}\right)+2\Gamma g_m v_g\left(\frac{F^*F+R^*R}{v_g}\right)=\frac{I}{qL} \qquad (7.79)$$

where

$$\frac{N}{\tau_r} \simeq \frac{N\{B_n(N/N_{tr})+C_n(N/N_{tr})^2\}}{\tau_{ro}} \qquad (7.80)$$

The optical density (photons per unit length) is given from $(F^*F+R^*R)/v_g$ as explained above. In eqn. 7.80 the recombination really has three terms:

(i) a linear term AN (partly attributed to nonradiative recombination) but $A=0$ is assumed here for simplicity;
(ii) a quadratic or 'bimolecular' term $BNP \sim N B_n(N/N_{tr})$ (attributed to radiative recombination); and
(iii) a 'cubic term' $(C'N^2P+C''NP^2) \sim NC_n(N/N_{tr})^2$ (attributed to nonradiative Auger recombination which does not contribute to the lasing field but removes carriers from the photonic interaction, so reducing the quantum efficiency).

The approximations refer to the assumption of electrical neutrality $N \sim P$. The effective electron recombination time $\tau_{ro} \sim 10^{-9}$ s, at the normalising carrier density $N_{tr} \sim 10^{24}$ m^{-3}, should be contrasted with the photon lifetime $\sim 10^{-12}$ s.

Normalisation is now made in eqn. 7.79, writing $N/N_{tr} \to$ Ne with this, in the MATLAB code, going to the array **Ne** varying from step to step. With uniform material then N_{tr} is uniform throughout the device but could be allowed to vary. The effective cross-sectional area for electronic/photonic interaction is $\Gamma \mathcal{A}$ so that $v_g N_{tr} \Gamma \mathcal{A}$ gives a normalising electron flow in numbers per second:

$$\frac{L}{v_g}\frac{d(Ne)}{dt}+\frac{L}{v_g}\frac{(Ne)}{\tau_r}+\frac{L}{v_g\tau_{ro}}2\Gamma g_M L\left\{\frac{(F^*F+R^*R)\tau_{ro}}{(N_{tr}\Gamma \mathcal{A}L)}\right\}$$

$$=\frac{L}{v_g\tau_{ro}}\left\{\frac{I\tau_{ro}}{q(N_{tr}\Gamma \mathcal{A}L)}\right\} \qquad (7.81)$$

where τ_r is now a function of **Ne** with τ_{ro} the value when **Ne**=1.

Knowing that the operational electron density will typically be a few times N_{tr} so that Ne ~ 2–4 provides an order of magnitude for the threshold drive current $I \sim NeqN_{tr}\Gamma \mathcal{A} L/\tau_{r0}$. Such rough order-of-magnitude checks can be reassuring in initial computations. Taking $L \sim 100$ μm, $v_g \sim 10^{14}$ μm/s and $\tau_r \sim 10^{-9}$ s, this normalisation indicates that $(L/v_g \, \tau_r) \sim 10^{-3}$ which is more than three orders of magnitude smaller than (L/s) (~ 30) where s is the spatial-step length for the computation. Hence the electron density in general changes on a much slower time scale than changes in the photonic fields. As indicated earlier, this relatively slow variation in eqn. 7.80 allows several key approximations to be made in the modelling:

(a) The electron density is approximately constant over two or three time steps even though the optical field may change significantly in these time steps.

(b) A forward finite-difference scheme has adequate accuracy for the electron equation (eqn. 7.80) which is composed of entirely real quantities. For the more rapidly changing complex optical fields in eqns. 7.77 and 7.78, a second-order central-difference scheme is used giving greater accuracy.

(c) Note that in the instantaneous stimulated emission current given from the real term proportional to $g(F^*F+R^*R)$, the instantaneous gain g is fundamentally real. Provided the laser is operating around a reasonably flat peak gain, then gain variations caused by the output being at different frequencies need not be added to the time-domain equations and indeed such routines have not been added in this book. However, if the laser is deliberately designed to operate away from the gain peak this gain has to be reduced appropriately.

While iteration would work, one recognises that the real stimulated recombination is more correctly given from the terms

$$\{(gF)F^* + (gF)^*F + (gR)R^* + (gR)^*R\} \quad (7.82)$$

The method of the complex Lorentzian filter can then be used on the complex terms (gF), $(gF)^*$, (gR), and $(gR)^*$. Alternatively, the methods of eqn A.5.30 could also be extended by writing

$$g \rightarrow [g(\omega_0) - j(\partial g/\partial \omega)_0 \partial/\partial t - \tfrac{1}{2}(\partial^2 g/\partial \omega^2)_0 \partial^2/\partial t^2] \quad (7.83)$$

Now, with ω_0 giving the peak gain taking $(\partial g/\partial \omega)_0 = 0$, consider eqns 7.83 and 7.82 and integrate by partial integration over one cycle of time. The partially integrated term and its conjugate are approximately zero, giving a real stimulated current (which is subsequently restricted to be non-negative) from

246 *Distributed feedback semiconductor lasers*

$$g(F^*F+R^*R) \rightarrow \{g(\omega_0)\,(F^*F+R^*R)$$
$$+\tfrac{1}{2}(\partial^2 g/\partial\omega^2)_0\,[\partial F/\partial t^*\partial F/\partial t+\partial R/\partial t^*\partial R/\partial t]\} \quad (7.84)$$

With negative gain curvature at the peak gain, any modulation of the fields will reduce the effective gain and reduce the stimulated emission current as the rate of modulation (i.e. deviation of frequency from the gain peak) increases.

7.10.4 Numerical programs

Program **dfbuni** in directory **dfb** brings these results together for a uniform DFB laser with real coupling constant κ. A vectorised system of equations helps to shorten the code and is well suited to MATLAB with the 'guts' of the program in the subroutine **dfbrun**. This program demonstrates that, at first switching on the drive to the laser, the electron density first has to increase, and to shorten the duration of the turn-on process, the electron density is started close to its transparency value. The first couple of runs simply show spontaneous noise appearing until a threshold electron density is reached when oscillations appear in the output waveform. Because the perfectly uniform DFB cannot discriminate the thresholds of the upper or lower mode close to the stop bands, both modes initially appear and which one finally wins depends upon the value of κL and the spontaneous emission. The reader may explore this directly but also may prefer to start with a program such as that for the phase-shifted DFB that gives the 'same' results on each run.

The program **dfb14** is a very similar DFB but now with a $\lambda_m/4$ phase shift in its centre. Figure 7.11 gives examples of the typical output that one expects to find from this program. Here there is strong variation with distance of both the electron density (spatial-hole burning) and the optical-intensity profile within the laser. This laser now switches on in one clear mode which is in the middle of the conventional stopband for the uniform DFB.

The program **dfb218** is again a similar DFB, as previously, but now with two phase shifts of $\lambda_m/8$, one on either side and offset from the laser's centre. This program shows that the lower frequency is the favoured mode of lasing with considerably less spatial-hole burning than for the single central phase shift. The program **dfb238** simply changes these phase shifts from $\lambda_m/8$ to $3\lambda_m/8$, and now the higher frequency is preferred. As discussed in Chapter 5, the detailed position of the phase shifts affects the performance and 'dfb238' lasers are generally less stable with a higher drive level than 'dfb218' lasers.

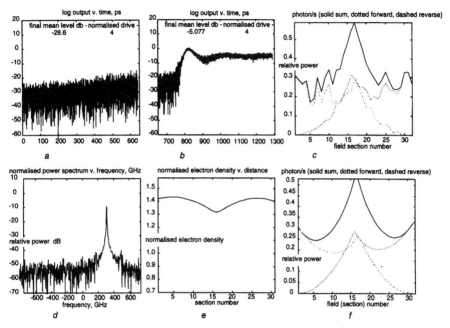

Figure 7.11   Typical output for dfb14 (a DFB with one central $\lambda_m/4$ phase shift)

a initial run; b continuation; c field profile with spontaneous emission; d spectrum; e electron density-distance; f field profile without spontaneous emission

Recent changes in technology have made devices with gain coupling possible. Here the value of κ is complex and the sign of the imaginary component determines whether the laser prefers the upper or the lower frequency around the stopband of the DFB. Strictly, the value of the imaginary component should change with electron density. However, here a constant complex $\kappa = \kappa_{real} + j\kappa_{gain}$ is inserted in a program called dfbgain. By changing the sign of κ_{gain}, the modal-selection properties of gain gratings can be demonstrated. The program would be very similar for a second-order grating but in a second-order grating there would be additional attenuation in the unscattered fields.

The program has the ability to run with bandwidth-limited (i.e. filtered) spontaneous emission. This slows down the program and has little practical effect because the gain filter is included as standard and, as stated earlier, once the gain becomes attenuation at the Nyquist-band edges the detailed shape of the spontaneous emission has little effect. The laser can also be run without the spontaneous emission provided that lasing is well established. This can provide 'ideal' theoretical field profiles. To operate this facility, wait until the laser has settled and stopped between runs, go into the keyboard and type spont0 and return. Appendix 12 summarises the programs that are available, while 1readme.txt files, provided within each directory, give more operational detail.

7.11 Summary

The chapter started with discussions on the numerical solution of an ordinary differential equation with complex rates of growth, highlighting the importance of the straightforward central-difference method. The advection equations or travelling-wave equations were considered, focusing on a central-difference technique with appropriate steps in space and time. Gain and phase changes were introduced along with coupled reflections. At each stage, numerical stability was considered and model test programs given. A uniform Bragg-grating amplifier was considered and the numerical model of a forward travelling wave inside a Bragg grating was implemented in MATLAB and checked against an analytic model.

The processes of spontaneous emission are modelled by uncorrelated Gaussian excitations injected into each section in the laser, and tutorial models of the spontaneous-emission processes were made available in MATLAB code.

The process of discretisation leads to limitations, and these were briefly discussed, particularly in relation to the fast Fourier transform. Strategies for Lorentzian spectral filters were outlined and then implemented to ensure that the gain inside the laser varied with frequency in an appropriate manner. The use of stochastic excitation to investigate broadband behaviour of a filter was demonstrated. Finally, a basic program for a DFB laser operating around the peak gain with spontaneous emission was discussed and implemented in MATLAB code.

It is a pleasure to acknowledge that many of the ideas in this chapter owe a considerable debt to research students Chi Tsang [39] who participated in the European COST-240 exercise [40] , Mark Nowell [41] and Barry Flanigan [42], Si Fung Yu [43] and David Jones [44] and especially to research associates Li Ming Zhang [43,45] (see also [12]) and Dominique Marcenac [45–47]. Their work has been pulled together with a more consistent use of central-difference methods for laser-diode DFB analysis than appears to have been previously established. It is hoped that the cut-down programs for use with student MATLAB will also encourage the reader to gain a better understanding of the numerical processes and some of their limitations, looking on fast numerical processing as an enjoyable experimental process.

7.12 References

1 AGRAWAL, G.P., and DUTTA, N.K.: 'Semiconductor lasers' (Van Nostrand Reinhold, New York, 1993), 2nd ed., Section 6.6.1
2 THOMPSON, G.H.B.: 'Physics of semiconductor laser devices' (J. Wiley, Chichester, 1980), Section 7.3.2
3 CARROLL, J.E.: 'Rate equations in semiconductor electronics' (Cambridge University Press, Cambridge, 1985), Section 6.4.2
4 NAGARAJAN, R., ISHIKAWA, M., FUKUSHIMA, T., GEELS, R.S., BOWERS, J.E., and COLDREN, L.A.: 'High speed quantum well lasers and carrier transport effects', *IEEE J. Quantum Electron.*, 1992, **28**, pp. 1990–2007
5 OHTSU, M.: 'Highly coherent semiconductor lasers' (Artech House, Boston, 1991), chap. 2
6 PETERMANN, K.: 'Laser diode modulation and noise' (Kluwer Academic Publishers, Dordrecht, 1991)
7 VANKWILEBERGE, P., MORTHIER, G., and BAETS, R. 'CLADISS—a longitudinal multimode model for the analysis of static and stochastic behaviour of diode lasers with distributed feedback', *IEEE J. Quantum Electron.*, 1988, **24**, pp. 2160–2169
8 MORTHIER, G., and VANKWIKELBERGE, P.: 'Handbook of distributed feedback lasers' (Artech, Boston, 1997)
9 BUUS, J.: 'Principles of semiconductor laser modelling', *IEE Proc.—J.*, 1985, **132**, pp. 42–51
10 COLDREN, L.A., and CORZINE, S.W.: 'Diode lasers and photonic integrated circuits (J. Wiley, New York, 1995), Section 3.5.2
11 BJORK, G., and NILSSON, O.: 'A new and efficient numerical matrix theory of complicated laser structures. Properties of asymmetric phase shifted DFB-laser' *J. Lightwave Technol.*, 1987, **5**, pp. 140–146
12 ZHANG, L.M., and CARROLL, J.E.: 'Large signal dynamic model of the DFB laser', *IEEE J. Quantum Electron.*, 1992, **28**, pp. 604–611
13 LOWERY, A.J.: 'Transmission line modelling of semiconductor lasers: the transmission-line laser model', *Int. J. Numer. Modell.*, 1989, **2**, pp. 249–265
14 LOWERY, A.J.: 'A qualitative comparison between two semiconductor laser amplifier equivalent circuit models', *IEEE. J. Quantum Electron.*, 1990, **26**, pp. 1369–1375
15 LOWERY, A.J., GURNEY, P.C.R., WANG, X-H., NGUYEN, L.V.T., CHAN, Y.C., and PREMARATNE, M.: 'Time-domain simulation of photonic devices, circuits and systems'. Proceedings of *Physics and Simulation of Optoelectronic Devices, SPIE Photonics West*, 1996, **2693**, pp. 624–635

16 HOEFER, W.J.R.: 'The transmission-line matrix method—theory and applications', *IEEE Trans.*, 1987, **MTT-35**, pp. 370–377
17 LOWERY, A.J.: 'A new dynamic semiconductor laser model based on the transmission line laser model', *IEE Proc.—J.*, 1988, **135**, pp. 126–132
18 WONG, Y.L, and CARROLL, J.E.: 'A travelling wave rate equation analysis for semiconductor lasers', *Solid-State Electron.*, 1987, **30**, pp. 13–19
19 LOWERY, A.J.: 'A new dynamic semiconductor laser model based on the transmission line modelling method', *IEE Proc.—J.*, 1983, **134**, pp. 281–289
20 LOWERY, A.J.: 'Transmission-line modelling of semiconductor lasers: the transmission-line laser model', *Int. J. Numer. Modell.*, 1990, **2**, pp. 249–265
21 COHEN, A.M.: 'Numerical analysis' (McGraw Hill, London, 1973)
22 ISERLES, A.: 'Numerical analysis of differential equations' (Cambridge University Press, 1996)
23 LAPIDUS, L., and PINDER, G.F.: 'Numerical solution of partial differential equations in science and engineering' (Wiley, New York, 1982)
24 PRESS, W.H., TEUKOLSKY, S.A., VETTERLING, W.T., and FLANNERY, B.P.: 'Numerical recipes in C' (Cambridge University Press, 1992), 2nd ed.
25 'The student edition of MATLAB' (Prentice Hall, Englewood Cliffs, 1995)
26 ISERLES, A.: 'Numerical analysis of differential equations' (Cambridge University Press, 1996), chap. 5
27 LAPIDUS, L., and PINDER, G.F.: 'Numerical solution of partial differential equations in science and engineering' (J. Wiley, New York, 1982), p. 566
28 IFEACHOR, E.C., and JERVIS, B.W.: 'Digital signal processing' (Addison-Wesley, Wokingham and Reading, 1993)
29 ABRAMOWITZ, M., and STEGUN, I.A.: 'Handbook of mathematical functions' (Dover, New York, 1965), p. 70
30 ISERLES, A.: 'Numerical analysis of differential equations' (Cambridge University Press, 1996), p. 312
31 CLOUDE, S.: 'Electromagnetic book' (UCL Press, London, 1995)
32 CUNNINGHAM, E.P.: 'Digital filtering' (Houghton Miflin, Boston, 1992)
33 COOLEY, J.W., and TUKEY J.W.: 'An algorithm for machine computation of complex fourier series', *Math. Comput.*, 1965 **19**, pp. 297–301
34 WESTBROOK, L.D.: 'Measurement of dg/dN and dn/dN and their dependence on photon energy in $\lambda=1.5$ μm InGaAsP laser diodes', *IEE Proc.—J.* 1986, **133**, pp. 135–142
35 NGUYEN, L.V.T., LOWERY, A.J., GURNEY, P.C.R., NOVAK, D., and MURTONEN, C.N.: 'Efficient material gain models for the transmission-line laser model', *Int. J. Numer. Modell. Electron. Netw., Devices Fields*, 1995, **8**, pp. 315–330
36 HENRY, C.H.: 'Theory of linewidth of semiconductor lasers', *IEEE J. Quantum Electron.*, 1982, **18**, pp. 259–264
37 HENRY, C.H.: 'Phase noise in semiconductor lasers', *J. Lightwave Technol.*, 1986, **4**, pp. 298–311

38 OSINSKI, M., and BUUS, J.: 'Linewidth broadening factor in semiconductor lasers—an overview', *IEEE Quantum Electron.*, 1987, **23**, pp. 9–28
39 TSANG, C.F.: 'Dynamics of distributed feedback lasers'. PhD dissertation, Cambridge, 1993
40 MORTHIER, G., BAETS, R., TSIGOPOULOS, A., SPHICOPOULOS, T., TSANG, C.F., CARROLL, J.E., WENZEL, H., MECOZZI, A., SAPIA, A., CORREC, P., HANSMANN, S., BURKHARD, H., PARADISI, A., MONTROSSET, I., OLESEN, H., LASSEN, H.E., and SCHATZ, R.: 'Comparison of different dfb laser models within the European cost-240 collaboration'. *IEE Proc.—Optoelectron.*, 1994, **141**(2), pp. 82–88
41 NOWELL, M.C.: 'Push-pull directly modulated laser diodes'. PhD dissertation, University of Cambridge, 1994
42 FLANIGAN, B.J.: 'Advances in push-pull modulation of lasers'. PhD dissertation, University of Cambridge, 1996
43 ZHANG, L.M., YU, S.F., NOWELL, M.C., MARCENAC, D.D., CARROLL, J.E., and PLUMB, R.G.S.: 'Dynamic analysis of radiation and side mode suppression in second order DFB laser using time-domain large signal travelling wave model', *IEEE J. Quantum Electron.*, 1994, **30**, pp. 1389–1395
44 JONES, D.J,. ZHANG, L.M., CARROLL, J.E., and MARCENAC, D.D.: 'Dynamics of monolithic passively mode-locked semiconductor lasers', *IEEE J. Quantum Electron.*, 1995, **31**, pp. 1051–1058
45 TSANG, C.F. MARCENAC, D.D., CARROLL, J.E., and ZHANG, L.M.: 'Comparison between 'power matrix model' and 'time domain model' in modelling large signal response of DFB lasers', *IEE Proc. Optoelectron.*, 1994, **141**, pp. 89–96
46 MARCENAC, D.D.: 'Fundamentals of laser modelling'. PhD dissertation, University of Cambridge, 1993
47 MARCENAC, D.D., and CARROLL, J.E.: 'Quantum mechanical model for realistic Fabry–Perot lasers', *IEE Proc.—J.*, 1993, **140**, pp. 151–171

Chapter 8
Future devices, modelling and systems analysis

8.1 Introduction

Computer-aided design (CAD) has been a vital initial ingredient in the realisation of very large-scale integrated (VLSI) electronic circuits. Now in optoelectronics projects, mathematical modelling, although at present lacking the sophistication of CAD in VLSI, is being recognised as a key component in the design of devices and systems [1]. One thrust of the book has been to lay out the physical- and mathematical-modelling techniques for the electromagnetic and electronic interactions within distributed feedback lasers to give better explanations and to facilitate new designs. This chapter outlines further areas in optoelectronics where such modelling will be at the forefront in the design of more effective prototypes. Optical systems where arrays of devices may be interconnected are one such important area and are discussed in Section 8.2. Novel concepts such as that of the push–pull laser (Section 8.3), tunable lasers with Bragg gratings (Section 8.4), and surface-emitting lasers (Section 8.5) are all candidate areas for applications of the modelling principles given in this text. The future for mathematical modelling, coupled with a sound physical understanding, is bright, extensive and assured.

8.2 Systems analysis

8.2.1 Introduction

The time-domain techniques (Chapter 7) for solving the electromagnetic interactions in lasers have led to powerful and efficient routines which are capable of solving a wide range of (single- and) multimodal problems covering Fabry–Perot lasers, uniform and phase-shifted DFB lasers, all of which may be driven with digital or analogue modulation. Similar routines have also been built using transmission-line laser modelling [2–4], an alternative time-domain technique with many of the virtues of the techniques discussed in Chapter 7. Considerable effort has been put into developing a general optical-system simulator based on embedded-device models that are interconnected with passive components so as to mimic system performance [5]. The laser simulator can, for example, model high-speed digital-modulation performance, fast pulse generation, passive mode locking, and dynamic instabilities. Extending such time-domain analyses to optical subsystems and systems containing many components will provide excellent insights into the interactions between these components over a wide range of frequencies. This section gives examples of some initial applications of such an optical-systems simulator.

8.2.2 Component modelling

The first ingredient for a 'systems simulator' is to have numerical models for all the components within the system. The basic numerical techniques described in Chapter 7 can be applied to Fabry–Perot and DFB lasers, and also to semiconductor optical amplifiers (SOA) which are essentially Fabry–Perot lasers with weak (or zero) reflection at their facets. Some geometrically complicated components may require techniques such as beam-propagation [6] or 'finite elements' [7] to solve the electromagnetic fields but these lie outside the scope of this work. The numerical techniques can be applied to both the electronic drives and the optical outputs/inputs. With an armoury of modelling methods, one can analyse combinations of lasers, electro-absorption modulators, ideal optical transmitters , y-junctions, lengths of dispersive fibre and so on, enabling complex systems to be tested numerically with a high degree of confidence in subsequent experimental trials.

Key features incorporated into such opto-electronic device/system modelling are:

(i) spatial-hole burning in the density of the charge carriers along the length of devices as the optical intensity varies;
(ii) time delays in injecting carriers from contact to active lasing regions (carrier-transport effects via three-level models, for example);

(iii) nonlinear gain (ϵ) as the photon density changes (introduces damping into the response of a laser to modulation drive);

(iv) wide varieties of carrier recombination (linear, bimolecular and Auger recombination and general terms of the form DN^P);

(v) multiple-mode operation (time-domain techniques automatically take many modes into account while frequency-domain techniques using, for example, transfer matrices often require one to track each mode separately);

(vi) losses can be fixed or dependent on charge-carrier densities (e.g. waveguide absorption and scattering, and intervalence band-absorption, respectively);

(vii) realistic gain variations with carrier density and wavelength (linear/log, peak-gain against carrier-density relationships, and changes of wavelength for the peak gain with carrier density);

(viii) realistic spontaneous-emission spectra with wavelength (changes of wavelength for the peak spontaneous emission as carrier density changes);

(ix) submounts with parasitic capacitances and inductances appropriate for the design of modulated laser/amplifier/electroabsorption modulators;

(x) arbitrary drive-voltage waveforms into multicontact optoelectronic devices;

(xi) reflections and radiation loss between sections of laser/amplifier/passive devices (note that full time-domain modelling of lasers attached to many kilometres of nonlinear fibre is difficult because lasers typically need subpicosecond time steps requiring billions of steps to model a millisecond delay fully); and

(xii) relative-intensity noise (RIN) due to inclusion of spontaneous emission [evaluating RIN in the kilohertz region also poses challenges for time-domain techniques for the same reasons as in (xi)].

Various devices (e.g. the electroabsorption modulator) will need data about absorption and refractive index variations and such variations can be modelled using high-order (e.g. fifth-order) polynomial fits to experimental or theoretical data at discrete sets of bias voltage, frequency etc. and then stored in look-up tables as part of the model, with interpolation providing data at intermediate points.

One useful component is the 'ideal optical transmitter' which generates optical waveforms with no amplitude patterning or intrinsic wavelength chirp though, of course, it contains the fundamental linewidth broadening caused by the information modulation (Fourier broadening). Such an ideal transmitter aids investigations of distortion, introduced for example by an SOA, without the complicating effects of an optical input having a nonideal waveform. Investigating the propagation of 'information-bandwidth-limited signals' along dispersive fibre links is another example where the ideal transmitter is of benefit: a realistic chirped source initially just confuses the issues. To model assemblages of subsystems and systems, it is essential to add y-junctions, attenuators, filters, partial reflectors, fibre links and four-port couplers. All these can be modelled at a variety of levels of complexity but many of these devices have linear input–output transfer functions with well defined spectral characteristics over the range of required power levels. It is then sometimes better to convert the time-domain input, using the FFT, to the frequency domain where transfer-matrix operations are performed to give the output which is then transformed via the inverse FFT back to the time domain. For linear elements this is a good straightforward technique with wide validity, however, considerable care is needed if the element is within a loop. The inclusion of fibre gratings, modelled like a Bragg grating in a laser, will be another component which leads to new and interesting applications in systems.

8.2.3 System modelling

A systems model will have a library of default and bespoke models of all the required devices which can then be selected and interconnected in arbitrary configurations to form the optical subsystems and systems. Such libraries will contain routines for interconnecting devices and subsystems and the number of these interconnects will be limited by the data storage available, remembering that each device/subsystem will need to store such entities as drive voltages/currents, input/output fields etc. for evaluation and for use in a later computation. Typically, one will examine at various points in the system:

(a) optical spectra and waveforms with selected time windows and, if required, effects of dispersion would be included;
(b) baseband spectra and filtered waveforms using RC, Butterworth or Bessel filters, typically with 1–7 poles;
(c) eye diagrams giving a qualitative indication of the system performance;
(d) instantaneous frequency against time (from rate of change of phase of complex field);

256 *Distributed feedback semiconductor lasers*

(*e*) instantaneous and mean powers over selected time windows; and
(*f*) data-transfer techniques for concatenating simulations and transfer to other computer systems and output hardware.

To find wide acceptance within any research/development organisation, such model libraries need to ensure that the software is very 'user friendly' and does not need significant prior knowledge of programming languages or complex physics. The difficult problem of the user interface is not addressed here but programs like MATLAB 5 are coming to the rescue with more user-friendly methods for data input and output.

In this section, results are presented from:

(i) the modelling of SOA post-amplification of output from high-speed digitally modulated optical sources;
(ii) DFB laser integrated with electroabsorption modulator allowing for residual reflections from modulator-output facet;
(iii) cross-gain and four-wave-mixing wavelength conversion in a semiconductor optical amplifier; and
(iv) cross-phase wavelength conversion in a Mach–Zehnder interferometer structure incorporating two semiconductor optical amplifiers.

8.2.4 10 Gbit/s power amplification

In the first example, an ideal optical transmitter injects a 10 Gbit/s waveform with no patterning or wavelength chirp into the SOA. The simulator generates the output from the SOA which includes the amplitude patterning and wavelength chirp induced by the amplifier-saturation process. The signal then is dispersed by a standard fibre and the received eye diagram of this dispersed waveform is examined as a function of the input optical power to the SOA by post-processing the stored waveforms.

Figure 8.1a shows the unfiltered transmitted eye diagram at 10 Gbit/s over one period with a pseudorandom sequence taken over an interval of 4.1 ns. The output is plotted every 4.1 ps although the time step in the computation is 0.05 ps. The optical output is then dispersed by 70 km of standard fibre detected and filtered (fifth-order Bessel, 7 GHz bandwidth) with Figure 8.1c showing the received eye diagram and Figure 8.1b showing the instantaneous frequency of the SOA output against time, for a 1 mW optical input to the SOA.

There is little amplitude patterning on the output from the SOA for this design, and the received eye diagram is symmetric with an

Future devices, modelling and systems analysis 257

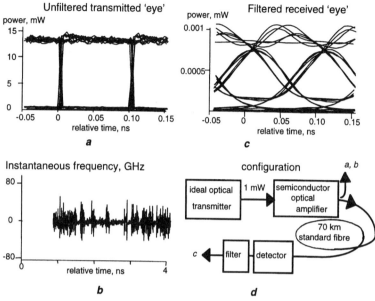

Figure 8.1 Simulated amplifier performance for 1 mW peak optical input power

adequate opening. The plot (*b*) of 'instantaneous frequency' or instantaneous rate of change of phase shows fluctuations which are found to be linked to the modulation. There is a high amplitude of frequency noise in the 'zeros' and a lower amplitude of frequency noise in the 'ones'. This effect is caused by the increased significance of the spontaneous emission noise and hence an enhanced phase noise when the input power to the SOA is low. Further, the gain of the SOA is larger in the 'zeros' and therefore also enhances the phase noise.

The same plots are shown in Figure 8.2 but now with a 5 mW optical input to the SOA. The SOA was not designed to operate at these power levels, and more amplitude patterning is clearly seen with saturation occurring over the duration of the pulse. There is also an increased instantaneous frequency excursion at turn-on and turn-off with a negative-frequency chirp at the leading edge of the 'ones'. This chirp is associated with the gain saturation process in the SOA. The filtered received eye diagram now shows an asymmetry because the isolated 'ones' are delayed relative to sequences of 'ones'. This is in agreement with experimental work* and the model was used to investigate its cause. The time constant associated with the gain-saturation process in

---

* COLLAR, A.J.: Private communication, 1966

258 *Distributed feedback semiconductor lasers*

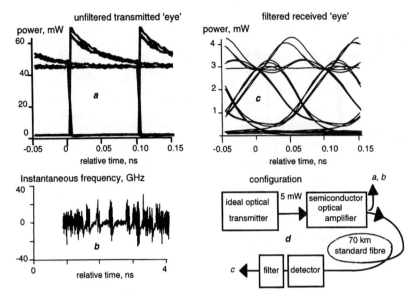

Figure 8.2 Simulated amplifier performance for 5 mW peak optical input power

SOAs causes the isolated 'ones' to propagate through a reducing average carrier density and increasing refractive index, resulting in a dynamic chirp to longer wavelengths, unlike 'ones' in a sequence of 'ones'. In a dispersive standard fibre system, a temporal delay relative to the rest of the waveform is found. In addition, the information bandwidth broadens the spectrum which, in the presence of dispersion, changes the pulse shape.

8.2.5 Direct modulation: recapitulation

The next system to be discussed is the laser with an external modulator instead of using direct modulation. From the earliest days, direct modulation of the current driving a semiconductor laser has been recognised as *the* low cost way of modulating the optical output from a laser [8] but there are limitations which are well recognised. The first of these is dynamic 'chirp'. To increase (reduce) the amplitude of the optical output from a laser, the gain has to be increased (reduced). As noted in Section 2.5, the linewidth-broadening factor α_H relates such changes of gain with necessary changes in the refractive index and resulting changes in the frequency. As seen in Section 6.2.5, this dynamic 'chirp' can increase the spread of frequencies to ~100 GHz for 10 Gbit/s modulation—well beyond the spectral broadening required from Fourier analysis of any conventionally modulated signal. This chirp, if not limited, leads (as discussed in Section 1.5) to spreading of the 'ones' into the 'zeros' giving rise to a loss of signal— a *dispersion penalty* that can be unacceptable even over a few kilometres.

Besides the dynamic chirp, there is also a fundamental limit to the bit rate for such direct modulation of a laser because of the photon–electron resonance discussed in Chapter 3. This limiting maximum modulation frequency $f_{M\,max}$, combined with package limitations, is usually significantly below 50 GHz depending on the differential optical gain for the material. For modulation frequencies $f_M > f_{M\,max}$, the optical output-modulation efficiency typically falls off at over 12 dB per octave increase in f_M.

The limitation imposed by both dispersion and photon–electron resonance may be mitigated by using a continuously operated DFB laser but integrated with a modulator after the laser [9–12] which turns the light on and off in a quite separate process. Such a combination, often with an amplifier between the laser and modulator, can also provide 'prechirp' which can so arrange the amplitude and phase of the frequency components forming the modulated optical pulse that the changes in group velocity with frequency within the fibre cause the pulse shape to compress along standard fibre before the pulse disperses again. This dispersion compensation combined with the removal of the photon–electron resonance offers excellent prospects for communication systems over standard fibre up to and beyond 10 Gbit/s with distances in excess of 100 km or more [13–16]. This option is currently preferred for high-quality systems, in spite of the extra cost and complexity.

8.2.6 Simulation of integrated DFB laser and electroabsorption modulator

As mentioned above, external modulators have the potential benefit of amplitude modulating a laser signal either without the large 'chirp' that is caused by α_H in directly modulated lasers or with a modified chirp formed to compensate for dispersion, but this latter is not considered here. Integration of the laser and (electroabsorption) modulator should give potential benefits of lower cost, higher output power and ease of packaging. However, the problem with external modulators that is exposed in this section is that, even with <1% reflections from an antireflection-coated output facet, the modulator can still cause complications by passing such reflections back through

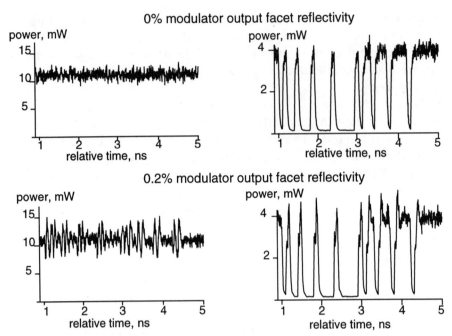

Figure 8.3 Simulation of effect of reflections from output facet of an integrated DFB laser and electroabsorption modulator on dynamic performance of source

Time step for the computation is 0.0956 ps while the curves are plotted every 4.11 ps

the modulator and interacting with the laser resulting in transient oscillations which lead to amplitude and phase variations being superimposed on the laser's nominally CW output field. Figure 8.3 illustrates the effects of changing the reflections from no reflection at the output facet of the modulator to 0.2% facet reflection where significant effects in terms of noisy power fluctuations and ringing on the 10 Gbit/s-modulated waveforms can be observed. However, the noise in the modulator output is less intrusive in the presence of modulation but the modelling clearly explains why reflections <0.1% are often specified.

8.2.7 Cross-gain and four-wave-mixing wavelength conversion in an SOA

A good example of the use of the time-domain-modelling method using multiple devices with multiple frequencies is given by considering wavelength conversion using an SOA subsystem. Figure 8.4 shows an ideal optical transmitter launching a 10 mW CW optical

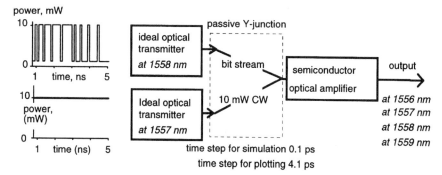

Figure 8.4 Four-wave mixing for frequency shifting

power at 1557 nm into a 350 μm-long SOA and a second ideal transmitter at 1558 nm launching a 10 Gbit/s '10110111011110111110100100010000100000' digitally modulated signal with 10 mW peak and 1 mW minimum optical power. The output spectrum shows signals at 1557 and 1558 nm, as expected, and also the four-wave-mixing products at 1556 and 1559 nm, as indicated in the spectrum of Figure 8.5.

The output spectrum is then optically filtered and Figure 8.6 shows the output power against time centred on the wavelengths of 1556, 1557, 1558 and 1559 nm using passbands each with a total width of 1 nm. The modulated signal at 1558 nm emerges amplified but with superimposed amplitude patterning associated with the SOA gain-saturation process. The gain recovery in the zeros can also be seen. The cross-gain saturation causes the CW signal at 1557 nm now to have superimposed on it an inverted version of the data but with a poor extinction ratio. The gain saturation leads to a completely different shape to the waveform for the inverted signal as the modulation is now driven entirely by the gain-saturation process, and so even an instantaneous change in the data leads to a delayed response in the inverted signal.

The four-wave-mixing products at 1556 and 1559 nm are shown in Figure 8.6 and are typically over a factor of 100 weaker than the main signals at 1557 and 1558 nm. Both mixing products (1556 and 1559 nm) show the data in noninverted form, with the 1556 nm signal being the stronger of the two and exhibiting less distortion but having a less satisfactory extinction ratio. It must be emphasised that these are not optimised results but simply indicative of the abilities of time-domain modelling to handle multiple frequencies.

The modelled example displays four-wave mixing as a result of the fluctuations induced in the carrier density and hence variations in gain

262 *Distributed feedback semiconductor lasers*

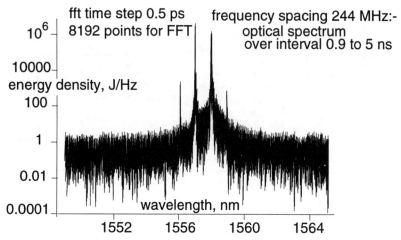

Figure 8.5 Simulated output spectrum from amplifier

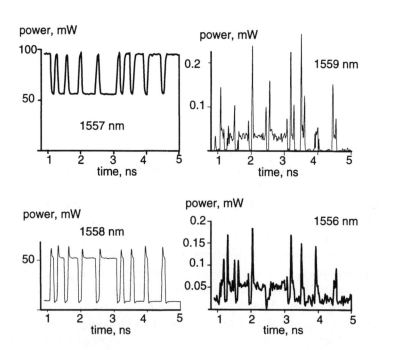

Figure 8.6 Simulated cross-gain and four-wave-mixing wavelength-converted signals at amplifier output

caused by the beating of the two optical-input signals. From the conventional analysis, the modulation of the carrier density decreases rapidly with frequencies above the photon–electron resonance. However, spectral-hole burning and carrier heating (see Section 2.4.4) may contribute to modulation of the gain up to much larger wavelength separations, and it is the gain modulation, for example, which could give the reported conversion efficiencies of up to 0 dB for 16 nm of wavelength separation (corresponding to 2 THz) [17, 18]. Two-photon absorption and the Kerr effect may also contribute to this remarkable result, but the numerical models presented here would need augmenting appropriately.

8.2.8 Simulation of cross-phase wavelength conversion in a Mach–Zehnder interferometer incorporating two SOAs

Signals can be sent to different locations via different routes (space-division multiplexing) using different time slots (time-division multiplexing or TDM) and via different wavelengths (wavelength-division multiplexing or WDM). One can then see that a wavelength converter will be a key component for future optical systems. Semiconductor-amplifier devices, combined with optical filters, offer significant potential in making such a device. Again, because there are wide ranges of frequencies and strong nonlinearities to be considered, time-domain modelling makes good sense. Preliminary results are shown here of simulating the performance of a cross-phase wavelength converter operating at 5 Gbit/s. The Mach–Zehnder interferometer contains an SOA in each arm and is injected with a CW optical input at 1557 nm and data at 1558 nm using ideal optical transmitters, as shown in Figure 8.7. There are four passive y-junctions as indicated which are here assumed to be lossless. In order to make the conversion process non-inverting a phase shift of π radians is added in the upper arm of the interferometer. This ensures that when injecting a '1' on the data, which is designed to induce a π phase shift in the lower SOA, the interferometer becomes balanced which in turns results in maximum output power at 1557 nm.

The optical data input at 1558 nm consists of a '1011011101001000' sequence with 4.6 mW in the 'ones' and 0 mW in the 'zeros'. Figure 8.8 presents the simulated output waveform at 1557 nm which has the inverted data superimposed by cross-gain saturation. The chirp introduced by the lower SOA gives negative-wavelength chirp on the leading edge of output pulses, which helps propagation in dispersive fibre, with Figure 8.8d showing the predicted filtered received eye

264 *Distributed feedback semiconductor lasers*

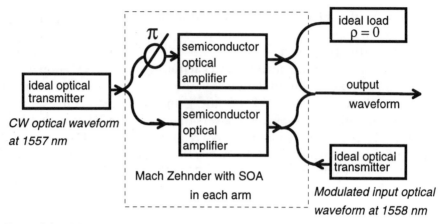

Figure 8.7 *Schematic diagram of cross-phase wavelength-conversion system*

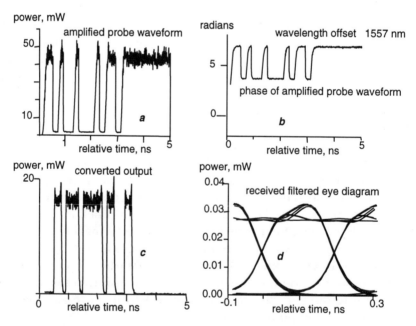

Figure 8.8 *Mach-Zehnder wavelength converter*

    a Simulated output waveform from lower amplifier at 1557 nm modulated by counter-propagating 1558 nm signal via cross-gain saturation
    b Phase shift imposed on this signal by the amplifier saturation process
    c The output from the wavelength converter
    d Eye diagram after 70 km of standard fibre giving dispersion of 17 ps/nm km and dispersion slope of 0 ps/nm^2 km with a loss of 0.2 dB/km

diagram after propagation through 70 km of fibre with 17 ps/nm km of dispersion.

These examples illustrate the power of a general optoelectronic simulator which has inclusive software to enable it to be applied to these very different applications. Such tools will mature over the next decade to the state now achieved by CAD for VLSI.

8.3 The push–pull laser

8.3.1 Introduction: push–pull electronics

Push–pull operation is well known in electronics where, for example (Figure 8.9), a pair of transistors T_1 and T_2 are coupled at their emitters (marked E). Because of a sufficiently large resistance R_t ('the long tail') there is a constant current I flowing equally between the two transistors in the steady state. On driving the input (the base of T_1) positive, the current in T_1 is enhanced by δI but, because of the constant-current source at the coupled emitters, the current into T_2 is reduced by δI. The base at transistor T_2 is shown to be fixed at 'ground' potential but could equally well have been driven by a second input provided that the polarity was opposite to that at the base input of T_1. The 'push–pull' outputs are then capable of driving another similar circuit. This circuit is a remarkably high-speed circuit because both transistors are operated with either unity voltage gain or unity current gain which then gives the potential for working at the limit of their gain–bandwidth product signified by the transition frequency f_t. The emitter E is the output for the first transistor (T_1) for driving transistor T_2. Consequently, the transistor T_1 is effectively operated as an emitter follower with unity voltage gain. The transistor T_2 is effectively operated in a common- (grounded-) base mode which then has unity current gain. Although each transistor is operating at unity voltage or current gain, the net power gain can still be in excess of 10 dB because of the change from a low input impedance to a higher output impedance. This emitter-coupled pair (or *long-tail pair*) provides the classic circuit for achieving the highest modulation speeds in transistors. It is typically used, with appropriate transistors, as an output stage for electronic drivers for lasers even up to 20 or 40 Gbit/s. With such drivers having two outputs O_1 and O_2 in antiphase, there is a natural speculation as to whether two contacts on a single laser operated in a push–pull mode could create benefits for high-bit-rate laser sources.

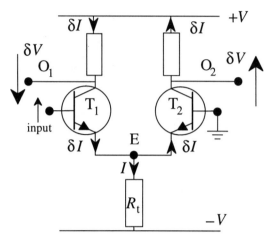

Figure 8.9 *Emitter-coupled amplifier giving push–pull output*

The output from T_1 which drives T_2 is the emitter E. Consequently, T_1 is driven as an emitter follower, while T_2 is driven as a grounded-base transistor. The (near) constant current I ensures that increment δI in the current flowing in T_1 is a decrement δI in the current flowing in T_2. Hence outputs O_1 and O_2 are in antiphase or push–pull. The configuration is also known as a 'long-tail pair'

8.3.2 Symmetrical push–pull DFB laser

The concept of push–pull laser operation is that modulation of the optical output is achieved by shifting light to and fro from facet to facet rather than increasing and reducing stored energy. Moving energy to achieve modulation is well known in the mode-locked Fabry–Perot laser where picosecond pulses of energy propagate up and down the laser, appearing at the facets with repetition rates in excess of 100 GHz [19,20] and with optical spectra close to the Fourier limits (so-called *bandwidth-limited spectra*). Bistable DFB lasers also can similarly give modulation at many tens of gigahertz by moving the energy back and forth from one facet to the other facet [21]. Unfortunately, neither mode locking nor bistability can give high-bit-rate pulses on demand and this is the contribution of push–pull modulation of split-contact lasers. It has been shown that the concept cannot work satisfactorily with conventional Fabry–Perot lasers because localised gain does not lead to localised optical intensity in these devices.* However, in a DFB laser the distributed feedback tends to localise the optical intensity to regions of optical gain and there is considerable potential [22,23], as discussed briefly here.

---

* YU, S.F.: Private communication, 1990.

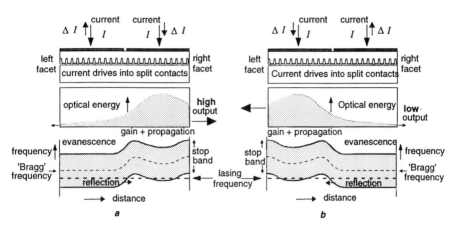

Figure 8.10 Stopband and energy/distance diagrams for push–pull DFB operation

a Right output high; b Right output low

Consider a symmetrical split-contact DFB laser as in Figure 8.10. The modulation of this laser is achieved by increasing I_R and simultaneously decreasing I_L so that the stored energy moves from the left to the right. The digital output changes from low (zero) to high (one). To change back to the low state, one decreases I_R and simultaneously increases I_L. From the symmetry, one can see that an output train of information Q at the right-hand facet is matched by a complementary output $\bar{Q}$ at the left-hand facet. Again, from the symmetry it follows that the steady frequency in the 'zero' state must be identical to that in the 'one' state. The dynamic chirp or change of frequency is limited to the transient region where the energy actually moves, and it will be seen shortly that there are ways of optimising this transient change.

To understand how the push–pull laser works without delving into deep mathematics, use can be made of the stopband/distance diagrams introduced in Section 5.6.2 and one can consider how these diagrams change as the modulation changes. Figure 8.10 shows the laser structure and the optical energy/distance and stopband/distance diagrams schematically. When the greatest current is driven into the right-hand contact as in Figure 8.10a, the gain is greatest in the right-hand section and light generated in the middle of this section propagates to the left and right but, because of the Bragg grating, the light evanesces on the left leaving very little light emerging from the left-hand facet. More light emerges from the nearer right-hand facet. The laser grating is designed so that the laser naturally prefers to lase in the -1 mode (the lower-frequency mode in the stopband diagram).

This selection of the mode can, in principle, be accomplished by having either an appropriate second-order grating [24] or an appropriate gain grating [25]. The preference for the -1 mode is further enhanced by the reflection from the left-hand section caused by the Bragg grating within that section being operated within its stopband (see the stopband diagram). The right-hand Bragg section provides adequate gain and reflection combined with phase shifting caused by the propagation to give a unity round-trip (complex) gain. The spatial-hole burning in the carrier density created by the optical energy within this half section will also aid the modal selection of the -1 lasing mode given, say, that $\kappa(L/2) \sim 2$.

Now switch the current in an appropriate mirror image as in Figure 8.10 and the steady-state condition is the same mirror image (Figure 8.10b), of the situation in (a). The right-hand Bragg section now ceases to provide enough gain for sustained lasing within that section. The stopband diagram readjusts to the changes of refractive index as the electron density is reduced. There is then a reflection from this right-hand Bragg section as there was previously from the left-hand section and this change of reflection actually helps the optical stored energy to move from the right-hand over to the left-hand section. The fact that the two states are mirror images must mean that they operate at the same lasing frequency. The chirp between the on and off states which is so evident in a conventional laser should therefore be significantly reduced except when there are switching transients.

Both modelling and experimental work indicate that it helps to have three sections rather than two, with a steady-current drive in the central section providing a source of stored photonic energy [26] which helps to reduce the chirp further. Typically, with such push–pull laser designs, one will find that the total length is over twice that of a conventional laser so that the end-modulated sections (of length L_s and operated in push–pull) each have a value of κL_s that is similar to the value of κL for the conventionally operated laser of length L. Figure 8.11 (after Flanigan [27]) shows the qualitative characteristic differences that are measured between a conventionally operated DFB laser and an unoptimised symmetrical push–pull-operated DFB laser highlighted by the amplitude-modulated waveforms and the localised frequency as a function of time. This latter is measured by an electronically tunable Fabry–Perot etalon which can be tuned to show the frequency for maximum energy through this etalon at any particular time during the output pulse [28]. The results for the conventionally modulated laser demonstrate the close link between

Future devices, modelling and systems analysis 269

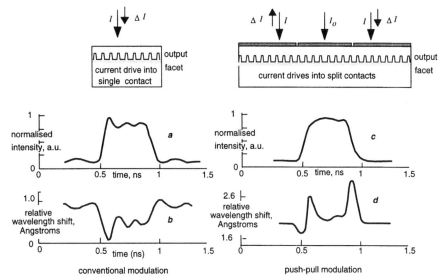

Figure 8.11 Output intensity and time-resolved chirp from commercial single-contact MQW DFB laser contrasted with unoptimised 3-contact push–pull bulk DFB laser

Pattern is 010 at 2.5 Gbit/s. The total bias level for single contact is 60 mA compared with 90 mA for push–pull and the modulation depth is 2 V_{p-p} (data after Flanigan [27])

the amplitude modulation (including the photon–electron resonance) and the dynamic chirp. The push–pull laser demonstrates that the photon–electron resonance has markedly changed the chirp with it now most significant only during the transients between the ones and zeros where it has the same direction at both turn-on and turn-off, in contrast to the chirp arising from conventional modulation.

8.3.3 Asymmetry and the push–pull DFB laser

The switching transients in the frequency output of the laser can be changed by changing the details of the drive to the push–pull laser. There are at least four ways of modifying the drive to a push–pull laser so as to influence the dynamic chirp and improve/change the laser's transmission performance over long distances. Figure 8.12 shows these schematically:

(i) modulation-depth asymmetry,
(ii) single-ended modulation which is a limiting case of modulation-depth asymmetry where only one contact (typically at the emitting end) is modulated,

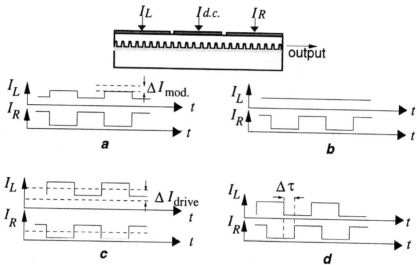

Figure 8.12 Schematic diagram illustrating various asymmetrical modulation schemes (after Flanigan [27])

a ΔI_{mod} negative-modulation-depth asymmetry *b* Single-ended modulation *c* ΔI_{drive} positive-DC-bias-level asymmetry *d* $\Delta \tau$ positive temporal asymmetry

(iii) DC bias level asymmetry, and
(iv) temporal asymmetry.

One could also consider building spatial asymmetry into the laser but this has not been investigated extensively.

The results on the time-resolved chirp for two such asymmetries are shown in Figure 8.13 which shows the effects of asymmetry in the modulation-current amplitudes. Note that a flat time-resolved chirp implies that the fundamental spectral spread in frequencies caused by Fourier's theorem is the only spread in frequencies that remains. The time-resolved chirp measures the excess spread in the frequency.

Experimental work has confirmed that the push–pull configuration for operating DFB lasers can achieve useful results. Pakulski has achieved a 0.8 dB penalty for symmetrical push–pull operating at 2.5 Gbit/s over a 370 km link using quantum-well devices [29]. When single-ended modulation was applied, the penalty could be reduced to 0.3 dB, indicating that asymmetry may help in reducing dispersion penalty. However, temporal asymmetry offers new possibilities and has yet to be tried at the time of writing. This is discussed briefly next.

Figure 8.14 considers temporal offset in an ideal symmetrical push–pull laser. The frequency changes at the leading and trailing edges of

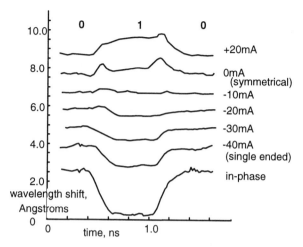

Figure 8.13 Effect of modulation-depth asymmetry *on the experimentally measured time-resolved chirp for a 010 1.5 Gbit/s pattern, using a push–pull modulated three-contact bulk-material device*

The fact that a slight negative asymmetry drive appears to give the 'best' results may indicate constructional asymmetries within the laser. The asymmetry is as defined in Figure 8.12 (after Flanigan [27])

the pulse have great significance when it comes to propagation along dispersive fibre. Notice how, in Figure 8.14, the pulses with +60 ps and −60 ps have the opposite changes in the wavelength shift at the leading and trailing pulse edges. The significance of these shifts is shown well in modelled 'eye' diagrams (Figure 8.15) for propagation of pseudorandom pulse trains over standard fibre. With positive temporal asymmetry, the frequency shifts in the leading- and trailing-edge frequencies partly compensate for the fibre's dispersion by tending to control the speed of the pulse edges of the ones, such that these edges do not move into the zeros. This keeps the ones well defined and the eye clearly open up to 400 km, and even reasonably open at 800 km with this 2.5 Gbit/s modulation, indicating that this form of push–pull modulation might compete with similar compensation schemes using external modulators [14]. With a negative temporal asymmetry, these frequency shifts at the trailing and leading edges of the pulse, combined with the fibre's dispersion, tend to move these edges more rapidly into the zeros and so close the eye more rapidly. Then at 400 km the eye is virtually closed. Zero temporal asymmetry is not shown, but is somewhere in between these two conditions.

272 *Distributed feedback semiconductor lasers*

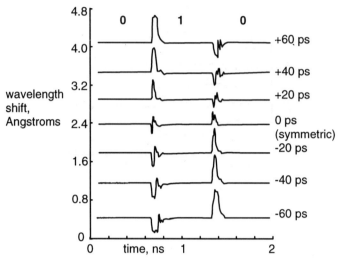

Figure 8.14 Modelled effect of temporal asymmetry on the time-resolved chirp for a three-contact MQW device push–pull modulated at 1.5 Gbit/s with a 010 pattern.

> The output was taken from the right facet, and a positive temporal asymmetry means that the modulation switch on the left contact leads the switch on the right contact. The DC bias levels on the end contacts were 22.5 mA, the centre contact bias was 50 mA. The modulation depth on the end contacts was 40 mA. The asymmetry is as defined in Figure 8.12 (after Flanigan [27]).

8.3.4 Speed of response for a push–pull DFB laser

Multigigabyte-per-second operation is required of directly modulated lasers for modern communication systems. While poor packaging can always limit the speed of response (Appendix 10), the internal physics of the laser creates the fundamental limits, for example the photon–electron resonance in a conventional laser even with negligible package parasitics. With the push–pull laser it is found that this resonance is not quite so important as the spectral separation between the -1 mode with its symmetrical fields and the -2 mode with its asymmetrical fields. Consider the split-contact laser *just at the point where switching occurs* with the drive approximately uniform. The field patterns within the laser have to be formed from a superposition of the modal patterns of fields created by modes within the laser (Figure 8.16). The -1 mode has a symmetrical envelope with distance along the laser and the frequency of the fields is f_{-1}. The -2 mode is asymmetrical and, relative to the -1 mode, beats at a frequency $f_{-1} - f_{-2}$, and so the net field pattern of these two modes will have

Future devices, modelling and systems analysis 273

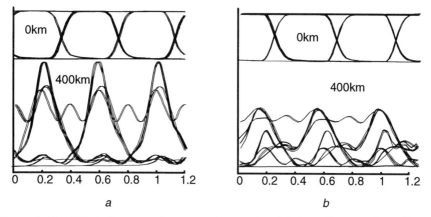

Figure 8.15 Effect of temporal asymmetry *on simulated eye diagrams for a three-contact push–pull-modulated MQW laser*

  a +60 ps modulation offset
  b −60 ps modulation offset (as defined in Figure 8.12)
  The bit rate is 2.5 Gbit/s. The steady bias is 22.5 mA on the end contacts and 50 mA on the centre contact. The modulation-depth current is ±40 mA on each end contact (after Flanigan [27]).

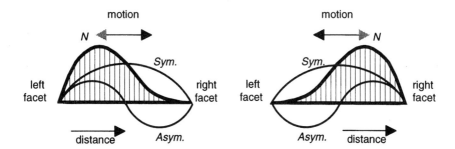

Figure 8.16 Beating of two modal patterns

  Symmetrical field envelope S contains fields at a frequency f_{-1} and asymmetrical field envelope A contains fields at a frequency f_{-2}. The net field pattern N then beats back and forth at a frequency $f_{-1} - f_{-2}$

energy concentrated either nearer the right-hand facet or nearer to the left-hand facet. The speed at which this energy can move from one facet to the other is therefore limited by $1/(f_{-1} - f_{-2})$, the reciprocal of this beat frequency. This does not mean that one excites the -2 mode into lasing, because that mode is always well below cutoff, but of course it means that, if one attempts to modulate the laser at

$f_M > (f_{-1} - f_{-2})$, the spread of frequencies extends to the -2 mode. Detailed numerical modelling confirms this result [30]. Although the modelling indicates the potential for modulation of these types of laser at 100 Gbit/s given optical material with sufficient gain, it is worth pointing out that, with standard fibre, dispersion rapidly closes the eye after a few kilometres even at 50 Gbit/s.

In conclusion, the push–pull laser is a device which requires extensive modelling to design and optimise. At the time of writing it is not clear that it will compete with the external modulator although it addresses the same issues of giving high-bit-rate modulation while helping to shape the spectra to compensate for the fibre's dispersion. A significant drawback comes when there is a long string of ones (or zeros) which gives a consequential change of the temperature between the two ends of the laser and so causes a small but troublesome temperature-induced change of frequency.

8.4 Tunable lasers with distributed feedback

8.4.1 Introduction

The emphasis in this book has been on single mode laser emitters with narrow and stable linewidths. However, future high-capacity optical networks need to employ tunable sources and detectors to increase capacity and flexibility [31], and for this purpose trial systems have been built with 80 or more channels in a 30 nm window centred around 1515 nm [32] using erbium-doped fibre amplifiers. Tunable lasers that are capable of wide tuning ranges are complex, expensive and have variable tuning ranges because of sensitivity to manufacturing tolerances. Sensing and measuring instruments can already accept the expense of currently available devices, so that high-volume communication applications should promote the improvement and cost reduction of tunable lasers.

In Chapter 2 it was pointed out that the gain spectra typically covered a frequency range $\Delta f \sim kT/h$, giving wavelength ranges to half-maximum gain of $\Delta \lambda \sim \lambda^2 kT/hc$, or in other words, to a first order, $\Delta \lambda \sim 50$ nm centred on 1500 nm, reducing to $\Delta \lambda \sim 15$ nm centred at 850 nm. These ranges can be extended to ~ 80 nm around 1550 nm central wavelength by pumping harder to increase the width of the gain spectrum and making use of low-loss material with adequate feedback. Figure 8.17 indicates schematically how quantum-well material, at high drive, can significantly increase the bandwidth for

Future devices, modelling and systems analysis 275

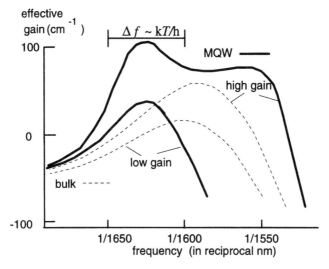

Figure 8.17 Schematic spectral widths of quantum-well and bulk materials

Greater spectral broadening with increasing drive is possible with quantum-well material than with bulk material because the reduced density of states for the lowest energy transitions forces the higher energy levels to be used—see Figure 8.18

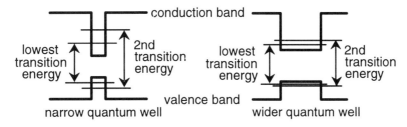

Figure 8.18 Gain-bandwidth broadening

The narrower the well, the greater the spectral separation between the first transition and the second transition. Narrow quantum wells favour a wide gain spectrum at sufficiently high drive levels. Careful design is required, though, in relation to saturation of the levels leading to gain saturation

gain compared with that of conventional bulk material. At low gain levels when only the lowest level is filled, the range of carrier energies is *lower* than in bulk material; but if the material is pumped strongly, the next quantised energy level begins to be filled and the energy range can be *greater* than in bulk material. Tuning ranges of over 110 nm have been reported using such quantum-well material [33].

Two points should be made about designing quantum-well structures for a *large* gain bandwidth. First, as shown schematically in Figure 8.18, deep narrow wells are more desirable than wide shallow ones because of the wider energy (frequency) separation between the first and subsequent transition energies. Secondly, gain saturation occurs at the wavelength corresponding to the first transition energy because filling of these quantum-well states is an essential mechanism of gain broadening. The design of lasers using this effect is therefore critical, with relatively restricted ranges of operation with respect to both temperature and output power.

Obtaining a material with a wide gain spectrum centred on the desired lasing frequency is only part of the problem of obtaining a tunable laser. The round-trip complex gain has to be unity at the lasing frequency; i.e. amplitude and phase from the feedback filter have to be matched independently at the desired lasing frequency, but not at any other frequency so as to avoid spurious or multiwavelength operation. An external grating (Section 1.4.2) has been one extremely successful method of achieving wideband tunability [34] and here the frequency of peak amplitude response is determined by rotating the grating, but to achieve a continuous tuning, a phase adjustment is also provided by means of slight changes to the grating–laser separation. However, the emphasis in this section is not on mechanically tunable or switched lasers but on monolithic electronically tuned DFB lasers where the principal methods of tuning are from changing the refractive index by:

(i) temperature (using, for example, electrical heating);
(ii) carrier density (via the plasma effect, see Appendix 7); and
(iii) electric field (via the Franz–Keldysh [35] or quantum-confined Stark effects [36,37]).

Temperature tuning [38,39] can, of course, be applied to any laser. The temperature coefficient of wavelength for InP-based single frequency lasers at 1300 nm and 1500 nm is about 0.09 nm/degC, and, given a temperature range of operation ΔT of at least 40 degC (e.g. $+10°C$ to $+50°C$), a tuning range of 3 nm or more is easily available. Appropriately designed DFB lasers are capable of operating at from -40 to $+80°C$, which leads to a tuning range of around 10 nm for a nominal 1500 nm laser. This approach is simple, predictable and reliable, but requires a Peltier heater/cooler to control the temperature, and the stabilisation to a new temperature then takes tens of milliseconds or more. At high temperatures the optical gain for a given carrier density is decreased, so the threshold current increases and, in

Future devices, modelling and systems analysis 277

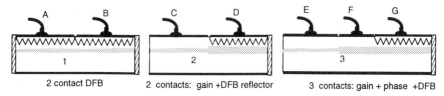

Figure 8.19 Some developments of multicontact DFB lasers

The narrow guide indicates material operating at the lasing wavelength while the wider guide indicates a guide made from material with a wider bandgap so as to avoid absorption losses

general, the power output and modulation efficiency decrease. At low temperatures the gain spectrum narrows (kT is decreased) and it becomes difficult to keep the correct mode, determined by the Bragg grating, aligned in frequency with the gain peak. A mode jump or multimode operation may well occur at some temperatures.

If the mean carrier density in a laser is increased, the mean refractive index *decreases* through the plasma effect and *shortens* the wavelength of the lasing modes; however, in most lasers, it is observed that the carrier densities are approximately clamped around some threshold value, and spatial variations generated by multiple contacts have to be employed to achieve significant tuning by this means: about 1% variation is possible. Smaller ranges than often expected are found in practice because the current, injected to increase the carrier density, will heat the laser and hence *lengthen* the wavelength, as noted in Section 2.6. Because one effect is electronic and the other thermal, they have very different associated time constants; this is observed in the FM response of DFB lasers [40] which changes sign at some modulation frequency in the megahertz range because at higher frequencies the thermal effects are too slow to follow the modulation.

8.4.2 Simple multicontact tunable lasers

When DFB lasers were introduced in the 1980s, it was immediately noted that a measure of electrically controlled tuning was desirable, and several laboratories subsequently demonstrated some tuning by simply taking a DFB laser and splitting its top contact into two or three sections. Since then a host of variations on this theme has been reported (e.g. [41–43]) and a brief summary is given below.

Figure 8.19 shows a series of increasingly complex tunable lasers incorporating a Bragg grating. The simplest design, laser 1, with only two contacts and a uniform grating, is not the easiest to understand.

When contacts A and B are driven identically, the device behaves like a conventional DFB laser. One way of making a tunable laser drives one contact (say A) alone until the laser reaches threshold and then, on increasing the current through contact B, tuning occurs. The important feature is that the asymmetry along the laser's length in the drive leads to an associated asymmetry in the effective linewidth-broadening factor α_H and also asymmetry in the gain saturation enabling the gain and phase tuning to be partly decoupled to different sections of the laser. In this device it is helpful to use quantum-well active regions so that, on the side which is close to gain saturation, the α_H factor is increased, thereby increasing the change of frequency for a given change in the carrier density. At 1500 nm, reductions in wavelength of 2–3 nm are possible by increasing the current in contact B before heating causes the refractive index, and thus the wavelength, to increase again. The device is easy to make but the tuning range is limited if relatively predictable.

Laser 2 may be regarded as a Fabry–Perot cavity (contact C) with feedback determined by one single-facet mirror, and one reflecting stack or Bragg grating (contact D) which is primary in determining the FP mode that is selected. The net round-trip gain from the reflector and the FP section must be appropriate but the important thing is that, *at approximately the lasing wavelength,* increasing current into D directly reduces the refractive index of the waveguide via the plasma effect, and hence reduces the wavelength of the peak reflection from the Bragg grating, thereby tuning the laser. To avoid the difficulties of interaction between gain and frequency in the Bragg section, this section can be fabricated in a wide-bandgap waveguide which neither absorbs nor gives gain to the lasing wavelength. Any spontaneous emission from this guide lies outside the frequencies of operational significance. In general with only two contacts, sufficient change in wavelength will lead to a jump to an adjacent Fabry–Perot mode ('mode hopping') so as to maintain the required round-trip phase matching.

This drawback of mode hopping is avoided in laser 3, where the Fabry–Perot section of laser 2 is now divided into two distinct regions. The region with contact F is a section of plain waveguide made from material with a bandgap which is too large to absorb or give gain at the lasing wavelength but can have its refractive index changed through the plasma effect and so provide the correct round-trip phase as the lasing frequency changes without creating a mode hop. The region with contact E provides the required round-trip gain (with some effect on the phase) as the Bragg reflector is tuned via the plasma effect as in laser 2.

Variations can be created, for example, by having the region, with contact F, formed from an intrinsic layer sandwiched within a P–N junction (forming a P–I–N junction). Then with a variable reverse bias with strong enough electric fields (>10 V/μm) the conduction and valence-band edges within the I-region shift so as to increase the bandgap energy. This is known as the Franz–Keldysh effect [35] in bulk materials, or the quantum-confined Stark effect [36,37] in quantum-well materials, and is primarily employed in electroabsorption modulators for high-speed low-chirp applications. The change in absorption also implies (from the Kramers–Kronig relationships) that there are changes in the refractive index which may be employed for tuning a laser. However, with those structures and voltages which are compatible with laser operation, the total refractive-index change available is typically less than 0.1% with a fractional tuning range of the same order. Four-section devices exist, similar to laser 3 but with tunable Bragg reflectors at both ends and phase- and gain-compensating regions.

8.4.3 Wide-tuning-range lasers with nonuniform gratings

Wider tuning ranges than those of Section 8.4.2 may be obtained in a relatively complex laser by use of interrupted [44,45] or superstructure [46–48] gratings. It is possible to show that, if one inserts periodic plain sections into an otherwise uniform grating then one can generate a comb of narrow reflections, with each reflection 'line width' $\Delta\lambda_r$ comparable with those from a single grating $\Delta\lambda_{stop}$, but with the comb covering a much wider range of frequencies $\Delta\lambda_{tune}$ (Figure 8.20). Spatial Fourier analysis of short grating 'patches' of length L_g gives the principal comb-envelope width $\Delta\lambda_{tune} \simeq \lambda_0 / N_g = \lambda_0^2 / 2L_g$ where N_g is the number of grating lines in each patch. The comb spacing is determined from $\Delta\lambda_{cmb} \simeq \lambda_0^2 / 2L_r$ where L_r is the periodicity of the interrupted grating structure. These approximate relationships may be derived from similar arguments of wave interference which were used for determining the fundamental reflection from a Bragg grating (Section 1.6). In order to obtain a wide comb with many lines, $L_g/L_r \ll 1$, but this fraction is also the factor by which the peak reflection per unit length (or effective-coupling coefficient) of the grating is reduced from its value when the grating is uniform. Consequently, long interrupted gratings are required to get adequately high reflectivity for laser operation but, as a recompense for

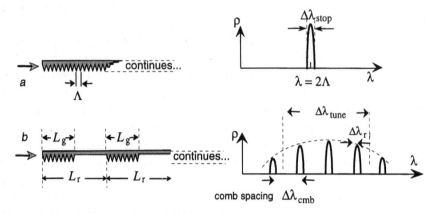

Figure 8.20   Reflections from modified gratings

the long length and low effective-coupling coefficient, narrow linewidths are achieved which can be well under 1 MHz [49].

Instead of the periodic interruption of the grating it is also possible to generate reflection combs from continuous *superstructure gratings*, with a periodically varying pitch. The interrupted gratings may be regarded as having a spatial periodic amplitude modulation of the reflections per unit length while a varying pitch is equivalent to a spatial periodic *phase* modulation of the reflections per unit length. Superstructures therefore have the advantage that the coupling coefficient remains higher than for interrupted gratings and consequently shorter reflectors are possible for the required feedback, thereby leading to less scattering loss and less free-carrier absorption in the waveguide when carriers are injected for tuning. The disadvantage is that superstructures are more difficult to fabricate than interrupted gratings.

These structures are used with *vernier tuning*, as illustrated in Figures 8.21 and 8.22. Here the device is a DBR type of laser with interrupted gratings with different pitches at either end. The central FP section of this device provides the gain. Adequate feedback will occur only when a line in *both* reflection combs aligns, as shown at λ_0 (Figure 8.21). Shifting the refractive index of grating 1 slightly will shift its whole reflection comb, and eventually lines at λ_{+1} will line up (or λ_{-1}) in the other direction. A small shift for the whole comb, created by the plasma effect and carrier injection, results in a larger shift for the lasing wavelength, with 'shift-multiplication ratio' $\approx \Delta\lambda_1/(\Delta\lambda_1 - \Delta\lambda_2)$. This multiplication ratio cannot be increased indefinitely in order to get larger tuning ranges, because of the finite comb linewidths $\Delta\lambda_r$.

Future devices, modelling and systems analysis 281

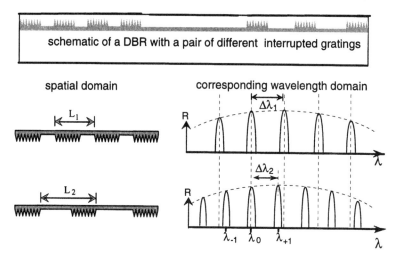

Figure 8.21   Pairs of periodic gratings and their reflection combs

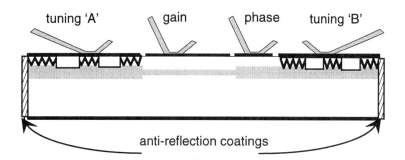

Figure 8.22   Four-section vernier-tuned laser

(i) Interrupted grating mirror A
(ii) Gain section
(iii) Phase-adjusting section
(iv) Interrupted grating mirror B

If $\Delta\lambda_1$ and $\Delta\lambda_2$ are too similar, there is insufficient discrimination between the adjacent comb lines and the laser becomes multimoded or has an inadequate sidemode-suppression ratio, especially near the extremes of the tuning range.

Figure 8.22 shows schematically a laser structure incorporating a pair of interrupted gratings. For the same reasons as in laser 3 (Figure 8.19) a phase-adjusting section has been added to the gain section to ensure phase matching at the desired wavelength and to reduce mode hopping. As with other tunable lasers, there are several currents that must be precisely controlled over the lifetime of the laser.

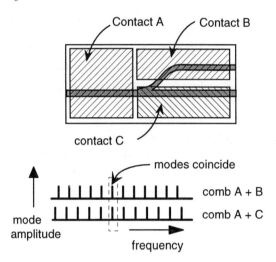

Figure 8.23 An asymmetric Y laser (schematic)

The two sets of Fabry–Perot modes formed by the two branches of arm A have slightly different frequency spacings for the modes. Only if both modes align in frequency will lasing occur, and then a small change in the refractive index in one arm permits a significant shift in frequency through the vernier effect [50,51]

8.4.4 Other tunable-laser structures

Another laser employing the vernier tuning action of Section 8.4.3 is the 'Y' laser [50,51]. These lasers contain a Y-branched waveguide so that two coupled sets of Fabry-Perot modes exist, and strong lasing occurs where the modes coincide, as indicated in Figure 8.23. They were introduced originally as single-frequency lasers for similar applications to DFB lasers; their construction is relatively easy using well tested technology of Fabry-Perot lasers. When properly adjusted, their performance is good, but multimode operation can occur at drive currents which are not exactly the correct design values, which in any event vary with temperature. However, in principle they have given the order of 50 nm tuning range by vernier action obtained by adjusting the relative currents into contacts A, B and C in Figure 8.23. The simplest concept of operation occurs when the waveguides under contacts B and C are transparent, so that their refractive indices may be changed by carrier injection without much gain change; however, operation is still possible even with gain in all the waveguides. Using the same arguments as in Section. 8.4.3, the combs shown in Figure 8.23 may have too little separation of frequency to allow sufficient frequency discrimination, and multimode operation would probably occur.

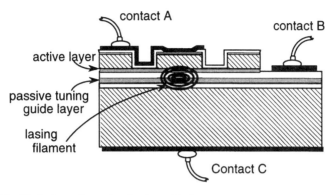

Figure 8.24 End-view schematic of twin-guide laser

With only two branches, modal discrimination can be a problem but excellent tuning has been demonstrated by extending the principle of operation by adding one (or more) further Y-junctions [52], so that at least three sets of modes have to coincide for lasing to occur, but this requires temperature control and four precisely controlled drive currents!

A related concept is the grating-assisted vertical-coupler filter (GAVCF) laser which gives a wide tuning range in principle by adjusting only one current [53,54]. Here a pair of waveguides with slightly different effective indices is 'phase matched' at one wavelength by a coarse grating (e.g. pitch 20 μm), so that power couples from one guide to the other. Small changes in the index of one guide cause much larger changes in the phase-matching wavelength. Unfortunately, such GAVCF devices have an insufficiently narrow passband to ensure single-mode operation so that a further auxiliary and adjustable filter must be added, again adding complexity to both manufacture and control.

A structure which gives a smaller tuning range, but is relatively easy to control, is the tunable twin-guide structure shown in Figure 8.24. Tuning ranges of over 10 nm in the 1500 nm band are possible [55], which is adequate for some applications. Referring to Figure 8.24, lasing gain is provided in the upper waveguide shown by injecting current from contact A to B, whereas *separately* the lower waveguide has its refractive index changed by carrier injection using a material composition so as to ensure transparency. Upper and lower guides are closely coupled so that the pair act as a single composite guide with adjustable index; problems with injection current and heating still exist, but the structure is well suited to optoelectronic integration [56].

In summary, while there are many candidates for the role of *the* tunable semiconductor laser, the really broadband devices are all difficult to characterise and require complex arrangements for the control of several drive currents. These features combined with the thermal limitations are a recurring problem of such tunable lasers and become particularly acute if the laser degrades with time. The control problem will need to be solved before such lasers become suitable for mass applications.

8.4.5 Tunable-laser linewidth and modulation

Tunable DFB lasers tend to use long cavities and low effective-coupling coefficients, which can result in very satisfactory linewidths of the order of 1 MHz for nominal 1500 nm lasers [57,58]. In cases where tuning and phase controls are not optimally adjusted or the vernier methods give inadequate gain and phase margins, the sidemode-suppression ratios are inadequate and multimode or noisy operation may occur. For critical systems using tunable lasers, it is essential to monitor the laser output for wavelength, spectral purity and sidemode suppression.

Direct *frequency* modulation, with moderate deviations, at constant amplitude is relatively straightforward for most monolithic tunable lasers by modulating the current (~1 GHz/mA is achievable) but direct *amplitude* modulation is more difficult. This is because systems require an amplitude-modulation depth of 10:1 or greater but, to achieve this modulation depth, large and rapid changes in the gain are required which also change the phase characteristics of any gain section. Changes of frequency even as low as 1 part in 10^4 give unacceptable dispersion penalties in the transmission of gigabyte-per-second information over 100 km fibre systems. For exacting applications, the answer is to use an external modulator along with an optical isolator to prevent changing feedback into the laser. For less critical applications, it may be that some hybrid of a two-section tunable DFB laser and a two-section push–pull laser could achieve an adequate result.

8.4.6 Modelling tunable semiconductor lasers

In principle, the time-domain model described in detail in Chapter 7 will work as well for tunable semiconductor lasers as for the fixed-wavelength grating-based lasers on which it was developed. However, there are several extra problems:

(i) Lasers with interrupted gratings need many sections to characterise them fully, so that well over 100 sections are likely to be required (at least one section for each 'plain' section, and one section for each grating section throughout the reflectors). Probably the easiest way to deal with this is to take the section lengths in the reflectors to the nearest integral number of distance/time steps, and then add a phase-shifting function to every section. This will result in a relatively complex program which will be outside the array-handling capabilities of the Student edition of MATLAB used in this book.

(ii) Thermal effects are frequently important in tunable lasers. The time-domain model is not at present suited to simulation of thermal transients lasting milliseconds or more, since with the number of sections and time steps the computer running time could become excessive. In some cases, this can be circumvented by precalculating the thermally dependent parameters with other modelling techniques, and then inserting these parameters into the normal time-domain model at accelerated time intervals, but care is needed when using this technique.

If only static tunable-laser performance is required, the characteristics of complex-interrupted-grating reflectors can be calculated by any of the usual methods employed for multilayer dielectric filters. Laser performance can then be calculated using a transfer-matrix or power-matrix model.

8.4.7 Multiple DFB lasers with optical couplers for WDM

The strength of the tunable laser lies in being able to tune to many wavelengths, but this also leads to weaknesses for some applications: the careful balance of control currents required to set the appropriate wavelengths can be upset by aging or temperature changes; the low threshold discrimination that is often necessary to obtain the widest tuning ranges means that dynamic increases in gain can excite unwanted modes; and feedback from external sources with spurious wavelengths may cause drastic changes of output by exciting near-threshold modes. An alternative for WDM systems is to have an array of fixed- but different-wavelength lasers which are switched in or out as required.

Electron-beam lithography, for example, may be used to write a set of gratings side-by-side but each with slightly different pitches, forming an array (~eight devices) of edge-emitting lasers on a single chip with predetermined wavelength spacing (~1–2 nm) [59–61]. The outputs from the different spatial positions are then combined externally, or integrated on the chip. For applications where eight or so wavelengths are required, such a device is attractive, because wavelength control is straightforward, spurious outputs are unlikely, and it is also possible to drive and modulate two or more lasers at different wavelengths simultaneously and independently for WDM applications.

The straightforward and popular approach is to combine the wavelengths in a 1:N star coupler or some integrated equivalent but such a device without frequency selectivity necessarily has a 1/N power loss which has to be made up by amplification either before or after the coupler. Amplification before the coupler may be preferred because, unless the amplification is strictly linear after the coupler, the nonlinearity or saturation can cause too much crosstalk between the channels. Frequency-selective couplers need not have the same loss as star couplers but are challenging to make typically using integrated two-dimensional optics to permit precise dimensioning of guides, couplers and gratings and with present technology such integrated eight-way couplers can still have around 7 dB of loss [62]. There is room for novel work in this area [63, 64], but it will be difficult to beat the straightforward methods.

Optical interference and crosstalk between such lasers need not be a problem, even for less than 2 nm wavelength spacing [65], while thermal crosstalk [66] between adjacent lasers will probably be measurable, but not serious in most applications. For applications where several tens or even hundreds of wavelengths are required, the fully tunable laser remains the more attractive option [67] and the compact nature of such semiconductor lasers contrasts well with other types of tunable laser [68].

8.5 Surface-emitting lasers

8.5.1 Introduction to surface-emitting lasers

The DFB structures so far discussed emit from their ends (edge emitters) rather than from their top or bottom surfaces, but there is a steadily increasing motivation to design lasers which emit from the chip's surface [69–71] for the following reasons:

(a) Distributing the optical output over a larger area of surface can reduce the peak power density at the emission facets and so facilitate yet higher-power lasers without the present difficulties of facet damage. With appropriate designs, one can also envisage improved heat sinking for high-power lasers leading to yet higher power outputs.

(b) The divergence of a laser beam is inversely proportional to the emission area (solid angle of emission $\sim \Omega = 4\pi \; \lambda_o^2/\mathscr{A}$) so that an increased emission area $\mathscr{A}$ can give narrow beams which are useful for applications such as free-space optical communication. Appropriate circular far-field patterns from surface emitters, as compared with asymmetric patterns from edge emitters, should also increase the coupling efficiency to circular optical fibres.

(c) Surface emitters can readily be arranged into two-dimensional (N × M) arrays (Figure 1.23) which could be useful for optical data switching and processing, whereas edge emitters can easily form only linear (1 × N) arrays.

Figure 8.25 shows two main different approaches to forming surface emitters. The first approach (Figure 8.25a and b) uses one form or another of edge emitter to feed either a second-order grating (Sections 1.6.1 and 4.4.3) or a 45° mirror so as to give surface emission [72–76]. High-power lasers have successfully been made by these techniques [77] and the angle of emission can also be steered over a limited range [78]. The second and very different approach is the vertical-cavity surface-emitting laser (VCSEL) [79,80,81], where the lasing filament is perpendicular to both the chip surface and the epitaxially grown layers of the structure.

VCSELs have become an exceedingly important new development, well deserving a research monograph [82] which contains the appropriate historical perspective and far more technical detail than can be accomplished here. The work here is limited to introductory material highlighting some of the significant challenges for the future to make VCSELs fulfil their potential.

8.5.2 Operating parameters of VCSELs compared with edge emitters

It is instructive to begin by comparing a conventional edge-emitting Fabry–Perot laser (having a cavity length $L_{edge} \sim 240$ μm and facet power reflectivity $R_{edge} \sim 30\%$) with a VCSEL (cavity length $L_{vcsel} \sim 1$ μm and facet power reflectivities R_{vcsel}, to be determined) as shown in Figure 8.26.

Assume to start with that the net material field gains per unit length (after scattering losses are subtracted) are g_{edge} and g_{vcsel} for the

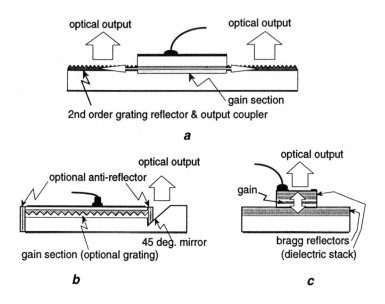

Figure 8.25 Grating-coupled DBR, mirror-coupled DFB and VCSEL

conventional laser and VCSEL, respectively, with confinement factors of $\Gamma_{edge} \sim 0.5$ and $\Gamma_{vcsel} \sim 1$. The lasing power-balance equations are

$$R_{edge}^2 \exp(4\, \Gamma_{edge}\, g_{edge}\, L_{edge}) = 1 \qquad R_{vcsel}^2 \exp(4\, \Gamma_{vcsel}\, g_{vcsel}\, L_{vcsel}) = 1 \qquad (8.1)$$

or, on rearranging to give the power gain explicitly,

$$2\, g_{edge} = (1/\, \Gamma_{edge}\, L_{edge})\, \ln\,(1/R_{edge}) \sim 100\ \text{cm}^{-1}$$

$$2\, g_{vcsel} = 10\,000\, \ln\,(1/R_{vcsel})\ \text{cm}^{-1} \qquad (8.2)$$

If VCSELs are to operate at high temperatures or need the long operating lives of conventional lasers, the gain for lasing and the resulting carrier density in the active region must not be significantly higher than in conventional edge-emitting lasers. For comparable power gains around 100 cm^{-1} (i.e. field gains of 50 cm^{-1}), it can be seen that the vertical-cavity surface emitter requires both top and bottom power reflectivities to be greater than 99%. However, these values cannot readily be achieved by metal reflectors or single dielectric interfaces but require Bragg reflectors formed from multiple $\lambda/4$ layers of high- and low-index semiconductor or vacuum-deposited dielectric stacks whose design and fabrication needs discussion.

Given a value of 99% for reflectivity of a Bragg grating, Figure 8.27 gives a rough estimate of the minimum number of quarter-wavelength

Future devices, modelling and systems analysis 289

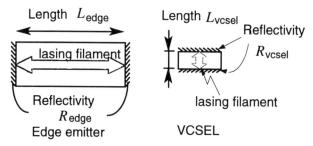

Figure 8.26   Schematic edge emitter and VCSEL

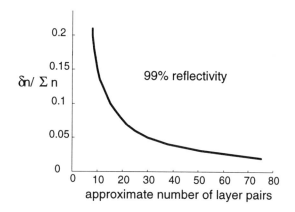

Figure 8.27   *Approximate number of layer pairs*

   Required to achieve 99% power reflectivity as a function of $\delta n/\Sigma n = (n_1 - n_2)/(n_1 + n_2)$. The analysis here does not account for significantly different refractive index of any terminating layer, nor for optical loss

layer pairs that will be required to achieve this reflectivity, assuming that there is a change in refractive index between each layer in any pair of $\delta n = (n_1 - n_2)$ with $(n_1 + n_2) = \Sigma n$. The calculation (see eqns. 5.35–5.37) neglects any differences in the refractive indices of the incident and final layers and assumes no optical loss. Optical loss limits the reflectivity, no matter how many layer pairs are used.

Figure 8.27 shows that when the adjacent layers have a normalised step in the refractive index $(\delta n/\Sigma n)$ falling below, say, 0.05, then 30–40 layer pairs are required to obtain adequate reflectivity, and this is a relevant problem at present for materials such as InGaAsP/InP forming appropriate combinations for emission at 1.3 or 1.5 μm communication wavelengths. The large number of layers of semiconductor material gives optical losses, further significantly limiting the available reflectivity. By contrast, Bragg stacks formed from high-index-step dielectrics such as amorphous silicon and silicon dioxide where $n_{\alpha\text{-}Si} \sim 3.4$ and $n_{SiO_2} \sim 1.9$ will give adequate reflectivity with around five or so pairs and also have sufficiently low optical absorption so that the reflectivity is not affected at this level. Bragg coupling coefficients range from $\kappa \sim 2000$ cm^{-1} for the dielectric stacks to $\kappa \sim 200$ cm^{-1} for the semiconductor stacks, and from such values one estimates the associated stopband, or spectral width of high reflection, to ranges from tens of nanometres for the dielectric stacks to a few nanometres for the semiconductor stacks, so that the material gain peak and the reflection peak have to be aligned more carefully than in edge emitters because of this need for high gain.

It might be thought that VCSELs would have a much faster optical-response time to direct-current modulation than edge emitters, simply because of their short length and low volume. However the high reflectivity of the facets means that the photon lifetimes are, in fact, comparable with those of edge emitters and more careful discussion is required. The small geometry of VCSELs makes them almost free from longitudinal spatial-hole burning and consequently (provided that the radial dimensions are small enough not to permit strong radial-hole burning) the dynamic behaviour can often be adequately dealt with by the standard rate equations of Section 4.1 provided that appropriate changes are made. The first such change notes that the dielectric mirrors store significant optical energy without any optical gain so that now a longitudinal confinement factor (of order 0.2) has to be introduced in addition to the transverse confinement factor, but this latter is normally high in VCSELs. The net confinement factor is then comparable with that for an edge emitter. A second notable change has to be made in the spontaneous-coupling coefficient β_{sp} which can be significantly larger ($>10^{-2}$) than for edge emitters [83–86] because the short, low volume cavity increases the lasing linewidth and the high reflectivity mirrors reduce the spontaneous linewidth compared with an edge emitter. Novel microcavity VCSEL structures are candidates for increasing β_{sp} and reducing the threshold current towards the so-called 'thresholdless laser' mentioned in Section 4.2.2. A third important change is that Auger recombination needs greater emphasis in the modelling. Auger recombination is already a significant factor in determining the threshold currents in edge-emitting long-wavelength lasers, and may well be even more serious with VCSELs unless these lasers can operate with lower threshold-current densities. The strong temperature dependence of the nonradiative Auger recombination will also be of concern.

Making allowance for such changes, analysis and experiment does confirm that there is an overall increase in the speed of response [87] for the following reasons:

(i) VCSELS tend to use high-gain material, at high carrier densities ($\sim 5 \times 10^{18}$ cm^{-3}), which then reduces the stimulated and spontaneous carrier lifetime, though it restricts the maximum temperature of operation;
(ii) the small dimensions of VCSELs give small electrical capacitances which are easier to charge and discharge; and
(iii) the resonance frequency increases as $\sqrt{(I-I_{th})}$ (Appendix 4) and the low threshold currents in VCSELs means that one can drive the laser at, say, 10 times threshold whereas edge emitters are rarely driven at more than three times threshold.

If one uses time-domain modelling for VCSELs, some changes from the techniques used for edge-emitting DFB/DBR lasers are advised. The significant refractive-index steps in the Bragg reflectors, combined with their length, suggests that the appropriate space step for the computation is that of a single layer or perhaps layer pair within the stack compared with the tens of grating periods used previously. Using such small step lengths gives excellent results for computing the reflection spectra of the stack mirrors, but also requires correspondingly smaller time steps $\sim 10^{-15}$ s, rather than $\sim 10^{-13}$ s for conventional DFBs.

8.5.3 Construction of VCSELs

VCSELs are fabricated using virtually identical technologies to those developed for edge emitters but one critical process (facet cleaving) is replaced by a different critical process: reflector stack fabrication. Figure 8.28 indicates some varieties. Testing is potentially much easier than with edge emitters since it may be done without dicing chips from their wafer; this is a very considerable advantage.

Formation of dielectric-mirror stacks by vacuum evaporation of materials such as SiO_2, Mg_2F, TiO_2 etc. is a well established and reliable technology. *In-situ* monitoring of reflectivity during deposition is often employed, so that drifts in deposition rates, optical densities etc. can

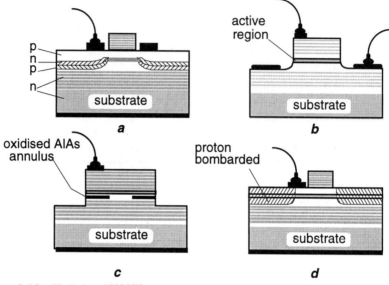

Figure 8.28 Varieties of VCSEL

All devices have circular emitting facets the order of 5–10 μm diameter
(a) Deposited top mirror, bottom grown, buried-heterostructure current confinement
(b) Grown mirrors, mesa current confinement
(c) Grown mirrors, large mesa with oxidised current confining annulus
(d) Deposited top mirror, grown bottom, proton-bombarded current confinement

be compensated, and high accuracy for the spectral characteristics of the mirror is assured. Good lasers may be fabricated [88], but injecting the current into the VCSEL junction with insulating mirrors remains a problem, with solutions such as annular contacts being employed (Figure 8.28c).

Conductive stacks grown from alternating layers of high- and low-index semiconductor [89] (all doped n-type or all p-type) make for a relatively simple laser structure (as in Figure 8.28b). However, there are associated difficulties caused by the stack of heterojunctions where, without careful design [90,91], the heterobarriers at the many interfaces inhibit conduction, with lasers needing 10 V of drive or more to achieve threshold. This problem can be avoided by doping the whole mirror stack very heavily, but high optical losses can then result. An elegant solution is either to grade the interfaces, or to employ 'delta doping', i.e. heavy doping in very thin layers at the interfaces [92, 93].

Although metal organic vapour-phase epitaxial (MOVPE) growth is a preferred growth technique for large-scale production, the stringent material tolerances required for VCSEL devices lead to the preference for growing VCSEL mirrors by molecular-beam-epitaxy (MBE) where *in-situ* monitoring of the growth process and optical reflectivity is easier [94,95], but unfortunately the difference in volatility of In and P makes MBE less easy to use for InP-based devices. The detailed choice of method(s) depends on the material and device design [96–100], but the difficulties of achieving the right compromises have made it easier to fabricate devices, with the GaAlAs/GaAs and GaInAs/GaAs systems giving outputs around 0.8 µm and 1.0 µm, respectively, where the step-index variations are sufficient to permit about 25–30 layer pairs to be used. Significantly more layers are usually required for InGaAsP/InP devices centred around the 1300 nm and 1550 nm bands. Although a higher step in the refractive index is possible at 1550 nm than at 1300 nm, the maximum gain tends to be lower and low-loss mirrors are a major difficulty at 1550 nm. Sophisticated techniques may provide a solution such as diffusion bonding a thin 'film' of VCSEL-active regions onto a substrate of an entirely different material in which low-loss mirrors can be formed. Even if a layer of dislocations is formed at such a bond, it need not have over-serious consequences on the operating lifetime of the laser provided that the dislocation layer and active region are sufficiently far apart. Two-dimensional arrays of VCSELs fabricated on the top of standard silicon integrated circuits would be especially attractive for optical signal processing, combining the best of Si VLSI and VCSEL technology. Progress towards these goals is being made using interfacial matching layers or diffusion bonding [101–104].

Because of the short length of the VCSEL, it is essential to have high optical gain which is best obtained by using quantum-well material. However, the position of any quantum wells is now more critical than in an edge-emitting laser. The precision required is illustrated by Figure 8.29, where it can be seen that the active layer (quantum well) now lies parallel to the wave fronts in the laser, unlike the edge-emitting laser. This gain region then should ideally be situated at a where the optical standing wave, at the desired longitudinal modal frequency, provides the strongest electric field. If the layer is at the node b, then only a weak interaction can occur, giving an excessive threshold current. Multiple wells should ideally be close to an appropriate antinode for maximum gain. The deposition of the layers then has to be significantly more accurate than $\lambda_m/4$—the quarter-wavelength inside the material.

294   *Distributed feedback semiconductor lasers*

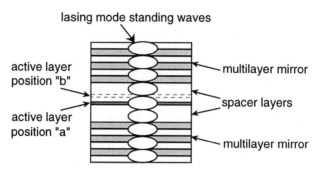

Figure 8.29   Active-layer position in relation to the standing-electrical-field wave

8.5.4 *Additional features of VCSELs*

The slab-waveguide model outlined in Chapter 3 is inadequate for the circular VCSEL devices other than to give an indication that the lateral dimensions are likely to permit more than one lateral mode, even though the cavity dimensions and selective mirrors ensure only one longitudinal mode. If a number of lateral modes is present, the far-field patterns are complicated [105]. The diameter of the active core ranges from 5 to 50 μm with the result that waveguiding for the principal mode is weak and guiding is susceptible to perturbations: the drive current can heat regions leading to thermal waveguiding; radial spatial-hole burning in the junction plane can depress the carrier density along the axis of the device. All this can cause significant changes in the mode guiding, especially if the laser is driven well above its threshold level [106–108] and provides the future challenges.

The polarisation is not usually well determined in VCSELs, which is a serious drawback for long-haul optical-communication systems [109]. Slight radial asymmetries can provide positive discrimination for polarisation; for example anisotropic strain generated by growing on misoriented substrates has had some success [110,111].

8.6 Summary

Four areas have been covered in this final chapter. The first has been systems modelling where advanced numerical models can predict the performance of assemblages of devices all of which are modelled separately but can be brought together. It is believed that, as device modelling improves in accuracy, this area will grow extensively and it will reduce the costs of trying out new systems and speed up their design. The second area discussed is a specialist area where numerical modelling has led to a new device: the push–pull laser. Even if this device does not become a production device, it highlights many mechanisms of importance within DFB lasers and illustrates how numerical modelling can aid invention.

The third and fourth areas are of growing importance. Tunable lasers or switchable laser systems will give WDM systems the versatility and flexibility they require. Which particular device or technique will win and become the industry standard for a WDM source is, at the time of writing, an open question. The final area of discussion is that of surface-emitting lasers and particularly VCSELs where the potential for growth in data processing and communications is vast. However, one recognises that there are serious challenges for designing and fabricating 1.55 μm wavelength VCSELs where the desirable materials at present give too small a refractive-index step and the nonradiative Auger recombination can inhibit lasing unless low enough current densities are achieved. It is possible that full and detailed three-dimensional dynamic models of VCSELs may be helpful in their optimisation, which may well be more critical than in edge emitters. This shows up future modelling requirements, where techniques of finite elements, for example, combined with the increasing power in personal computers, will make such computation and design as easy as the current time-domain models for edge-emitting DFBs.

8.7 References

1 LOWERY, A.J.: 'Computer-aided photonics design', *IEEE Spectrum*, 1997, **34**, pp. 26–31
2 LOWERY, A. J.: 'Semiconductor device and lightwave systems performance modelling'. Technical digest of *Optical Fiber Communication, 97 (OFC'97)*, OSA Technical Digest Series, 1997, **6**, paper ThG6, pp. 267–268
3 LOWERY, A.J.: 'Transmission line modelling of semiconductor lasers: the transmission line laser model', *Int. J. Numer. Model.*, 1989, **2**, pp. 249–265
4 LOWERY, A.J.: 'Transmission-line laser modelling of semiconductor laser amplified optical communications systems', *IEE Proc. J*, 1992, **139**, pp. 180–188
5 LOWERY, A.J., GURNEY, P.C.R., WANG, X–H., NGUYEN, L.V.T., CHAN, Y.C., and PREMARATNE, M.: 'Time–domain simulation of photonic devices, circuits and systems', Proceedings of *Physics and Simulation of Optoelectronic Devices*, at SPIE Photonics West, 1996, **2693**, pp. 624–635
6 YOUNG, T.P.: 'CAD tools for optoelectronic subsystems', *GEC J. Res.*, 1994, **11**, pp. 110–121

7 SILVESTER, P.P., and FERRARI, R.L.: 'Finite elements for electrical engineers' (Cambridge University Press, 1966), 3rd edn.
8 PAOLI, T., AND RIPPER, J.E.: 'Direct modulation of semiconductor lasers', *Proc. IEE*, 1970, **58**, pp. 1457–1465
9 OJALA, P., PETTERSSON, C., STOLTZ, B., MORNER, A.C., JANSON, M., and SAHLEN, O.: 'DFB laser monolithically integrated with an absorption modulator with low residual reflectance and small chirp', *Electron. Lett.*, 1993, **29**, pp. 859–860
10 HASHIMOTO, J.I., NAKANO, Y., and TADA, K.: 'Influence of facet reflection on the performance of a DFB laser integrated with an optical amplifier/modulator', *IEEE J. Quantum Electron.*, 1992, **28**, pp. 594–603
11 ZHANG, L.M., and CARROLL, J.E.: 'Semiconductor 1.55 micron laser source with gigabit/second integrated electroabsorptive modulator', *IEEE J. Quantum Electron.*, 1994, **30**, pp. 2573–2577
12 MORITO, K., SAHARA, R., SATO, K., KOTAKI, Y., and SODA, H.: 'High power modulator integrated DFB laser incorporating strain compressed MQW and graded SCH modulator for 10 Gb/s transmission', *Electron. Lett.*, 1995, **31**, pp. 975–976
13 BINSMA, J.J.M., KUINDERSMA, P.I., VANGESTEL, P., VANDERHOFSTAD, G.L.A., CUIJPERS, G.P.J.M., PEETERS, R., VANDONGEN, T., and THIJS, P.J.A.: 'DFB lasers with integrated electroabsorption modulator', *Opt. Quantum Electron.*, 1996, **28**, pp. 455–462
14 GNAUCK, A.H., KOROTKY, S.K., VESELKA, J.J., NAGEL, J., KREMMERER, C.T., MINFORD, K.W., and MOSER, D.T.: 'Dispersion penalty reduction using an optical modulator with adjustable chirp', *IEEE Photonics Technol. Lett.*, 1991, **3**, pp. 916–918
15 REICHMANN, K.C., MAGILL, P., KOREN, U., MILLER, B.I., YOUNG, M., NEWKIRK, M., and CHEN, M.D.: '2.5 Gb/s transmission over 674 km at multiple wavelengths using a tunable DBR laser with an integrated electroabsorption modulator,' *IEEE Photonics Technol. Lett.*, 1993, 5, pp. 1098–1100
16 PARK, Y.K., NGUYEN, T.V., MORTON, P.A., JOHNSON, J.E., MIZUHURA, O., JEONG, J., TZENG, L.D. et al.: 'Dispersion penalty free transmission over 130 km standard fiber using a 1.55 mm 10 Gb/s integrated EA/DFB laser with low extinction ratio and negative chirp,' *IEEE Photonics Technol. Lett.*, 1995, **8**, pp. 1255–1257
17 D'OTTAVI, A., MARTELLI, F., SPANO, P., MECOZZI, A., SCOTTI, S., DALL'ARA, R., ECKNER, J., and GUEKOS, G.: 'Very high efficiency four-wave mixing in a single semiconductor traveling wave amplifier', *Appl. Phys. Lett.* 1996, **68**, pp. 2186–2188
18 D'OTTAVI, A., IANNONE, E., MECOZZI, A., SCOTTI, S., SPANO, P., DALL'ARA, R., GUEKOS, G., and ECKNER, J.: 'Ultrafast gain and refractive index dynamics of semiconductor amplifiers measured by four-wave mixing', Conference Digest, *14th IEEE International Semiconductor Laser Conference*, 1994, paper T2.3, pp. 63–64
19 CHENG, Y.-K., and WU, M.: 'Monolithic colliding–pulse mode–locked quantum well lasers', *IEEE J. Quantum Electron.*, 1992, **28**, pp. 2176–2185

20 MARTINS-FILHO, J.F., and IRONSIDE, C.N.: 'Multiple colliding-pulse mode-locked operation of a semiconductor laser', *Appl. Phys. Lett,* 1994, **65,** pp. 1894-1896
21 MARCENAC, D.D.: 'Comparison of self-pulsation mechanisms in DFB lasers', Presented at *IEEE Lasers and Electro-Optics Society 7th Annual meeting,* Baltimore 1994, Paper SL8.5
22 NOWELL, M.C., ZHANG, L.M., CARROLL, J.E., and FICE, M.J.: 'Chirp reduction using push-pull modulation of three-contact lasers', *IEEE Photonics Technol. Lett.,* 1993, **5,** pp. 1368-1371
23 NOWELL, M.C., CARROLL, J.E., PLUMB, R.G.S., MARCENAC, D.D., ROBERTSON, M.J., WICKES, H., and ZHANG, L.M.: 'Low chirp and enhanced resonant frequency by direct push-pull modulation of DFB lasers', *IEEE J. Sel. Topics Quantum Electron.,* 1995, **1,** pp. 433-441
24 KAZARINOV, R.F., and HENRY, C.H.: 'Second order distributed feedback lasers with mode selection provided by first order radiation losses', *IEEE J. Quantum Electron.,* 1985, **21,** pp. 144-150
25 MAKINO, T.: 'Mechanisms for high single mode stability of gain-coupled DFB lasers with periodically etched quantum wells', *Electron. Lett.,* 1995, **31,** pp. 1579-1580
26 GRIFFIN, P.S., WILLIAMS, K.A., WHITE, I.H., and FICE, M.J.: 'Substantial chirp reduction of dynamic wavelength chirp of 2 Gbit/s pseudorandom optical signal using 3-contact InGaAsP/InP multi-quantum well distributed feedback laser.' *Electron. Lett.,* 1992, **28,** pp. 2045-2046
27 FLANIGAN, B.J.: 'Advances in push-pull modulation of lasers', PhD dissertation, University of Cambridge, 1996
28 OLSEN, C.M., and OLESEN, H.: 'Time-resolving wavelength chirp with Fabry-Perot etalons and gratings: a theoretical approach,' *J. Lightwave Technol.,* 1991, **9,** pp. 436-441
29 PAKULSKI, G., BLAAUW, C., CLEROUX, M., and KIM, H.B.: 'Adjustable chirp in three-electrode directly modulated distributed feedback lasers'. Proceedings of *IEEE Lasers and Electro-Optics Society 8th annual meeting,* San Francisco, 1995, **1,** paper SCL 5.2
30 MARCENAC, D.D., NOWELL, M.C., and CARROLL, J.E.: 'Theory of enhanced amplitude modulation bandwidth in push-pull modulated DFB lasers,' *IEEE Photonics Technol. Lett.,* 1994, **30,** pp. 2064-2072
31 HAWKER, I., TANDON, V., COTTER, D., and HILL, A.: 'New network infrastructures for the 21st century. Part 1: Ultra high capacity wavelength routed networks', *Br. Telecomm. Eng.,* 1995, **13,** pp. 103-111
32 YOUNG, M.G.: 'InP-based components for wavelength division multiplexing', *Proc. SPIE,* 1994, **2153,** pp. 251-258
33 ISHII, H., TOHMORI, Y., YAMAMOTO, M., TAMAMURA, T., and YOSHIKUNI, Y.: 'Modified multiple phase shift superstructure grating DBR lasers for broad wavelength tuning', *Electron. Lett.,* 1994, **30,** pp. 1146-1147
34 TABUCHI, H., and ISHIKAWA, H.: 'External grating tunable MQW laser with wide tuning range of 240 nm', *Electron. Lett.,* 1990, **26,** pp. 742-743

35 PANKOVE, J.I.: 'Optical processes in semiconductors' (Dover, New York, 1971)
36 MILLER, D.A.B., CHEMLA, D.S., DAMEN, T.C., GOSSARD, A.C., WIEGMANN, W., WOOD, T.H., and BURRUS, C.A.: 'Electric field dependence of optical absorption near the band gap of quantum well structures', *Phys. Rev. B*, 1985, **32**, pp. 1043–1060
37 CAI, B., SEEDS, A.J., and ROBERTS, J.S.: 'MQW tuned semiconductor lasers with uniform frequency response', *IEEE Photonics Technol. Lett.*, 1994, **6**, pp. 496–498
38 WOODWARD, S.L., KOREN, U., MILLER, B.I., YOUNG, M.G., NEWKIRK, M.A., and BURRUS, C.A.: 'A DBR laser tunable by resistive heating', *IEEE Photonics Technol. Lett.*, 1992, **4**, pp. 1330–1332
39 DUTTA, N.K., HOBSON, W.S., LOPATA, J., and ZYDZIK, G.: 'Tunable InGaAs/GaAs/InGaP laser', *Appl. Phys. Lett.*, 1997, **70**, pp. 1219–1220
40 PANDIAN, G.S., and DILWALI, S.: 'On the thermal FM response of a semiconductor-laser diode', *IEEE Photonics Technol. Lett.*, 1992, **4**, pp. 130–133
41 GRIFFEL, G., and CHEN, C.H.: 'Static and dynamic analysis of tunable 2-section high-speed distributed-feedback laser utilizing the gain lever effect', *IEEE J. Quantum Electron.*, 1996, **32**, pp. 61–68
42 TOHYAMA, M., ONOMURA, M., FUNEMIZU, M., and SUZUKI, N.: 'Wavelength tuning mechanism in three electrode DFB lasers', *IEEE Photonics Technol. Lett.*, 1993, **5**, pp. 616–618
43 HILLMER, H., ZHU, H.L., and BURKHARD, H.: 'Enhanced tunability of asymmetric three section InGaAsP/InP distributed feedback lasers', *J. Appl. Phys.*, 1993, **73**, pp. 1035–1038
44 JAYARAMAN, V., HEINBUCH, M.E., COLDREN, L.A., and DENBAARS, S.P.: 'Widely tunable continuous wave InGaAsP/InP sampled grating lasers', *Electron. Lett.*, 1994, **30**, pp. 1492–1494
45 OUGIER, C., TALNEAU, A., DELORME, F., RAFFLE, Y., LANDREAU, J., and MATHOORASING, D.: 'Sampled-grating DBR lasers with 80 addressable wavelengths over 33 nm for 2.5 Gbit/s WDM applications', *Electron. Lett.*, 1996, **32**, (17), pp. 1592–1593
46 ISHII, H., TOHMORI, Y., YOSHIKUNI, Y., TAMAMURA, T., and KONDO, Y.: 'Multiple phase shift superstructure grating DBR lasers for broad wavelength tuning', *IEEE Photonics Technol. Lett.*, 1993, **5**, pp. 613–615
47 ISHII, H., TOHMORI, Y., YAMAMOTO, M., TAMAMURA, T., and YOSHIKUNI, Y.: 'Modified multiple phase shift superstructure grating DBR lasers for broad wavelength tuning', *Electron. Lett.*, 1994, **30**, pp. 1146–1147
48 KANO, F., ISHII, H., TOHMORI, Y., and YOSHIKUNI, Y.: 'Characteristics of superstructure grating (SSG) DBR lasers under broad range wavelength tuning', *IEEE Photonics Technol. Lett.*, 1993, **5**, pp. 611–613
49 TOHMORI, Y., ANO, F.K., ISHII, H., YOSHIKUNI, Y., and KONDO, Y.: 'Wide tuning with narrow linewidth in DFB laser with superstructure grating (SSG)', *Electron. Lett.*, 1993, **29**, pp. 1350–1352

50 DUTTING, K., HILDEBRAND, O., BAUMS, D., IDLER, W., SCHILLING, M., AND WUNSTEL, K.: 'Analysis and simple tuning scheme of asymmetric Y lasers', *IEEE J. Quantum Electron.*, 1994, **30**, pp. 654–659
51 KUZNETSOU, M., VERLANGIERI, P., DENTAI, A.G., JOYNER, C.H., BURRUS, C.A.: 'Asymmetric Y–branch tunable semiconductor laser with 1.0 THz tuning range', *IEEE Photonics Technol. Lett.*, 1992, **4,** pp. 1093–1095
52 KUZNETSOV, M., VERLANGIERI, P., DENTAI, D.G., JOYNER, C.H., and BURRUS, C.A.: 'Widely tunable (45 nm, 5.6 thz) multi quantum well three branch $Y^β$ lasers for WDM networks', *IEEE Photonics Technol. Lett.*, 1993, **5,** pp. 879–882
53 OBERG, M., NILSSON, S., STREUBEL, K., WALLIN, J., BACKBON, L., and KLINGA, T.: '74 nm wavelength tuning range of an InGaAsP/InP vertical grating assisted codirectional coupler laser with rear sampled grating reflector', *IEEE Photonics Technol. Lett.*, 1993, **5,** pp. 735–737
54 KIM, I., ALFERNESS, R.C., KOREN, U, BUHL, L.L., MILLER, B.I., YOUNG, M.G., CHIEN, M.D., KOCH, T.L., PRESBY, H.M., RAYBON, G., and BURRUS, C.A.: 'Broadly tunable vertical coupler filtered tensile strained InGaAs/InGaAsP multiple quantum well laser', *Appl. Phys. Lett.*, 1994, **64,** pp. 2767–2769
55 WOLF, T., ILLEK, S., RIEGER, J., BORCHERT, B., and AMMAN, M.C.: 'Tunable twin guide (TTG) distributed feedback (DFB) laser with over 10 nm continuous tuning range', *Electron. Lett.*, 1993, **29,** pp. 2124–2125
56 THULKE, W., ACHATZ, V., BORCHERT, B., ILLEK, S., and WOLF, T.: 'Integratable tunable twin guide laser on semi–insulating substrate', *Jpn. J. Appl. Phys. 2, Lett.*, 1993, **32,** pp. L914–L916
57 DIECKMANN, A.: 'Linewidth broadening in wavelength tunable laser diode due to absorption and parasitic currents', *Jpn. J. Appl. Phys. 2, Lett.*, 1994, **33,** pp. L1222–L1223
58 AMANN, M.C., HAKIMI, R., BORCHERT, B., and ILLEK, S.: 'Linewidth broadening by 1/f noise in wavelength tunable laser diodes', *Appl. Phys. Lett.*, 1997, **70,** pp. 1512–1514
59 ZAH, C.E., PATHAK, B., FAVIRE, F.J., ANDREAKIS, N.C., BHAT, R., CANEAU, C., CURTIS, L., MAHONEY, D.D., YOUNG., W.C., and LEE, T.P.: 'Monolithic integrated multiwavelength laser arrays for WDM lightwave systems', *Optoelectron. Devices Technol.*, 1994, **9,** pp. 143–166
60 YOUNG, M.G., KOREN, U., MILLER, B.I., CHIEN, M., KOCH, T.L., TENNANT, D.M., FEDER, K., DREYER, K., and RAYBON, G.: '6–Wavelength laser array with integrated amplifier and modulator', *Electron. Lett.*, 1995, **31,** pp. 1835–1836
61 KATOH, Y., KUNII, T., MATSUI, Y., and KAMIJOH, T.: '4 wavelength DBR laser array with waveguide couplers fabricated using selective MOVPE growth', *Opt. Quantum Electron.* 1996, **28,** pp. 533–540
62 OKAMOTO, K., TAKIGUCHI, R., and OHMORI, Y.: '8–channel flat spectral response arrayed-waveguide multiplexer with asymmetrical Mach–Zehnder filters', *IEEE Photonics Technol. Lett.*, 1996, **8,** pp. 373–374

63 PARK, C.Y., KIM, D.B., YOON, T.H., KIM, J.S., OH, K.R., LEE, S.W., LEE, S.M., AHN, J.H., KIM, H.M., and PYUN, K.E.: 'Fabrication of wavelength-tunable InGaAsP/InP grating-assisted codirectional coupler filter with very narrow bandwidth', *Electron. Lett.*, 1997, **33,** pp. 773–774

64 ASGHARI, M., ZHU, B., WHITE, I.H., SELTZER, C.P., NICE, C., HENNING, I.D., BURNESS, A.L., and THOMPSON, G.H.B.: 'Demonstration of an integrated multichannel grating cavity laser for WDM applications', *Electron. Lett.*, 1994, **30,** pp. 1674–1675

65 MILLER, B.I., DREYER, K., BEHRINGER, R.E., KOREN, M.G., CHIEN, M., RAYBON, G., and CAPIK, R.J.: 'Low chirp wavelength selectable 1 x 6 arrays suitable for WDM applications', *15th IEEE International Semiconductor Laser Conference*, 1996, paper WI.3, pp. 129–130

66 SATO, K., and MURAKAMI, M. 'Experimental investigation of thermal crosstalk in a distributed feedback laser array', *IEEE Photonics Technol. Lett.*, 1991, **3,** pp. 501–503

67 AMMAN, M.C.: 'Recent progress on wavelength tunable laser diodes', *Inst. Phys. Conf. Ser.*, 1994, **136,** pp. 257–264

68 MOLLENAUER, L.F., and WHITE, J.C. (Eds.): 'Tunable lasers' (Springer Verlag, 1987)

69 OKUDA, H., SODA, H., MORIKI, K., MOTEGI, Y., and IGA, K.: 'GaInAsP–InP surface emitting injection laser with buried heterostructures', *Jpn. J. Appl. Phys.*, 1981, **20,** pp. L563–L566

70 MOTEGI, Y., SODA, H., and IGA, K.: 'Surface emitting GaInAsP–InP injection laser with short cavity length', *Electron. Lett.*, 1982, **18,** pp. 461–463

71 IGA, K., ISHIKAWA, S., OHKOUCHI, S., and NISHIMURA, T.: 'Room temperature pulsed operation of GaAlAs–GaAs surface emitting injection laser', *Appl. Phys. Lett.*, 1984, **45,** pp. 348–350

72 OU, S.S., YANG, J.J., and JANSEN, M.: '635 nm GaInP/GaAlInP surface emitting laser diodes', *Appl. Phys. Lett.*, 1993, **63,** pp. 3262–3264

73 TAKAMORI, T., COLDREN, L.A., and MERZ, J.L.: 'Folded cavity transverse junction stripe surface emitting laser', *Appl. Phys. Lett.*, 1989, **55,** pp. 1053–1055

74 SAITO, H., and KONDO, Y.: '4 x 4 surface emitting 1.55 µm InGaAsP/ InP laser arrays with microcoated reflectors fabricated by reactive ion etching', *Jpn. J. Appl. Phys. Lett.*, 1991, **30,** pp. L599–L601

75 GOODHUE, W.D., DONNELLY, J.P., WANG, C.A., LINCOLN, G.A., RAUSCHENBACH, K., BAILEY, R.J., and JOHNSON, G.D.: 'Monolithic 2-dimensional InGaAs/AlGaAs and AlInGaAs/AlGaAs diode laser arrays with over 50% differential quantum efficiencies', *Appl. Phys. Lett.*, 1991, **59,** pp. 632–634

76 STEGMULLER, B., WESTERMEIER, H., THULKE, W., FRANZ, G., and SACHER, D.: 'Surface emitting InGaAsP/InP distributed feedback laser diode with monolithic integrated microlens', *IEEE Photonics Technol. Lett.*, 1991, **3,** pp. 776–778

77 EVANS, G.A., CARLSON, N.W., BOUR, D.P., LURIE, M., DEFREEZ, R.K., and BOSSERT, D.J.: '14-W peak power grating surface–emitting laser array', *Electron. Lett.*, 1990, **26,** pp. 1380–1381

78 YU, S.F., PLUMB, R.G.S., and CARROLL, J.E.: 'Spatial active optical switching by using grating coupled surface emitting DFB lasers', *Electron. Lett.*, 1993, **29,** pp. 1147–1148
79 JEWELL, J.L., HUANG, K.F., TAI, K., LEE, Y.H., FISCHER, R.J., MCCALL, S.L., and CHO, A.Y.: 'Vertical cavity single quantum well laser', *Appl. Phys. Lett.*, 1989, **55,** pp. 424–426
80 JEWELL, J.L., HARBISON, J.P., SCHERER, A., LEE, Y.H., and FLOREZ, L.T.: 'Vertical-cavity surface-emitting lasers—design, growth, fabrication, characterization', *IEEE J. Quantum Electron.*, 1991, **27,** pp. 1332–1346
81 IGA, K., ISHIKAWA, S., OHKOUCHI, S., and NISHIMURA, T.: 'Room temperature pulsed operation of GaAlAs–GaAs surface emitting injection laser', *Appl. Phys. Lett.*, 1984, **45,** pp. 348–350
82 SALES, T.E., 'Vertical cavity surface emitting lasers', (Wiley, New York, 1995)
83 SHTENGEL, G., TEMKIN, H., BRUSENBACH, P., UCHIDA, T., KIM, M., PARSONS, C., QUINN, W.E., and SWIRHUN, S.E.: 'High–speed vertical-cavity surface-emitting laser', *IEEE Photonics Technol. Lett.*, 1993, **5,** pp. 1359–1362
84 BABA, T., HAMANO, T., KOYAMA, F., and IGA, K.: 'Spontaneous emission factor of a microcavity DBR surface emitting laser', *IEEE J. Quantum Electron.*, 1991, **27,** pp. 1347–1358
85 BABA, T., HAMANO, T., KOYAMA, F., and IGA, K.: 'Spontaneous emission factor of a microcavity DBR surface emitting laser, 2, effects of electron quantum confinements', *IEEE J. Quantum Electron.*, 1992, **28,** pp. 1310–1319
86 HAMANO, T., IGA, K., and BABA, T.: 'Spontaneous emission behavior and its injection level dependence in 3–dimensional microcavity surface-emitting lasers', *Electron. Commun. Jpn. 2, Electron.*, 1996, **79,** pp. 46–54
87 SANDUSKY, J.V., and BRUECK, S.R.J.: 'Observation of spontaneous emission microcavity effects in an external-cavity surface-emitting laser structure', *Appl. Phys. Lett.*, 1996, **69,** pp. 3993–3995
88 WATANABE, I., KOYAMA, F., and IGA, K.: 'GaInAsP–InP CBH surface emitting laser with a dielectric multilayer reflector', *Jpn. J. Appl. Phys. 1, Regul. Pap. Short Notes*, 1987, **26,** pp. 1598–1599
89 CHAILERTVANITKUL, A., IGA, K., and MORIKI, K.: 'GaInAsP–InP surface emitting laser ($\lambda = 1.4$ μm, 77 K) with heteromultilayer Bragg reflector', *Electron. Lett.*, 1985, **21,** pp. 303–304
90 HOUNG, Y.M., TAN, M.R.T., LIANG, B.W., WANG, S.Y., YANG, L., and MARS, D.E.: 'InGaAs(0.98 μm)/GaAs vertical cavity surface emitting laser grown by gas-source molecular beam epitaxy', *J. Crystal Growth*, 1994, **136,** pp. 216–220
91 SUGIMOTO, M., KOSAKA, H., KURIHARA, K., OGURA, I., NUMAI, T., and KASAHARA, K.: 'Very low threshold current density in vertical cavity surface emitting laser diodes with periodically doped distributed Bragg reflectors', *Electron. Lett.*, 1992, **28,** pp. 385–387
92 ZEEB, E., MOLLER, B., REINER, G., RIES, M., HACKBARTH, T., and EBELING, K. J.: 'Planar proton implanted VCSELs and fiber-coupled 2-D VCSEL arrays', *IEEE J. Select. Topics Quantum Electron.*, 1995, **1** , pp. 616–623

93 KIM, J.Y., HAYASHI, Y., MUKAIHARA, T., OHNOKI, N., KOYAMA, F., and IGA, K.: 'Improvement of current density-voltage characteristics of GaAs/AlAs distributed Bragg reflectors in InGaAs/AlGaAs surface-emitting lasers', *J. Korean Phys. Soc.*, 1995, **28**, pp. 495–498

94 IGA, K., NISHIMURA, T., YAGI, K., YAMAGUCHI, T., and NIINA, T.: 'Room temperature pulsed oscillation of GaAs/GaAs surface emitting junction laser grown by MBE', *Jpn. J. Appl. Phys. 1 Regul. Pap. Short Notes*, 1986, **25**, pp. 924–925

95 GOURLEY, P.L., and DRUMMOND, T.J.: 'Visible, room temperature, surface emitting laser using an epitaxial Fabry–Perot resonator with AlGaAs/AlAs quarter wave high reflectors and AlGaAs/GaAs multiple quantum wells', *Appl. Phys. Lett.*, 1987, **50**, pp. 1225–1227

96 KOYAMA, F., TOMOMATSU, K., and IGA, K.: 'GaAs surface emitting lasers with circular buried heterostructure grown by metalorganic chemical vapor deposition and two-dimensional laser array', *Appl. Phys. Lett.*, 1988, **52**, pp. 528–529

97 YANG, L., WU, M.C., TAI, K., TANBUNEK, T., and LOGAN, R.A.: 'InGaAsP (1.3 μm)/InP vertical cavity surface emitting laser grown by metalorganic vapor-phase epitaxy', *Appl. Phys. Lett.*, 1990, **56**, pp. 889–891

98 HIBBSBRENNER, M.K., SCHNEIDER, R.P., MORGAN, R.A., WALTERSON, R.A., LEHMAN, J.A., KALWEIT, E.L., LOTT, J.A., LEAR, K.L., CHOQUETTE, K.D., and JUERGENSEN, H.: 'Metalorganic vapor phase epitaxial growth of red and infrared vertical cavity surface emitting laser diodes', *Microelectron. J.*, 1994, **25**, pp. 747–755

99 LEI, C., ROGERS, T.J., DEPPE, D.G., and STREETMAN, B.G.: 'InGaAs–GaAs quantum well vertical cavity surface emitting laser using molecular beam epitaxial regrowth', *Appl. Phys. Lett.*, 1991, **58**, pp. 1122–1124

100 LEI, C., ROGERS, T.J., DEPPE, D.G., and STREETMAN, B.G.: 'ZnSe/CaF$_2$ quarter wave Bragg reflector for the vertical cavity surface emitting laser', *J. Appl. Phys.*, 1991, **69**, pp. 7430–7434

101 OHISO, Y., AMANO, C., ITOH, Y., TATENO, K., TADOKORO, T., TAKENOUCHI, H., and KUROKAWA, T., '1.55 μm vertical-cavity surface-emitting lasers with wafer-fused InGaAsP/InP-GaAs/AlAs DBRs', *Electron. Lett.*, 1996, **32**, pp. 1483–1484

102 CHUA, C.L., LIN, C.H., ZHU, Z.H., LO, Y.H., HONG, M., MANNAERTS, J.P., and BHAT, R.: 'Dielectrically-bonded long-wavelength vertical-cavity laser on GaAs substrates using strain-compensated multiple-quantum wells', *IEEE Photonics Technol. Lett.*, 1994, **6**, (12), pp. 1400–1402

103 EGAWA, T., HASEGAWA, Y., JIMBO, T., and UMENO, M.: 'Room temperature pulsed operation of AlGaAs/GaAs vertical-cavity surface-emitting laser-diode on Si substrate', *IEEE Photonics Technol. Lett.*, 1994, **6**, (6), pp. 681–683

104 SCHRAUD, G., GROTHE, H., and SCHRODER, S.: 'Substrateless singlemode vertical-cavity surface emitting GaAs/GaAlAs laser diode', *Electron. Lett.*, 1994, **30**, (3), pp. 238–239

105 DELLUNDE, J., VALLE, A., and SHORE, K.A.: 'Transverse mode selection in external cavity surface emitting laser diodes', *J. Opt. Soc. Am. B, Opt. Phys.*, 1996, **13**, (11), pp. 2477–2483

106 SCOTT, J.W., GEELS, R.S., CORZINE, S.W., and COLDREN, L.A.: 'Modelling temperature effects and spatial hole burning to optimise vertical cavity surface emitting laser performance', *IEEE J. Quantum Electron.*, 1993, **29,** pp. 1295–1308
107 WILSON, G.C., KUCHTA, D.M., WALKER, J.D., and SMITH, J.S.: 'Spatial hole-burning and self focusing in vertical cavity surface emitting laser diodes', *Appl. Phys. Lett.*, 1994, **64,** pp. 542–544
108 WIPIEJEWSKI, T., PETERS, M.G., THIBEAULT, B.J., YOUNG, D.B., and COLDREN, L.A.: 'Size dependent output power saturation of vertical cavity surface emitting laser diodes', *IEEE Photonics Technol. Lett.*, 1996, **8,** pp. 10–12
109 FIEDLER, U., REINER, G., SCHNITZER, P., and EBELING, K.J.: 'Top surface-emitting vertical cavity laser diodes for 10 Gbit/s data transmission', *IEEE Photonics Technol. Lett.*, 1996, **8,** pp. 746–748
110 NUMAI, T., KURIHARA, K., KUHN, K., KOSAKA, H., OGURA, I., KAJITA, M., SAITO, H., and KASAHARA, K.: 'Control of light output polarization for surface emitting laser type device by strained active layer grown on misoriented substrate', *IEEE J. Quantum Electron.*, 1995, **31,** pp. 636–642
111 TAKAHASHI, M., VACCARO, P., FUJITA, K., WATANABE, T., MUKAIHARA, T., KOYAMA, F., and IGA, K.: 'InGaAs–GaAs vertical cavity surface emitting laser grown on GaAs (311)A substrate having low threshold and stable polarization', *IEEE Photonics Technol. Lett.*, 1996, **8,** pp. 737–739

Appendix 1
Maxwell, plane waves and reflections

This appendix provides a summary of plane-wave interactions at dielectric interfaces and a summary of special cases which are of importance in laser-diode design, in particular, indicating one reason why TE modes are slightly more strongly reflected from a cleaved facet than TM modes.

A1.1 The wave equation

Start with Maxwell's two famous equations [1–3] for the electromagnetic fields:

$$\text{curl } \boldsymbol{E} = -\mu_r \mu_0 \partial \boldsymbol{H}/\partial t \qquad (\mu_r=1 \text{ in this book}) \qquad (A1.1)$$

$$\text{curl } \boldsymbol{H} = \varepsilon_r \varepsilon_0 \partial \boldsymbol{E}/\partial t \qquad (A1.2)$$

In this work, the optical waveguide is formed from uniform layered slabs of different relative permittivity $\varepsilon_r = n^2 = (n_r + j n_i)^2$, giving the material's complex refractive index n with n_r determining the real refractive index and n_i determining the optical gain (Appendix 5). Taking the notation of Figure 3.1, the normal to the optical slab guides is in the Oy direction and the TE (TM) waves may be classified as those modes with the $\boldsymbol{E}$ ($\boldsymbol{H}$) field in the plane of the slabs i.e. in the Ox direction. With a uniform permittivity within each slab, elimination of $\boldsymbol{H}$ from eqns. A1.1 and A1.2 leads to the wave equation

$$\nabla^2 \boldsymbol{E} - \mu_r \mu_0 \varepsilon_r \varepsilon_0 \, \partial^2 \boldsymbol{E}/\partial t^2 = 0 \qquad (A1.3)$$

A1.2 Linearly polarised plane waves (in a uniform 'infinite' material)

Taking an infinite uniform isotropic slab of material where there is no variation over the xy plane ($\partial/\partial x = \partial/\partial y = 0$) but only electric fields E_x

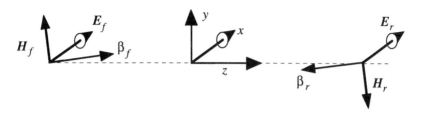

Figure A1.1 Right-handed sets for plane waves

When a plane wave is launched at an angle to the Oz axis with E parallel to Ox then note there is a component H_z and a component field H_y. This is typical in slab guides for TE modes where a pencil of different directions for $\boldsymbol{\beta}$ is possible with $\boldsymbol{\beta}$ not quite parallel to the Oz axis. See Appendix 2

with optical frequencies ω, these fields propagate in the Oz direction as:

$$E_x = F\exp\{j(\omega t - \beta z)\} \qquad (A1.4)$$

where, from eqn. A1.3,

$$\beta^2 = \omega^2 \mu_0 \varepsilon_r \varepsilon_0 = (n\omega/c)^2 \qquad (A1.5)$$

with $c = (1/\mu_o \varepsilon_o)^{1/2} = 3 \times 10^8$ m/s, the velocity of light in vacuum. The phase velocity of light within material is $c_m = 3 \times 10^8/n_r$ m/s which varies slightly with the optical frequency. In general, with optical gain or loss where n_i is nonzero, β is complex.

The H fields are given from eqn. A1.1 by putting $\partial/\partial x = \partial/\partial y = 0$ with $\partial/\partial t \rightarrow j\omega$ so that only a component H_y is found. For forward-propagating fields:

$$(\text{curl }\boldsymbol{H})_x = -\partial H_y/\partial z = j\beta H_y = \varepsilon_r \varepsilon_0 \partial E_x/\partial t = j\omega \varepsilon_r \varepsilon_0 E_x \qquad (A1.6)$$

$$H_y = (\varepsilon_r \varepsilon_0 / \mu_r \mu_0)^{1/2} E_x \rightarrow E_x/H_y = (377/n) \ \Omega \qquad (A1.7)$$

In isotropic material, the E field could lie along any direction but the orientation of fields with respect to the vector direction of propagation $\boldsymbol{\beta} = (\beta_x, \beta_y, \beta_z)$ must be maintained. For plane *linearly polarised waves*, the forward E field lies along one direction with $\boldsymbol{E}_f$, $\boldsymbol{H}_f$ and $\boldsymbol{\beta}_f$ forming a right-handed set for the forward wave as shown in Figure A1.1 and similarly for the reverse wave $\boldsymbol{E}_r$, $\boldsymbol{H}_r$ and $\boldsymbol{\beta}_r$ form a right-handed set.

A1.3 Snell's law and total internal reflection

When plane waves are incident on an interface, they give rise to transmitted and reflected plane waves, as indicated in a ray diagram

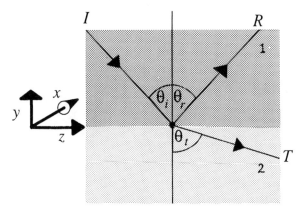

Figure A1.2   Notation for Snell's law

Total internal reflection occurs when $\sin\theta_t > 1$, i.e. θ_t is imaginary

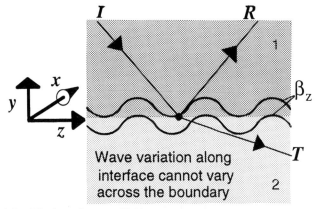

Figure A1.3   The boundary conditions

The continuity of any fields cannot be satisfied unless they have the same spatial periodicity actually over the interface

showing the choice of axes in Figure A1.2. Because all tangential fields have to be continuous, the periodicity of these three waves, resolved along the interface must be identical (Figure A1.3) to one another. This periodicity is determined from β_{zi}, the axial component of the propagation vector $(0, \beta_{yi}, \beta_{zi})$ of the incident wave, and similarly replacing the subscript i with r and t for the reflected and transmitted waves, respectively. Note that $\beta_{xi/r/t}=0$ shows that the light rays lie in the yz plane. Hence, from the equal periodicity condition,

$$\beta_{zi}=\beta_{zr}=\beta_{zt} \tag{A1.8}$$

The ray angles of the incident transmitted and reflected directions are

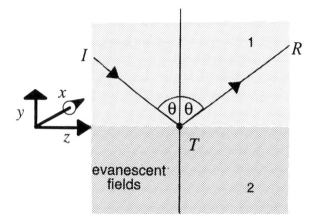

Figure A1.4   Total internal reflections

given from:

$$\tan \theta_i = \beta_{zi}/\beta_{yi} \qquad \tan \theta_r = \beta_{zr}/\beta_{yr} \qquad \tan \theta_t = \beta_{zt}/\beta_{yt} \qquad (A1.9)$$

From eqn. A1.5, one has for the i and r waves in region 1 and t wave in region 2, respectively,

$$(\beta_{yi/r}^2 + \beta_{zi/r}^2)^{1/2} = (\omega/c) n_1 = (\beta_{zi/r}/\sin \theta_{i/r});$$
$$(\beta_{yt}^2 + \beta_{zt}^2)^{1/2} = (\omega/c) n_2 = (\beta_{zi}/\sin \theta_t) \qquad (A1.10)$$

Hence from eqn. A1.8 one recovers Snell's law for reflection and refraction in lossless material with real refractive indices

$$\sin \theta_i = \sin \theta_r = (n_2/n_1) \sin \theta_t \qquad (A1.11)$$

The 'critical ray' has $\theta_t = 90°$ when $\sin \theta_i = (n_2/n_1) < 1$. When $\sin \theta_i > 1$ it is not possible to find any real solution to Snell's law of eqn. A1.11 but explicit expressions can still be found for the vertical 'propagation' in the region 2:

$$\beta_{yt}^2 = \{(\beta_0 n_2)^2 - (\beta_0 n_1)^2 \sin^2 \theta_i\} < 0 \qquad (A1.12)$$
$$\beta_{yt} = \pm j\{(\beta_0 n_1)^2 \sin^2 \theta_i - (\beta_0 n_2)^2\}^{1/2} = \pm j\gamma_{yt} \qquad (A1.13)$$

The 'y propagation coefficient' is now imaginary, giving '*evanescent*' variations $\exp(\pm \gamma_{yt} y)$ in the y direction. With the next boundary 'far away', only the decaying evanescent solution is physically possible in region 2 (Figure A1.4) with the evanescent fields storing energy only significantly within a wavelength from the boundary, but note that this energy does *not* propagate away from the boundary. Power flow in the y direction is solely in region 1. This is the physics of total internal reflection (TIR).

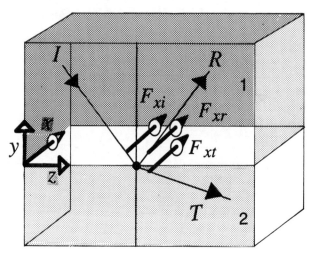

Figure A1.5   Reflection/transmission amplitudes at surfaces
$F = E$ for TE fields
$F = H$ for TM fields

A1.4 Reflection amplitudes at surfaces: TE fields

If the E field lies parallel to the interfacial plane as sketched in Figure A1.5, the field excites dipoles in the dielectric which in turn excite E-fields in the same direction in both region 1 and 2. The TE fields are therefore maintained throughout the slab. The H_z fields are provided by (taking $\mu_r = 1$) and using eqn. A1.1:

$$(\text{curl } \boldsymbol{E})_z = -\partial E_x/\partial y = j\beta_y E_x = -\mu_r\mu_0 \partial H_z/\partial t = -j\omega\mu_0 H_z \qquad (A1.14)$$

giving

$$377\,(\omega/c)H_z = -\beta_y E_x \qquad (A1.15)$$

The continuity of the tangential-field components E_x and H_z requires

$$E_{xi} + E_{xr} = E_{xt} \qquad \beta_{yi}E_{xi} + \beta_{yr}E_{xr} = \beta_{yt}E_{xt} \qquad (A1.16)$$

Noting that $\beta_{yi/r} = \pm(\omega n_1/c)\cos\theta_i$; $\beta_{yt} = -(\omega n_2/c)\cos\theta_t$ and $\theta_i = \theta_r$

$$n_1\cos\theta_i E_{xi} - n_1\cos\theta_r E_{xr} = n_2\cos\theta_t E_{xt} \qquad (A1.17)$$

Eliminating appropriately from eqns. A1.15 and A1.17 leads, after a little work, to reflection and transmission coefficients:

$$\rho = E_{xr}/E_{xi} = \{(n_1\cos\theta_i - n_2\cos\theta_t)/(n_1\cos\theta_i + n_2\cos\theta_t)\} \qquad (A1.18)$$

$$T = E_{xt}/E_{xi} = \{2n_1\cos\theta_i/(n_1\cos\theta_i + n_2\cos\theta_t)\} \qquad (A1.19)$$

A1.5 TE reflection amplitudes: three special cases

(i) *Normal incidence:* $\theta_i = \theta_t = 0$ gives the so-called 'normal' incidence where the amplitude reflection coefficient $\rho = E_{xr}/E_{xi}$ is given from eqn. A1.18 by

$$\rho_0 = (n_1 - n_2)/(n_1 + n_2) \qquad (A1.20)$$

This is used in considering Bragg gratings.

(ii) *Total internal reflection:* $\cos \theta_t$ is purely imaginary, and $\rho = E_{xr}/E_{xi}$ is then of the form $\rho = (C - jK)/(C + jK)$ where C and K are appropriate real numbers; whence $|\rho| = 1$ confirming that there can be no power lost. Total internal reflection is fundamental to the understanding of guided waves along slab structures.

(iii) *Near-normal incidence:* For small angles close to the normal incidence $\cos \delta\theta_i \simeq 1 - \frac{1}{2}\delta\theta_i^2$ then, making use of Snell's law, $\cos \theta_t \simeq 1 - \frac{1}{2}(n_1/n_2)^2 \delta\theta_i^2$. The reflection coefficient for the TE modes may be written, again using eqn. A1.18, as

$$\rho_{TE} \simeq \rho_0 \{1 + \delta\theta_i^2 (n_1/n_2)\} \qquad (A1.21)$$

The near-normal incidence is typical of the wavefronts close to the facets of a laser with a range of $\delta\theta_i$.

A1.6 Reflection amplitudes at surfaces: TM fields

Significantly different results occur when the H field is parallel to the interface, taking $F = H$ in Figure A1.5. Again the arguments about dipole excitations of E fields ensure that TM field will be preserved across the boundary. The tangential fields which are continuous are now the H_x fields and the E_z fields which are related through eqn. A1.2:

$$(\text{curl } H)_z = -\partial H_x/\partial y = j\beta_y H_x = \varepsilon_r \varepsilon_0 \partial E_z/\partial t = -j\omega\varepsilon_r\varepsilon_0 E_z \qquad (A1.22)$$

$$(1/377) n^2(\omega/c) E_z = \beta_y H_x \qquad (A1.23)$$

Similarly, the transverse electric field may also be found from eqn. A1.2:

$$(1/377) n^2(\omega/c) E_y = \beta_z H_x \qquad (A1.24)$$

Because the E field reflection coefficient for the TM mode will be required in order to compare with the E field reflection coefficient of the TE mode, the continuity equations for the tangential H fields and E fields are written as:

$$E_{zi}+E_{zr}=E_{zt}$$
$$n_1^2 E_{zi}/\beta_{yi}+n_1^2 E_{zr}/\beta_{yr}=n_2^2 E_{zt}/\beta_{yt}$$
(A1.25)

Noting again that $\beta_{yi/r}=\pm(\omega n_1/c)\cos\theta_i$; $\beta_{yi/r}=-(\omega n_2/c)\cos\theta_t$ and still $\theta_i=\theta_r$,

$$(n_1 E_{zi}/\cos\theta_i) - (n_1 E_{zr}/\cos\theta_i) = (n_2 E_{zt}/\cos\theta_t) \quad (A1.26)$$

Comparing with the TE case, one can make the appropriate changes to find

$$E_{zr}=E_{zi}[\{(n_1/\cos\theta_i) - (n_2/\cos\theta_t)\}/\{(n_1/\cos\theta_i) + (n_2/\cos\theta_t)\}] \quad (A1.27)$$

$$E_{zt}=E_{zi}[2(n_1/\cos\theta_i)/\{(n_1/\cos\theta_i) + (n_2/\cos\theta_t)\}] \quad (A1.28)$$

The E field reflection coefficient corresponding to $\rho=E_{xr}/E_{xi}$ in eqn. A1.18 should take the full magnitude of the E field, so that now

$$\rho=(E_{zr}/\cos\theta_r)/(E_{zi}/\cos\theta_i)$$
$$=\{(n_1\cos\theta_t - n_2\cos\theta_i)/(n_1\cos\theta_t + n_2\cos\theta_i)\} \quad (A1.29)$$

A1.7 TM reflection amplitudes at surfaces: four special cases

(i) *Normal incidence* and (ii) *Total internal reflection:* These remain in most essentials unaltered from the TE case. The reflection coefficient at normal incidence has to be identical because, with normal incidence, the E field is also parallel to the surface and there physically cannot be any difference. For the total internal reflection, the total amplitude-reflection coefficient remains of modulus unity, but there are different phase changes compared with the TE case. The similar results here indicate that waves can be guided and Bragg gratings will still operate in essentially the same manner for either TE or TM modes. There will be significant differences because of the boundary conditions and slight differences at near normal reflections as seen below.

(ii) *Near normal incidence:* The third special case of nearly normal incidence is important for laser structures because it helps to show why TE rather than TM modes are selected in structures where there are facets. When $\cos\delta\theta_i \approx 1-\tfrac{1}{2}\delta\theta_i^2$; $\cos\theta_t \approx 1-\tfrac{1}{2}(n_1/n_2)\,\delta\theta_i^2$ and now the reflection coefficient for the TM modes can be written as ρ_{TM} using $\rho_o = \{(n_1-n_2)/(n_1+n_2)\}$

$$\rho_{TM} = \rho_0\{1 - \delta\theta_i^2(n_1/n_2)\} \tag{A1.30}$$

giving

$$\rho_{TM}/\rho_{TE} \simeq \{1 - 2\,\delta\theta_i^2(n_1/n_2)\} \tag{A1.31}$$

(iv) *Brewster's angle:* A new special case arises in that it is now possible to have $\rho_{TM}=0$ at a critical angle: the 'Brewster' angle. This happens when

$$\{(n_1/\cos\theta_i) - (n_2/\cos\theta_t)\} = 0 \tag{A1.32}$$

Using Snell's law and writing $n_1^2(1-\sin^2\theta_t) = n_2^2(1-\sin^2\theta_i)$

$$\text{gives } \sin\theta_i = n_2/\sqrt{(n_1^2+n_2^2)} \quad \text{or} \quad \tan\theta_i = n_2/n_1 \tag{A1.33}$$

This is the Brewster angle of incidence where there is no reflection of the TM wave and only a TE wave is reflected. This is the principle on which Polaroid glasses help to cut out some of the glare from reflection. This angle is of special significance in gas lasers, and has also been used in semiconductor lasers to cut out facet reflections (at one polarisation) at the semiconductor/air interface. However, the steepness of the Brewster angle ($\theta_i \sim 70°$) for a typical semiconductor/air interface makes this unattractive compared with the present technology for low-reflectivity surfaces where multiple dielectric thin films are deposited with precision over the surface and can keep reflections below 0.1% for near normal incidence.

A1.8 Reflection for waveguide modes at facets

In many situations with an index-guided mode, the reflection at the facet may be calculated with sufficient accuracy from the effective relative refractive index and unity (air). However, it is important to be able to understand the significant difference between the observed reflectivities of TE and TM modes.

All waveguide modes can be regarded as having a pencil of rays covering a mean-square range of incident angles $\overline{\delta\theta_i^2}$ with this average taking into account the different weighting of each ray angle with the ray intensity. Hence from eqns. A1.21 and A1.30 it follows that the ratio of TM modes to TE-mode reflectance $\sim (1 - 2\,\overline{\delta\theta_i^2}\,n_{\mathit{eff}})$ where n_{eff} is the effective refractive index of the guide. The slightly higher reflectivity for the TE mode gives a lower threshold to a laser and this is one reason for the selectivity of certain laser structures in oscillating in TE rather than TM modes. However, the detailed calculation of reflection of the wave front within a waveguide is not entirely straightforward [4]: the wavefront will curve to some extent, especially with strong gain, and radiation modes can affect the outcome. When it comes to the TE gain margins for DFB lasers, there are further detailed problems concerned with the effects of the grating reflections in TE/TM mode reflections which this appendix has not addressed but are left for further reading [5].

The difficulty of precise calculation of reflection is compounded, of course, when multilayer antireflection or high-reflectivity layers are evaporated onto the facets or grown into the laser, as with VCSELs, where optical diffraction can reduce the reflectivity calculated from a pure-plane-wave analysis [6]. To gain low facet reflectivities for lasers, careful experimental monitoring is usually performed rather than using precise calculation.

A1.9 References

1 STRATTON, J.A.: 'Electromagnetic theory' (McGraw Hill, New York, 1941)
2 RAMO, S., WHINNERY, J.R. and VAN DUZER, T.: 'Fields and waves in communication electronics', (J.Wiley, New York, 1993), 3rd Ed.
3 LONGAIR, M.S.: 'Theoretical concepts in physics' (Cambridge University Press, 1984) (case study 2 gives an excellent historical perspective)
4 HERZINGER, C.M., LU, C.C., DETEMPLE, T.A., and CHEW, W.C.: 'The semiconductor waveguide facet problem', *IEEE J. Quantum Electron.*, 1993, **29,** pp. 2273–81
5 YU, B.M., and LIU, J.M.: 'Gain margin analysis of distributed feedback lasers for both transverse electric and magnetic-modes', *IEEE J. Quantum Electron.*, 1992, **28,** pp. 822–832
6 SALE, T.E.: 'Vertical cavity surface emitting lasers' (Research Studies Press, Taunton; Wiley, New York, 1995)

Appendix 2
Algorithms for the multilayer slab guide

A2.1 TE slab modes

This appendix gives a systematic approach to providing a program for solving propagation, confinement factor and far-field emission when electromagnetic waves are guided by slab waveguides. Almost arbitrary numbers of layers can be computed with complex refractive indices. Chapter 3 covers slabs with only three layers, where there are many fine texts that can provide back-up material [1,2]. Here, by examining a five-layer system, the way forward to a semi-automated method for an arbitrary number of layers with complex refractive indices is demonstrated. Multilayer guides are particularly relevant in discussing DFB lasers where the longitudinally periodic profile of the permittivity has lateral variations across the waveguide which can be taken into account by having a series of layers with different patterns of permittivity, giving an average permittivity and an average periodic component which can be calculated using the programs.

The classification of TE was given in Appendix 1, where TE plane waves had an H_z component because the associated set of orthogonal E, H and β vectors has β at a slight angle to the actual Oz as in Figure A1.1. This plane wave then 'sloms' between the slabs with β_z common to all layers but with varying vertical 'propagation' β_y from layer to layer. Five layers of arbitrary material will demonstrate the procedures. Here the top layer is labelled with the symbol T and other layers labelled from 1 to 4 (etc.) sequentially.

Now in each layer, by definition of a mode, one has to have the same value for the axial propagation β_z and for uniformity of notation one now has a vertical propagation as $\exp(-\gamma_r y)$ in the rth layer for the wave 'travelling' in the '$+y$' direction and varying as $\exp(+\gamma_r y)$ 'travelling' in the '$-y$' direction. Temporal variations are taken as exp($+j\omega t$) and the magnetic fields relate to the electric fields from $\mu_0 \partial H_z / \partial t = \partial E_x / \partial y$ that with the wave equation give

$$\gamma_r^2 = \beta_z^2 - n_r^2 k_0^2$$
$$Z_r = E_x / H_z = -j\mu_0 \omega / \gamma_r = -jZ_0(k_0/\gamma_r) \quad \text{(A2.1)}$$
$$Z_0 = (\mu_0/\varepsilon_0)^{1/2}$$

It is then observed that this problem is like a transmission-line problem [3] where there is, effectively, a terminating impedance Z_T for the top layer being fed from layers of different impedances and optical thicknesses. The impedance Z_r is the wave impedance for the layer r with Z_0 the characteristic wave impedance of free space. $Y_0 = 1/Z_0$ is the wave admittance of free space. Starting from the top layer labelled T, the wave in this layer T is evanescing as $\exp(-\gamma_T y)$. The wave admittance for this layer is $jY_0(\gamma_T/k_0)$ and this is the 'terminating' impedance for the layer 1. The sign of the real part of γ_T is important because the wave must evanesce away from the main guide and may not grow or propagate if the guide is to sustain a guided mode. The layer 1 is of thickness d_1 and one has a 'forward' wave $\exp(-\gamma_1 y)$ with a wave admittance of $jY_0(\gamma_1/k_0)$ and a 'reverse' wave varying as $\exp(\gamma_1 y)$ with a wave admittance of $-jY_0(\gamma_1/k_0)$. The notation here uses a subscripted $r+$ on the E fields for the 'forward' wave and a subscripted $r-$ on the 'reverse' wave. The interface in question is denoted by subscripts $(\ )_{pq}$ so that $(\ )_{T1}$ denotes the interface between the T and 1 layers etc. Then at the interface we have matching E fields and H fields:

$$(E_{x1+} + E_{x1-})_{T1} = (E_{xT+} + E_{xT-})_{T1} \qquad \gamma_1(E_{x1+} - E_{x1-})_{T1} = \gamma_T(E_{xT+} - E_{xT-})_{T1}$$
$$\text{(A2.2)}$$

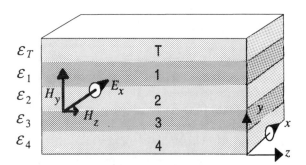

Figure A2.1 Multilayer slab guide

In the ideal theory, the slab is infinite in the x-direction with no x variations; in practice the x variations simply have to be slow enough

Eqn. A2.2 may be rewritten in a matrix form as

$$\gamma_1 \begin{bmatrix} E_{x1+} \\ E_{x1-} \end{bmatrix}_{T1} = \frac{1}{2} \begin{bmatrix} \gamma_1 + \gamma_T & \gamma_1 - \gamma_T \\ \gamma_1 - \gamma_T & \gamma_1 + \gamma_T \end{bmatrix} \begin{bmatrix} E_{xT+} \\ E_{xT-} \end{bmatrix}_{T1} \quad (A2.3)$$

Moving a distance $-d_1$ (i.e. back from the interface) to reach the interface 2/1, the propagation inside the layer is accounted for through

$$\begin{bmatrix} E_{x1+} \\ E_{x1-} \end{bmatrix}_{12} = \begin{bmatrix} \exp(\gamma_1 d_1) & 0 \\ 0 & \exp(-\gamma_1 d_1) \end{bmatrix} \begin{bmatrix} E_{x1+} \\ E_{x1-} \end{bmatrix}_{T1} \quad (A2.4)$$

Then at the 2/1 interface one has the fields at the start of region 2 given from

$$\gamma_2 \begin{bmatrix} E_{x2+} \\ E_{x2-} \end{bmatrix}_{12} = \frac{1}{2} \begin{bmatrix} \gamma_2 + \gamma_1 & \gamma_2 - \gamma_1 \\ \gamma_2 - \gamma_1 & \gamma_2 + \gamma_1 \end{bmatrix} \begin{bmatrix} E_{x1+} \\ E_{x1-} \end{bmatrix}_{12} \quad (A2.5)$$

Then at the 3/2 interface one has the fields at the end of region 2 given from

$$\begin{bmatrix} E_{x2+} \\ E_{x2-} \end{bmatrix}_{23} = \begin{bmatrix} \exp(\gamma_2 d_2) & 0 \\ 0 & \exp(-\gamma_2 d_2) \end{bmatrix} \begin{bmatrix} E_{x2+} \\ E_{x2-} \end{bmatrix}_{12} \quad (A2.6)$$

This process is repeated until the bottom layer is reached. For example, with the five layers as illustrated, define

$$E_r = \begin{bmatrix} \exp(\gamma_r d_r) & 0 \\ 0 & \exp(-\gamma_r d_r) \end{bmatrix} \quad D_{sr} = \frac{1}{2} \begin{bmatrix} \gamma_s + \gamma_r & \gamma_s - \gamma_r \\ \gamma_s - \gamma_r & \gamma_s + \gamma_r \end{bmatrix} \quad (A2.7)$$

$$M = D_{43} E_3 D_{32} E_2 D_{21} E_1 D_{1T} \quad (A2.8)$$

$$\gamma_1 \gamma_2 \gamma_3 \gamma_4 \begin{bmatrix} E_{x4+} \\ E_{x4-} \end{bmatrix}_{43} = M \begin{bmatrix} E_{xT+} \\ E_{xT-} \end{bmatrix}_{T1} \quad (A2.9)$$

316 Distributed feedback semiconductor lasers

It is again important to get the sign of γ_4 right because, just as for the top layer T, the bottom layer (here layer 4) has to have an evanescent wave with only E_{x4-} evanescing away from the guiding layer. One also notices that E_{x4+} is zero and consequently we have the situation that

$$\gamma_1\gamma_2\gamma_3\gamma_4 \begin{bmatrix} 0 \\ E_{x4-} \end{bmatrix}_{43} = \begin{bmatrix} M_{11} & M_{12} \\ M_{21} & M_{22} \end{bmatrix} \begin{bmatrix} E_{xT+} \\ 0 \end{bmatrix}_{T1} \qquad (A2.10)$$

leading to the determinantal equation for guiding:

$$M_{11} = 0 \qquad (A2.11)$$

Solving eqn. A2.11 numerically is not found to be the best technique because, especially with complex refractive indices, the convergence of search routines is not assured. The better method (see Figure 3.6) is to note that, with only the correct form of evanescent wave 'transmitted' at the bottom of the slab but an arbitrary propagation coefficient, in general there will be a 'reflected' evanescent wave at the top with the incorrect sign (i.e. growing as it moves away from the guide: $y \to \infty$) as well as the 'incident' evanescent component with the correct sign (i.e. decaying as it moves away from the guide). A 'reflection coefficient' is found by assuming that E_{xT-} at the top of the guide is not quite zero but $|\rho| = |E_{xT-}/E_{xT+}| = |M_{11}/M_{12}|$ and it is found that log $|\rho|$ displays clearer minima when one is close to a solution than any attempt to find where M_{11} goes to zero. The value of $|\rho| = 0$ ensures that there is only one evanescent wave at the top and, by construction, only one at the bottom of the slab and both have the correct sign to ensure confinement of the optical wave with the amplitudes decaying away from the guiding layers. Although at first sight it is an apparently minor mathematical alteration to consider $|M_{11}/M_{12}|$ rather than $|M_{11}|$, this method turns out to be a major step towards a more stable computational process with complex guides.

The numerical problem starts by making a series of estimates of the effective refractive index and calculates log $|\rho|$, noting those indices which give the lowest value. By alternately scanning the imaginary parts and the real parts (rather like balancing an old-fashioned impedance bridge in an electrical laboratory) one can converge rapidly on values for the effective refractive index which make $|\rho| < 0.001$, which ensures a reasonable accuracy to the effective refractive index and an extremely good accuracy for the intensity profile.

A2.2 TM slab modes

TM modes follow closely the scheme of solution for the TE mode. The same value of β_z holds for each layer, with vertical 'propagation' as $\exp(-\gamma_r y)$ in the rth layer for the wave 'travelling' in the '$+y$' direction and varying as $\exp(+\gamma_r y)$ 'travelling' in the '$-y$' direction. Temporal variations $\exp(+j\omega t)$ with $\varepsilon_r \varepsilon_0 \partial E_z/\partial t = -\partial H_x/\partial y$ along with the wave equation give

$$\gamma_r^2 = \beta_z^2 - n_r^2 k_0^2$$
$$Z_r = E_z/H_x = -j\gamma_r/\varepsilon_r\varepsilon_0\omega = -jZ_0(\gamma_r/\varepsilon_r k_0)$$
$$Z_0 = (\mu_0/\varepsilon_0)^{1/2} \quad \text{(A2.12)}$$

As before, one starts with the top layer T with a single evanescent wave varying as $\exp(-\gamma_T y)$ with a wave impedance of $-jZ_0(\gamma_r/\varepsilon_r k_0)$. The layer 1, of thickness d_1, has a 'forward' wave $\exp(-\gamma_1 y)$ with a wave impedance of $-jZ_0(\gamma_r/\varepsilon_r k_0)$ and a 'reverse' wave varying as $\exp(\gamma_1 y)$ with a wave impedance of $jZ_0(\gamma_r/\varepsilon_r k_0)$. As in the TE case, $r+$ and $r-$ indicate, respectively, the 'forward' and 'reverse' types of evanescent wave in the layer r. Then at the interface we have matching H fields and E fields, respectively:

$$(H_{x1+} + H_{x1-})_{T1} = (H_{xT+} + H_{xT-})_{T1}$$
$$(\gamma_1/\varepsilon_1)(H_{x1+} - H_{x1-})_{T1} = (\gamma_T/\varepsilon_T)(H_{xT+} - H_{xT-})_{T1} \quad \text{(A2.13)}$$

$$(\varepsilon_T\gamma_1)\begin{bmatrix} H_{x1+} \\ H_{x1-} \end{bmatrix}_{T1} = \frac{1}{2}\begin{bmatrix} (\varepsilon_T\gamma_1)+(\varepsilon_1\gamma_T) & (\varepsilon_T\gamma_1)-(\varepsilon_1\gamma_T) \\ (\varepsilon_T\gamma_1)-(\varepsilon_1\gamma_T) & (\varepsilon_T\gamma_1)+(\varepsilon_1\gamma_T) \end{bmatrix}\begin{bmatrix} H_{xT+} \\ H_{xT-} \end{bmatrix}_{T1}$$
$$\text{(A2.14)}$$

Moving a distance $-d_1$ (i.e. back from the interface) to reach the interface 2/1, one accounts for the propagation inside the layer using

$$\begin{bmatrix} H_{x1+} \\ H_{x1-} \end{bmatrix}_{12} = \begin{bmatrix} \exp(\gamma_1 d_1) & 0 \\ 0 & \exp(-\gamma_1 d_1) \end{bmatrix}\begin{bmatrix} H_{x1+} \\ H_{x1-} \end{bmatrix}_{T1} \quad \text{(A2.15)}$$

Then at the 2/1 interface one has the fields at the start of region 2 given from

$$(\varepsilon_1\gamma_2)\begin{bmatrix} H_{x2+} \\ H_{x2-} \end{bmatrix}_{12} = \frac{1}{2}\begin{bmatrix} (\varepsilon_1\gamma_2)+(\varepsilon_2\gamma_1) & (\varepsilon_1\gamma_2)-(\varepsilon_2\gamma_1) \\ (\varepsilon_1\gamma_2)-(\varepsilon_2\gamma_1) & (\varepsilon_1\gamma_2)+(\varepsilon_2\gamma_1) \end{bmatrix}\begin{bmatrix} H_{x1+} \\ H_{x1-} \end{bmatrix}_{12}$$
$$\text{(A2.16)}$$

Then at the 3/2 interface one has the fields at the end of region 2 given from

$$\begin{bmatrix} H_{x2+} \\ H_{x2-} \end{bmatrix}_{23} = \begin{bmatrix} \exp(\gamma_2 d_2) & 0 \\ 0 & \exp(-\gamma_2 d_2) \end{bmatrix} \begin{bmatrix} H_{x2+} \\ H_{x2-} \end{bmatrix}_{12} \quad (A2.17)$$

As for the TE mode, this process is repeated until the bottom layer is reached. For example, with the five layers as illustrated, define

$$\boldsymbol{H}_r = \begin{bmatrix} \exp(\gamma_r d_r) & 0 \\ 0 & \exp(-\gamma_r d_r) \end{bmatrix} \quad \boldsymbol{D}_{sr} = \tfrac{1}{2}\begin{bmatrix} \varepsilon_r \gamma_s + \varepsilon_s \gamma_r & \varepsilon_r \gamma_s - \varepsilon_s \gamma_r \\ \varepsilon_r \gamma_s - \varepsilon_s \gamma_r & \varepsilon_r \gamma_s + \varepsilon_s \gamma_r \end{bmatrix}$$

(A2.18)

$$\boldsymbol{M} = \boldsymbol{D}_{43}\boldsymbol{H}_3\boldsymbol{D}_{32}\boldsymbol{H}_2\boldsymbol{D}_{21}\boldsymbol{H}_1\boldsymbol{D}_{1T} \quad (A2.19)$$

$$\gamma_1\gamma_2\gamma_3\gamma_4\varepsilon_T\varepsilon_1\varepsilon_2\varepsilon_3 \begin{bmatrix} H_{x4+} \\ H_{x4-} \end{bmatrix}_{43} = \boldsymbol{M}\begin{bmatrix} H_{xT+} \\ H_{xT-} \end{bmatrix}_{T1}$$

$$\gamma_1\gamma_2\gamma_3\gamma_4\varepsilon_T\varepsilon_1\varepsilon_2\varepsilon_3 \begin{bmatrix} 0 \\ H_{x4-} \end{bmatrix}_{43} = \begin{bmatrix} M_{11} & M_{12} \\ M_{21} & M_{22} \end{bmatrix}\begin{bmatrix} H_{xT+} \\ 0 \end{bmatrix}_{T1}$$

(A2.20)

leading to the requirement for a solution, similar to that for the TE mode, that

$$|\rho| = |M_{11}/M_{12}| = 0 \quad (A2.21)$$

The solution technique is followed as for the TE mode.

The word of warning given elsewhere is repeated: it is the E fields which interact with the electronic dipoles in the material caused by electrons and holes recombining to give gain. Consequently it is the E fields which are important in determining the strength of the interaction and in determining the confinement factor in a laser. When discussing the TM modes, any computing programs constructed may well find it easiest initially to compute the H field but the final interaction should compute the total E fields to indicate the strength of the interaction with the electronic gain along the guide and to estimate the confinement factor using the E fields.

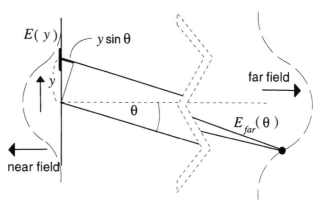

Figure A2.2   Far field contributions

A2.3 Far fields

Not only does one need to find the field intensities within the guide, but also one requires the far-field pattern. The first effect on the far field is given from the phase changes of all the contributions to the far field from the near field. Consider Figure A2.2 where a distribution of near fields in the y direction leads to a distribution of the far field as a function of the angle θ.

Each element of the near field contributes to the far field but with a phase shift caused by the additional distance $y \sin \theta$ with the associated phase shift $(2\pi y/\lambda) \sin \theta$. As seen in Figure A2.2, the far field is contributed to in the same proportion by each element of the near field apart from this phase shift which adds a further distance of travel so that fields can cancel or reinforce one another. On summing all the elemental contributions from the near field, one finds what is, in effect, a Fourier transform of the near field:

$$E_{far}(\theta) \propto \int_{-\infty}^{\infty} E_x(y) \exp(-jk_0 y \sin \theta) dy \qquad (A2.22)$$

This transform gives a first result for the far-field intensity as proportional to $|E_{far}(\theta)|^2$.

However, there is another significant effect. This is caused by the Huygens pattern of radiation from any aperture [4], shown in Figure A2.3. It can be shown that any elemental aperture, no matter how small, re-radiates into free space with a field polar diagram determined by $\frac{1}{2}(1+\cos \theta)$. One notices that there is some reradiation of the field in the reverse directions where $\frac{1}{2}\pi < |\theta| < \pi$. However, this standard

320 *Distributed feedback semiconductor lasers*

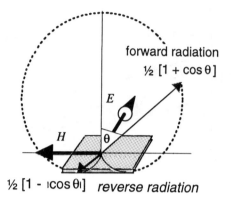

Figure A2.3   Huygens' re-radiation pattern

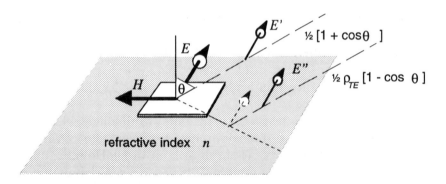

Figure A2.4   Radiation with E field normal to plane of emission

> The radiating aperture (white) with its known fields are placed just in front of the laser facet (shaded) to show more clearly that the reverse re-radiated light is reflected from the facet

pattern is valid only if there is no reflection of this reverse radiation.

With the semiconductor present, the material with refractive index $n > 1$ causes this reverse radiation to be reflected, as indicated in Figure A2.4 where one uses the reflection coefficients for TE and TM fields as given in Appendix A1. Hence, for a TE mode, where

$$\rho_{TE}(\theta) = \{\cos\theta - \sqrt{(n^2 - \sin^2\theta)}\}/\{\cos\theta + \sqrt{(n^2 - \sin^2\theta)}\}$$

the net far-field intensity is formed from the initial effect combining with a modified Huygens factor to give

$$I_{far} = |E_{far}(\theta)|^2 O_{TE}(\theta) \tag{A2.23}$$

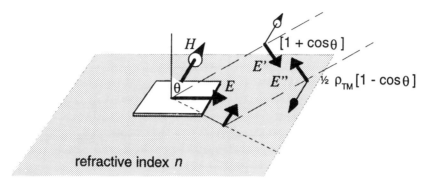

Figure A2.5 Radiation with H field normal to plane of emission

The radiating aperture (white) with its known fields are placed just in front of the laser facet (shaded) to show more clearly that the reverse reradiated light is reflected from the facet

where $O_{TE}(\theta)$ is a power *obliquity factor* given from the expression $|\{\frac{1}{2}(1+\cos\theta)+\rho_{TE}(\theta)\frac{1}{2}(1-\cos\theta)\}|^2$ and leads to

$$O_{TE}(\theta) = \cos^2\theta |\{1+\sqrt{(n^2-\sin^2\theta)}\}/\{\cos\theta+\sqrt{(n^2-\sin^2\theta)}\}|^2$$

$$\text{for } -\pi/2 < \theta < \pi/2 \quad (A2.24)$$

Now when the plane of the far field lies along the E field, one at first sight simply replaces $\rho_{TE}(\theta)$ by $\rho_{TM}(\theta)$ given by the expression $\{\sqrt{(n^2-\sin^2\theta)}-n^2\cos\theta\}/\{n^2\cos\theta+\sqrt{(n^2-\sin^2\theta)}\}$. However, a little more care is required, as shown in Figure A2.5. The E-field components, resolved tangential to the exciting E field, must lie in the same sense in both the reverse and forward directions, as seen in Figures A2.4 and A2.5. This is seen to force the direct and reflected E field components to be in antiphase for their far-field contributions. When the far-field components combine, one has to have a field-obliquity factor obtained by subtracting the reflected fields from the direct fields to obtain $\frac{1}{2}\{(1+\cos\theta)-\rho_{TM}(\theta)(1-\cos\theta)\}$.

The resulting far-field pattern is now given from

$$I_{far} = |E_{far}(\theta)|^2 O_{TM}(\theta) \quad (A2.25)$$

where $O_{TM}(\theta)$ is the obliquity factor given from

$$O_{TM}(\theta) = \cos^2\theta |\{n^2+\sqrt{(n^2-\sin^2\theta)}\}/\{n^2\cos\theta+\sqrt{(n^2-\sin^2\theta)}\}|^2$$

$$\text{for } -\pi/2 < \theta < \pi/2 \quad (A2.26)$$

Experimentally observed narrowing of the far fields is caused by these

322   *Distributed feedback semiconductor lasers*

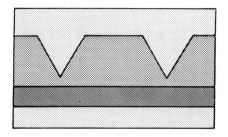 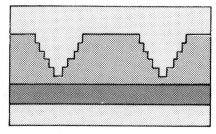

Figure A2.6   *Replacing etched/refilled thick layer with many thin layers for calculation of a slab waveguide which includes a Bragg grating*

Permittivity is then averaged along the length of the guide to find the field patterns

obliquity factors, as was noted by Hockham [5]. The treatment here is simpler than Hockham's and owes much to Lewin [6] who noted the significance of the reflection from the dielectric material. If both the result here and Hockham's results are normalised, the obliquity factors give practically identical results for the TE mode with either calculation, but it is hoped that the derivation here is more physically intuitive as well as being based on sound physics. The reader may also be interested in angled mirrors, and is referred to the literature for this more advanced material [7].

A2.4 Slab waveguide program

Included with the series of MATLAB 4.0 programs is a program **slabexec** in a directory **slab**. The **1readme** file will guide the reader for obtaining results with multislab guides with complex permittivities and selecting either the TE or TM mode. For quantum-well materials, one can have a succession of thin slabs modelling individually the quantum-well layers, the barriers and the separate confinement layers required to confine the light around the quantum wells. There is, in principle, no difficulty in having the large number of layers as required, other than the limitations imposed by a student edition of MATLAB 4.0. Field intensities, confinement factors and far-field patterns are calculated (see Figures 3.7 and 3.8). Bragg-grating calculations can also be taken into account by using a stack of thin layers with varying periodic patterns of permittivity, as shown schematically in Figure A2.6. The program takes the average permittivity over the length of each of the thin layers, calculates the transverse-field pattern and then estimates the coupling coefficient using the form of eqn. 4.37. However, the reader is advised to become familiar with running the slab-waveguide program without gratings before requesting the calculation for a Bragg grating.

A2.5 References

1 SNYDER, A.W., and LOVE, J.D.: 'Optical waveguide theory' (Chapman and Hall, London, 1983)
2 ADAMS, M.J.: 'Introduction to optical guided waves' (Wiley, Chichester, 1981)
3 ROZZI, T.E., and IN'TVELD, G.H.: 'Fields and network analysis of interacting step discontinuities in planar dielectric waveguides', *IEEE Trans.*, 1979, **MTT-27,** pp. 303–309
4 RAMO, S., WHINNERY, J.R., and VAN DUZER, T.: 'Fields and waves in communication electronics' (Wiley, 1994), 3rd ed., pp. 618–627
5 HOCKHAM, G.A.: 'Radiation from a solid state laser', *Electron. Lett.*, 1973, **9,** pp. 389–391
6 LEWIN, L.: 'Obliquity factor for radiation from solid state laser', *Electron. Lett.*, 1974, **10,** pp. 134–135
7 LAU, T., and BALLANTYNE, J.M.: 'Two-dimensional analysis of a dielectric waveguide mirror', *J. Lightwave Technol.*, 1997, **15,** pp. 551–558

Appendix 3
Group refractive index of laser waveguides

The group velocity is the velocity of a wave packet (i.e. the velocity of energy) that is centred on a central carrier frequency $f = \omega/2\pi$. The group velocity is different from the phase velocity for two reasons: (i) the waveguide changes the propagation coefficient as a function of frequency; and (ii) the material permittivity changes with frequency so that the propagation coefficient in the material changes with frequency. This appendix illustrates these two roles using a symmetric three-layer waveguide so that one can appreciate the physics through putting numbers into an analytic solution.

The analytic solution for the propagation coefficient as a function of frequency is well documented for the TE mode in a symmetrical three-slab guide [1–4] . The guide is illustrated in Figure 3.5, where the effect of weak transverse guiding is neglected and only the strong lateral guiding is considered. The thickness of the central layer is taken here to be d and it has a relative permittivity ε_{r1} and is surrounded by layers with relative permittivity ε_{r2}. The axial-propagation coefficient β_z determines, through the wave equation with an angular frequency ω, that the central region has a lateral propagation coefficient β_y:

$$\beta_y^2 = k_0^2 \varepsilon_{r1} - \beta_z^2 \tag{A3.1}$$

where $k_0 = \sqrt{(\omega^2 \varepsilon_0 \mu_0)} = (\omega/c) = 2\pi/\lambda$ with λ the free-space wavelength. In the outer regions there is an evanescent wave decaying spatially in y with a rate

$$\gamma_y^2 = \beta_z^2 - k_0^2 \varepsilon_{r2} \tag{A3.2}$$

There are several different notations used in the literature giving equivalent results, but the use of symmetry of the guide gives a straightforward result [3,4] that the field profile with continuity of the E fields must be of the form:

$$E_x = E_0 \cos(\beta_y y) \qquad [0<y<d/2] \qquad (A3.3)$$

$$= E_0 \cos\left(\frac{\beta_y d}{2}\right) \exp\{-\gamma_y(y-d/2)\} \qquad [d/2<y] \qquad (A3.4)$$

Continuity of the H field is equal to continuity of $\partial E_x/\partial y$ at the boundary where $y = d/2$, leading to the requirement that, for propagation along the guide, the lateral propagation and evanescent constants have to satisfy a transcendental equation:

$$\tan\left(\frac{\beta_y d}{2}\right) = \frac{\gamma_y}{\beta_y} \qquad (A3.5)$$

This equation may also be rewritten as

$$\cos\left(\frac{\beta_y d}{2}\right) = \sqrt{\frac{\beta_y^2}{\beta_y^2 + \gamma_y^2}} = \frac{\beta_y}{k_0 \sqrt{(\delta\varepsilon_r)}} \qquad (A3.6)$$

where $\delta\varepsilon_r = (\varepsilon_{r1} - \varepsilon_{r2})$. A normalised thickness for such a slab guide may be given in terms of the 'V parameter' where $V = (\omega d/c)\sqrt{(\delta\varepsilon_r)}$ and then, defining the optical phase change across half the width of the guide $\Theta = (\beta_y d/2)$, the equation which determines the guide's main mode of operation is:

$$\cos\Theta = 2\frac{\Theta}{V} \qquad (A3.7)$$

As $V \to 0$, so the guide width tends to zero but as Θ increases $(2\Theta/V)$ is limited to magnitude 1 and so V has to increase.

The phase velocity v_p along the guide for this zero-order mode is given from (ω/β_z) and this is found from

$$\frac{\omega d\sqrt{(\delta\varepsilon_r)}}{c} = V = 2\frac{\Theta}{\cos\Theta} \qquad (A3.8)$$

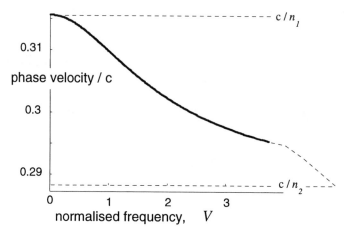

Figure A3.1 Phase velocity against normalised guide frequency

The outer layers of the slab guide have the lower refractive index $n_2 = \sqrt{\varepsilon_{r2}}$ and with thin enough guides the energy resides in the outer layers giving the higher phase velocity. As the optical thickness increases, V increases and the energy is stored in the central part of the guide. At high enough V, the phase velocity is c/n_1 where $n_1 = \sqrt{\varepsilon_{r1}}$ but at such high V numbers the guide is overmoded

$$\beta_z d = \sqrt{\left\{\frac{\omega^2 d^2 \varepsilon_{r1}}{c^2} - 4\Theta^2\right\}} = \sqrt{\left\{V^2\left(\frac{\varepsilon_{r1}}{\delta \varepsilon_r}\right) - 4\Theta^2\right\}} \qquad (A3.9)$$

$$v_p = \frac{\omega d}{\beta_z d} = c/\sqrt{\{\varepsilon_{r1} - \delta\varepsilon_r(\cos\Theta)^2\}} \qquad (A3.10)$$

From the equations, observe that when $\cos\Theta \sim 0$ then $v_p \sim c/\sqrt{(\varepsilon_{r1})}$. Physically, this arises because the optical width of the guide is so large that all the stored optical energy is, effectively, in the central region. However, when $\cos\Theta \sim 1$ where the optical width is small then all the energy is effectively in the cladding and so $v_p \sim c/\sqrt{(\varepsilon_{r2})}$. The fundamental mode only extends usefully up to V values of around π when the first higher-order mode starts to propagate so that Θ is only considered up to values around $\pi/4$ rather than the full range. Figure A3.1 plots the phase velocity as a function of the normalised frequency V for materials given in Table A3.1; these are typical materials for heterostructure DFB lasers. Typical V numbers for $d = 0.1$ to 0.2 μm-thick guiding layers at wavelengths around 1.55 μm are in the range 0.6–1.3.

To find the group velocity v_g, differentiate β_z with respect to ω to obtain

$$c\frac{d\beta_z}{d\omega} = \frac{c}{v_g} = \frac{c}{v_p} + \frac{v_p}{c}\delta\varepsilon_r \omega\Theta'\cos\Theta\sin\Theta + 0.5\frac{v_p}{c}\{\omega\varepsilon'_{r1} - \omega\delta\varepsilon'_r(\cos\Theta)^2\} \quad (A3.11)$$

where $\Theta' = d\Theta/d\omega$; $\varepsilon'_{r1} = d\varepsilon_{r1}/d\omega$; $\delta\varepsilon'_r = d(\delta\varepsilon_r)/d\omega$ and, from eqn. A3.8, calculate $d\omega/d\Theta$ to obtain

$$\omega\Theta' = \left(1 + 0.5\frac{\omega\delta\varepsilon'_r}{\delta\varepsilon_r}\right)\frac{\Theta}{1 + \Theta\tan\Theta} \quad (A3.12)$$

The equation for the reciprocal of the group velocity may be summarised as

$$\frac{c}{v_g} = \frac{c}{v_p} + \frac{v_p}{c}\delta\varepsilon_r\{1 + A(\omega)\}\frac{\Theta\sin\Theta\cos\Theta}{1 + \Theta\tan\Theta} + B(\omega)\varepsilon_{r1}\frac{v_p}{c} \quad (A3.13)$$

where

$$A(\omega) = 0.5\frac{\omega\delta\varepsilon'_r}{\delta\varepsilon_r}; \quad B(\omega) = 0.5\left\{\frac{\omega\varepsilon'_{r1}}{\varepsilon_{r1}} - \frac{\omega\delta\varepsilon'_r}{\varepsilon_{r1}}(\cos\Theta)^2\right\}$$

With no dependence of the material permittivities on frequency, $A(\omega) = B(\omega) = 0$. Under these conditions, one notices from the algebra that the group velocity equals the phase velocity at the end points where $\cos\Theta = 1$ or 0, again consistent with the concept that all the energy effectively resides in the cladding layer or the central layer, respectively. Then using Θ as a parameter one can plot the phase and group indices as a function of the normalised frequency parameter V, with and without material effects, by the inclusion or omission of $A(\omega)$ and $B(\omega)$.

The results are best considered for specific materials and cases. Table A3.1 gives refractive indices n and their rates of change with free-space wavelength λ at wavelengths around 1.55 μm for two typical laser materials. The connection between $dn/d\lambda$ and $\varepsilon'_r = d\varepsilon_r/d\omega$ is given from $n = \sqrt{\varepsilon_r}$ with $\omega\lambda = 2\pi c$ yielding $\omega\varepsilon'_r = -2n\lambda\, dn/d\lambda$. The reader will find it instructive to use a MATLAB program called **group** in directory **slab** to explore these effects.

328 Distributed feedback semiconductor lasers

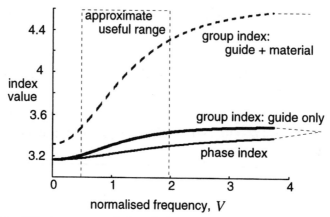

Figure A3.2   Effective group and phase indices against normalised frequency

Group and phase indices as appropriate for a guide formed from a cladding layer InP with a refractive index of 3.168 and a core layer of $In_{1-x}Ga_xAs_yP_{1-y}$ with a refractive index 3.531. For a core-layer thickness of 0.2 μm, one finds $V \sim 1.25$ for $\lambda \sim 1.55$ μm

Table A3.1   Refractive index/permittivity and differentials

Material	Refractive index n	Relative permittivity ε_r	$dn/d\lambda$ at $\lambda = 1.55$ μm	$\omega d\varepsilon_r/d\omega$ at $\lambda = 1.55$ μm
InP	3.168	10.0362	-9.8064×10^4	0.963067
$In_{1-x}Ga_xAs_yP_{1-y}$; $y=0.9$	3.531	12.4680	-7.4322×10^5	8.13536

A3.1 References

1 HAUSS, H.A.: 'Waves and fields in optoelectronics' (Prentice Hall, Englewood Cliffs, 1984), p. 175
2 KOGELNIK, H., and RAMASWAMY, V.: 'Scaling rules for thin-fiber optical waveguides', *Appl. Opt.*, 1974, **8**, p. 1857
3 COLDREN, L.A., and CORZEINE, S.W.: 'Diode lasers and photonic integrated circuits' (Wiley, 1995)
4 ANDERSON, W.W.: 'Mode confinement and gain in junction lasers', *IEEE J. Quantum Electron.*, 1965, **6**, pp. 228–236

Appendix 4
Small-signal analysis of single-mode laser

This appendix provides a more detailed account of the classic rate-equation analyses of appropriately uniform lasers and shows the first-order effects of carrier transport into the active quantum-well region, and the effects of spontaneous emission and gain saturation on damping of the photon–electron resonance. The appendix ends with large-signal rate equations showing the influence of four key parameters which shape the main features of the large-signal response.

A4.1 Rate equations: steady-state and small-signal

This appendix evaluates the small-signal responses of a laser diode to modulation of the drive current more comprehensively than in Chapter 4. The photonic equation (eqn. 4.1) is now written as

$$\frac{dS}{dt} = \Gamma G'_m(N - N_{tr})\frac{S}{1+\epsilon S} - \Gamma bNS - \frac{S}{\tau'_p} + \beta_{sp} BNP \qquad \text{(A4.1)}$$

The effective photon lifetime τ'_p has been used as in Chapter 4. The intervalence-band-absorption term bNS is added here for completeness and the spontaneous emission is recognised as being formed from bipolar recombination so that, when the holes and electrons are of equal density, the radiative spontaneous emission is proportional to N^2. The gain is linearised about the transparency density N_{tr} but gain saturation with optical density is included. The spontaneous-emission coupling β_{sp} is now assumed to have absorbed the confinement factor. The electronic eqn. 4.2 is also written more fully as

$$\frac{dN}{dt} = \frac{J}{qd} - AN - BN^2 - CN^3 - G'_m(N - N_{tr})\frac{S}{1+\epsilon S} \quad (A4.2)$$

where the effective recombination rate N/τ_r in eqn. 4.2 has been expanded to become

$$AN + BN^2 + CN^3 \quad (A4.3)$$

The first (linear) term is often identified with nonradiative recombinations and the final (cubic) term identified with Auger (nonradiative) recombinations. The current density into the active area is J and the active region thickness is d ($J/qd = I/q\Gamma\mathcal{V}$).

As is standard for a small-signal approach, the terms split into steady-state and dynamic terms varying at the modulation angular frequency ω_M:

$$S = S_0 + S_1 \exp(j\omega_M t)$$
$$N = N_0 + N_1 \exp(j\omega_M t) \quad (A4.4)$$
$$J = J_0 + J_1 \exp(j\omega_M t)$$

Substituting eqns. A4.4 into eqn. A4.1 and selecting the steady-state term,

$$S_0 = \beta'_{spont} BN_0^2 \Big/ \left\{ -\Gamma G'_m(N_0 - N_{tr}) + \left(\Gamma bN_0 + \frac{1}{\tau'_p}\right)(1+\epsilon S_0) \right\} > 0 \quad (A4.5)$$

If $\beta'_{spont} BN_0^2 \to 0$ but nevertheless S_0 remains significant, then the electron density N_0 is pinned to what is known as the threshold electron density N_{th}:

$$N_{th} = \left(G'_m N_{tr} + \frac{1+\epsilon S_0}{\Gamma\tau'_p} \right) \Big/ (G'_m - b - b\epsilon S_0) > 0 \quad (A4.6)$$

For small enough gain-saturation terms ϵS_0, the electron density is pinned and this is the value required to start the lasing. To see how the electron density approaches its threshold value, one looks at the steady-state conditions from eqn. A4.2:

$$\frac{J_0}{qd} = AN_0 + BN_0^2 + CN_0^3 + G'_m(N_0 - N_{tr})\frac{S_0}{1+\epsilon S_0} \quad (A4.7)$$

With S_0 zero, the electron density increases with the current input until $N=N_{th}$, when

$$\frac{J_{th}}{qd} = AN_{th} + BN_{th}^2 + CN_{th}^3 \tag{A4.8}$$

As J continues to increase, the optical output then increases almost linearly with drive current above threshold, provided that the gain saturation is negligible:

$$S_0 = \frac{J_0 - J_{th}}{qdG'_m(N_{th} - N_{tr})} \tag{A4.9}$$

The dynamic first-order equations are found by retaining only the first-order terms in eqns. A4.1 and A4.2 using eqns. A4.4. In particular, the small-signal version of eqn. A4.1 may be rewritten after some manipulation as

$$\left(j\omega_M + \frac{1}{\tau_{d1}}\right) S_1 = \Xi_1 N_1 \tag{A4.10}$$

where

$$\frac{1}{\tau_{d1}} = (\beta'_{sp} BN_0^2 / S_0) + \Gamma G'_m(N_0 - N_{tr}) \frac{\epsilon S_0}{(1 + \epsilon S_0)^2}$$

$$\Xi_1 = \left\{ \Gamma S_0 \left(G'_m \frac{1}{1 + \epsilon S_0} - b \right) + 2\beta_{sp} BN_0 \right\}$$

With negligible spontaneous emission $[(\beta_{sp} N_0 / S_0) \to 0]$ and also with negligible optical gain saturation, the changes in the electron density and changes in photon density are in quadrature (i.e. the greatest rates of increase/decrease of photons occur at the peaks and troughs of the electron-density changes with time). However, when the output is low enough just around threshold, so that $(\beta_{sp} N_0 / S_0)$ is no longer a negligible number, the spontaneous emission damps the response of the light to changes in electron density with a damping time constant $\tau_{d1} \sim S_0 / (\beta_{sp} BN_0^2)$. Note that too much intervalence-band absorption (b too large) kills the possibility of lasing by making Ξ_1 negative. Because, as an order of magnitude $\beta_{sp} \sim 10^{-5}$, the spontaneous emission is neglected in the remaining work which applies to above-threshold conditions but gain saturation adds the damping rate $1/\tau_{d1} \sim \{\Gamma G'_m(N_0 - N_{tr})\epsilon S_0\}/(1+\epsilon S_0)^2$.

The modulation changes caused by the changes in the drive current give

$$j\omega_M N_1 = \frac{J_1}{qd} - \frac{N_1}{\tau_r} - G'_m N_1 \frac{S_0}{1+\epsilon S_0} - G'_m(N_0 - N_{tr}) \frac{S_1}{(1+\epsilon S_0)^2} \quad (A4.11)$$

where

$$\frac{1}{\tau_r} = A + 2BN_0 + 3CN_0^2 \quad (A4.12)$$

This can be rearranged to give

$$\left(j\omega_M + \frac{1}{\tau_{d2}}\right) N_1 + \Xi_2 S_1 = \frac{J_1}{qd} \quad (A4.13)$$

where

$$\frac{1}{\tau_{d2}} = \frac{1}{\tau_r} + G'_m \frac{S_0}{1+\epsilon S_0} \quad \text{and} \quad \Xi_2 = G'_m(N_0 - N_{tr}) \frac{1}{(1+\epsilon S_0)^2}$$

The steady state gives

$$\Gamma G'_m(N_0 - N_{tr}) \frac{1}{1+\epsilon S_0} = \Gamma bN_0 + \frac{1}{\tau'_p} \quad (A4.14)$$

Rearranging to obtain the change in photon intensity S_1 as driven by the changes in the current density J_1,

$$(-\omega_M^2 + j\omega_M \gamma + \omega_r^2) S_1 = \frac{\Xi_1 J_1}{qd} \quad (A4.15)$$

where

$$\omega_r^2 = \Xi_1 \Xi_2 + \frac{1}{\tau_{d2}\tau_{d1}} \quad \text{and} \quad \gamma = \left(\frac{1}{\tau_{d2}} + \frac{1}{\tau_{d1}}\right)$$

Neglecting the IVBA term b gives the damping and resonance from

$$\gamma = \frac{1}{\tau_r} + \frac{G'_m S_0}{(1+\epsilon S_0)}\left(1+\frac{\epsilon}{G'_m \tau'_p}\right) \qquad \omega_r^2 = \frac{G'_m S_0}{(1+\epsilon S_0)\tau'_p}\left(1+\frac{\epsilon}{G'_m \tau_r}\right) \qquad (A4.16)$$

Eqn. A4.15 shows a classic damped resonant response with the overall damping rate determined principally by the effective recombination time but with increased damping at high output-power levels and with gain saturation. The resonance frequency increases as $\sqrt{S_0}$ [or alternatively as $\sqrt{(I_0 - I_{th})}$ when the steady current increases above the threshold value I_{th}, assuming that there is negligible gain saturation]. Notice that G'_m in eqn. A4.16 is the differential power gain per unit time so that knowledge of the resonance frequency as a function of intensity S_0 can yield an estimate of this differential gain. The most important consequence of gain saturation is that it can both lead to a marked reduction in the resonant frequency ω_r at high power levels and also give additional damping of the optical response to changes in current. Therefore at high enough optical-power levels, gain saturation slows down the dynamic response of a laser to direct modulation of the current drive.

At this point a little care is needed to note that S_1 already represents an optical power so that the -3 dB bandwidth of the optical-power modulation for the laser occurs when

$$|-\omega_M^2 + j\omega_M \gamma + \omega_r^2| = 2\omega_r^2 \qquad (A4.17)$$

Usually $\omega_M \gamma \ll \omega_r^2$, allowing the approximate -3 dB bandwidth to be defined as

$$\omega_{M-3dB} \sim (\sqrt{3})\,\omega_r \qquad (A4.18)$$

Measurements of the small-signal AM response are often normalised to the response at low frequency. The magnification at resonance is given by

$$\text{Magnification}|_{resonance} = \frac{\omega_r}{\gamma} = \frac{\left(\dfrac{G'_m S_0}{\tau'_p}\right)^{0.5}}{\left(\dfrac{1}{\tau_r} + G'_m S_0\right)} \qquad (A4.19)$$

which has a maximum when the photon density is given by

$$S_0 = \frac{1}{G'_m \tau_r} \qquad (A4.20)$$

334 *Distributed feedback semiconductor lasers*

Finally, it is worth noting that confusion can arise in interpreting experimental small-signal AM-response measurements. It is usual to plot the response in decibels normalised to the steady-state value. This is the ratio of the power dissipated in a resistive load on the detector to the power supplied to the laser. However, taking the load-resistor currents at the modulation frequency ω_M to be $I_1(\omega_M)$, which is proportional to the optical power $S_1(\omega_M)$, one has the optical frequency response given from $10\log_{10}\{S_1(\omega_M)/S_1(0)\}\text{dB} \propto 10\log_{10}\{I_1(\omega_M)/I_1(0)\}$. However, the electrical frequency response is given from $20\log_{10}\{I_1(\omega_M)/I_1(0)\}\text{dB}$. Consequently, when measuring frequency responses, $\text{dB}_{\text{electrical}} = 2\,\text{dB}_{\text{optical}}$.

A4.2 Carrier-transport effects

The high differential gain in multiple-quantum-well (MQW) lasers led to an initial expectation that they would offer improved high-speed performance. In practice, the initial devices were a disappointment until the mechanism responsible for the additional time constant was identified as the slow transport of carriers, particularly holes, through the separate optical-confinement layers into the active MQW region. This is outlined in Section 2.7.4. A simple but useful model can be developed which provides a good explanation of the basic effects.

The carriers driven into the laser waveguide are now referred to as N_{wg}, rather than N as previously. These carriers diffuse through the waveguide layer as before, possibly with some field assistance. However, there is an important difference from bulk material in that, before the electrons can reach the quantum wells, which are the real active regions, they have also to diffuse through the separate confinement-heterojunction layers of thickness d_{SCH}, as in Figure A11.1. This characteristic time to be captured by the quantum wells is taken to be a constant τ_{cap} and is a composite time constant taking into account both the diffusion time across the distance d_{SCH} and actual capture time, which may be only a picosecond. Diffusion is often the dominant factor. The thickness d_{SCH} provides then a limit to how low τ_{cap} can be made. The carriers within the quantum wells have a density N_{qw} but these can then escape through their thermal energy overcoming the barriers and diffuse back into the waveguide layer with another characteristic time constant τ_{esc}. The separate confinement region has the role of helping these carriers to be recaptured by the quantum wells, and this is another reason for keeping d_{SCH} low. The carrier rate equations describing this effect for the waveguide and active regions are

$$\frac{dN_{wg}}{dt} = \frac{J}{qd_{SCH}} + \frac{N_{qw}}{\tau_{esc}} r - \frac{N_{wg}}{\tau_{cap}} \quad (A4.21)$$

The ratio $r = d_{qw}/d_{SCH} < 1$ gives a measure of the concentration of the charge carriers as they fall into the confinement of the quantum wells which have an effective overall thickness d_{qw}. The equivalent small-signal equations can then be written as

$$\left(j\omega_M + \frac{1}{\tau_{cap}}\right) N_{1wg} = \frac{J_1}{qd_{SCH}} + \frac{N_{1qw}}{\tau_{esc}} r \quad (A4.22)$$

Eliminating N_{1wg} and retaining only first-order ω terms on the left-hand side yields an equation which is almost identical to the previous equation for N_1 but with the principal addition that there is a further delay

$$\left(j\omega_M R + \frac{1}{\tau_{eff}} + G'_m S_0\right) N_{1qw} + G'_m (N_{0qw} - N_t) S_1 \simeq \frac{J_1}{qd_{qw}} \frac{1}{(1 + j\omega_M \tau_{cap})} \quad (A4.23)$$

$$R = 1 + \frac{\tau_{cap}}{\tau_{esc}} \quad (A4.24)$$

The photon density is now driven by N_{qw}, whereas previously it had been driven by N:

$$\frac{dN_{qw}}{dt} = \frac{1}{r} \frac{N_{wg}}{\tau_{cap}} - \frac{N_{qw}}{\tau_{esc}} - AN_{qw} - BN_{qw}^2 - CN_{qw}^3 - G'_m(N_{qw} - N_{tr}) S \quad (A4.25)$$

The results, then, of the previous small-signal analysis are substantially the same except for the new feature of the extra delay which is created by the time constant τ_{cap} which summarises the effects of both diffusion and capture along with some numerical differences created by the factor R. The escape time does not figure so prominently as the capture time because it is has been assumed that electrons that escape from the wells back into the guides can be recaptured by the wells.

The conclusion of this simplified analysis is that, provided that the capture time for electrons to enter the quantum wells can be made short enough so that $\omega_r \tau_{cap} \ll 1$, it is the photon–electron resonance ω_r

which limits the amplitude response rather than the capture rate of charge carriers into the quantum wells. On the other hand, if it takes too long for the wells to capture the charge carriers, there is a significant first-order time constant which can limit the response well before the photon–electron resonance plays a role and this phenomenon exhibits itself principally by an additional roll-off in the small-signal AM response.

A4.3 Small-signal FM response of single-mode laser

In principle, rate equations for the balance between photons and charge carriers contain no information about the frequency of operation of the laser because they contain no phase information. However, the photon lifetime is usually in the picosecond range while the time scale for the change of drive current is measured in nanoseconds or hundreds of picoseconds, and it is therefore not surprising to find that the laser's frequency settles sufficiently rapidly with time so that, for uniform lasers where the frequency is determined by the optical length of, say, the Bragg pitch or Fabry–Perot length, the changes in frequency follow the average refractive index as given in eqn. 4.15 with λ_0 being the value of the free-space wavelength:

$$\Delta f / f \sim -\overline{\Delta n_r / n_r} \sim (\lambda_0 / 2\pi) \, \overline{(\alpha_H \Delta g / n_r)} \qquad (A4.26)$$

However, the differential optical-power-gain/unit-time G'_m used in the rate-equation analysis above, is linked to the differential field gain per unit distance from:

$$G'_m = 2 v_g \, dg/dN \qquad (A4.27)$$

Putting $n_r v_g \sim c$ gives the small-signal change in frequency with electron density from

$$\Delta f_1 \sim N_1 \alpha_H G'_m / 4\pi \qquad (A4.28)$$

The key to finding how the frequency/wavelength changes with drive is therefore linked to finding N_1 as a function of the drive current J_1. In the absence of any effects of charge transport, one has to rearrange eqns. A4.10 and A4.13:

$$N_1 = \left(j\omega_M + \frac{1}{\tau_{d1}} \right) \left(\frac{J_1}{qd} \right) \Big/ (-\omega_M^2 + j\omega_M \gamma + \omega_r^2) \qquad (A4.29)$$

The FM response can therefore be seen to follow closely that of the AM response with the same resonance and damping, but the FM rolls off as $1/\omega_M$ as ω_M increases above ω_r, whereas the AM response rolls off as $1/\omega_M^2$ as ω_M increases above ω_r.

A4.4 Small-signal FM response and carrier transport

The carrier-transport mechanism clearly affects the current injected into the active region, but in addition the presence of stored charge in the waveguide layer alters the local refractive index. This in turn changes the effective refractive index of the guided mode and leads to wavelength/frequency chirp for the lasing modes. These changes of refractive index are approximated as follows:

$$n_{eff} = n_{eff,0} + \Gamma_{wg} N_{wg} \frac{\mathrm{d}n}{\mathrm{d}N}\bigg|_{wg} + \Gamma_{qw} N_{qw} \frac{\mathrm{d}n}{\mathrm{d}N}\bigg|_{qw} \qquad (A4.30)$$

The detailed results are straightforward if 'messy' and shed little light on the physics other than to once again indicate that the key effect is to add an additional delay so that the form of the FM response has the general form

$$\frac{N_{qwl}}{(J_1/qd_{qw})} = \frac{\xi}{(1+j\omega_M \tau_{cap})(-\omega_M^2 \alpha + j\omega_M \gamma + \omega_r^2)} \qquad (A4.31)$$

where α, γ and ω_r^2 are similar to the parameters that previously would have been found for the AM response and ξ is again proportional to $j\omega_M$ as ω_M increases. The important point is that transport adds an additional roll-off term for the FM response just as it does for the AM response.

Figure A4.1 summarises, in its upper sketch, the schematic features of AM responses as found from the classic electron–photon rate equations using logarithmic scales. The damping factor has been chosen arbitrarily to give a peak magnification of 3 dB. It may be seen that the transport time constant needs to be such that $\omega_r \tau_{cap} < 1$ for transport effects to have negligible effects on the frequency response. The lower sketch in Figure A4.1 shows the FM response. If there were no nonlinear gain, the FM response would tend to zero as the frequency fell. Nonlinear gain flattens the response. Note that the

338 *Distributed feedback semiconductor lasers*

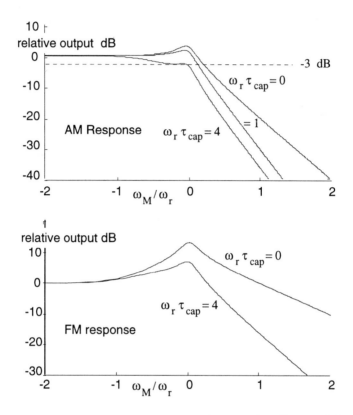

Figure A4.1 Schematic amplitude response $10 \log_{10} (P_{opt\,1}/J_{drive\,1})$ of laser-output intensity to current modulation at frequency ω_M and frequency response $10 \log_{10} (\Delta f/J_{drive\,1})$ with $J_{drive\,1}$ appropriately normalised

Values are normalised to those at low frequency. Parameters chosen to illustrate the schematic features of resonance and fall off of responses above resonance.

resonance frequency ω_r in both the AM and FM responses is the same and is proportional to $\sqrt{(I_0 - I_{th})}$. This is the same resonance as is found in the relative-intensity noise (Appendix 8). [1] gives a detailed discussion of the effects of carriers on the FM response.

In practical devices, the FM response below a few megahertz is significantly affected by the change in refractive index with temperature. If the current increases slowly, there are increases in the temperature which reduce the lasing frequency in opposition to the linewidth broadening or plasma effect, which increases the frequency with carrier density (and hence with current).

A4.5 Photonic and electronic equations for large-signal analysis

This section looks at grouping and normalising variables for large-signal analyses. The equations are as before except that now the spontaneous emission is assumed to have a single time constant. Gain saturation is still included. The pair of rate equations is

$$\frac{dS}{dt} = \Gamma G'_m (N - N_{tr}) \frac{S}{1 + \epsilon S} - \frac{S}{\tau'_p} + \beta_{sp} \frac{\Gamma N}{\tau_r} \qquad (A4.32)$$

$$\frac{dN}{dt} = \frac{1}{\Gamma}\left(\frac{dS}{dt} + \frac{S}{\tau'_p}\right) - \frac{N}{\tau_r}(1 - \beta_{sp}) + \frac{I}{q\Gamma V} \qquad (A4.33)$$

The nominal threshold electron density is then given from

$$N_{th} = N_{tr} + \frac{1}{\Gamma G'_m \tau'_p} \qquad (A4.34)$$

and this, along with the effective photon lifetime, is used to normalise parameters:

$$Q = \frac{\tau_r}{\tau'_p} \qquad a = \frac{N_{tr}}{N_{th}} \qquad T_n = \frac{t}{\tau'_p} \qquad e_s = \frac{\epsilon \Gamma N_{th}}{Q} \qquad (A4.35)$$

$$N_n = \frac{N}{N_{th}} \qquad L_n = \frac{Q}{\Gamma} \frac{S}{N_{th}} \qquad D = \frac{I \tau_r}{q \Gamma V N_{th}} \qquad (A4.36)$$

L_n is the normalised light output, N_n is the normalised electron density and D is the normalised drive with T_n the normalised time and e_s the normalised gain-compression factor so that:

$$\frac{dL_n}{dT_n} = \frac{(N_n - a)}{1 - a} \frac{L_n}{1 + e_s L_n} - L_n + \beta_{sp} N_n \qquad (A4.37)$$

$$Q \frac{dN_n}{dT_n} = -\left(\frac{dL_n}{dT_n} + L_n\right) - N_n(1 - \beta_{sp}) + D \qquad (A4.38)$$

In the steady state where $d/dT_n \to 0$,

$$N_n = \{L_n + e_s(1-a)L_n^2\}/\{L_n + (1-a)(1+e_sL_n)\beta_{sp}\} \qquad L_n = D - N_n(1-\beta_{sp})$$
(A4.39)

Change the variables to $M = N_n + (L_n/Q)$ and L_n:

$$\frac{dL_n}{dT_n} = \frac{\{M - (L_n/Q) - a\}}{1-a}\frac{L_n}{1+e_sL_n} - L_n + \beta_{sp}\{M - (L_n/Q)\}$$
(A4.40)

$$\frac{dM}{dT_n} = \frac{1}{Q}[-L_n - \{M - (L_n/Q)\}(1-\beta_{sp}) + D]$$
(A4.41)

Note how there are then four normalised parameters β_{sp}, a, Q and e_s which determine the normalised light/time characteristics for a given normalised drive D. Note also that, in general, the light changes at a much faster rate than the electrons because of the factor $1/Q$ in the normalised rate equation for the electrons. The electrons vary on a time scale governed mainly by the electron-recombination time, which is longer by a factor of Q (typically ~1000) than the photon lifetime which governs the rate of change of the photons. This is an important point in modelling because it is the photon lifetime which determines the temporal step that is required.

There are two MATLAB programs, fpstat and fpdyna, in directory fabpero. These allow exploration of the qualitative effects of spontaneous-emission parameters and gain saturation.

In summary, the rate-equation approach has been extremely successful in dealing with several major characteristics of Fabry–Perot lasers where the fields are sufficiently uniform and phase effects of fields are not important. However, considerable care is required in applying the results from such equations to distributed-feedback lasers where phasing, nonuniform fields, and the distribution of the feedback change the laser's behaviour in important ways.

A4.6 Reference

1 VANKWIKELBERGE, P., BUYTAERT, F., FRANCHOIS, A., BAETS R., KUINDERSMA, P.I., and FREDRIKSZ, C.W.: 'Analysis of the carrier induced FM response of DFB lasers', *IEEE J. Quantum Electron.*, 1989, **25**, pp. 2239–2254

Appendix 5
Electromagnetic energy exchange

There are three aims of this appendix:

(i) to give a reconciliation of the classical electromagnetic-field exchange of energy and the quantum-particle exchange of energy;
(ii) to show how the rate equations derived from the particle balance are consistent with Maxwell's equations; and
(iii) to show more formally why the group velocity appears in the travelling-field equations used in Chapter 3.

A5.1 Dielectric polarisation and energy exchange

A physical model to have in mind for 'classical' laser interactions between electrons and optical fields is that displaced charges can be treated as dipoles. Figure A5.1 shows the concept of the equilibrium charge and the same charge which has been displaced by an amount x_d because of the interaction with an electric field forcing the charge to move. Each displaced charge is then equivalent to a dipole being

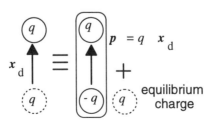

Figure A5.1 *Displaced charge*
   Equivalent to a dipole and equilibrium charge

superimposed on the equilibrium charge. With a density N of charge carriers, there is then a polarisation $\boldsymbol{P} = qN\boldsymbol{x}_d$, where there are two components to the displacement. The first component is dependent on the field strength $\boldsymbol{E}$ in a complex way which depends on the optical frequency and the details of the optical interaction, while the second component represents a random dipole fluctuation:

$$\boldsymbol{P} = \chi \varepsilon_0 \boldsymbol{E} + \boldsymbol{P}_{spont} \quad (A5.1)$$

The parameter χ is called the susceptibility with $1 + \chi = \varepsilon_r$ giving the relative permittivity. In electro-optic material, it is more correctly considered as a tensor so that the polarisation is not necessarily in exactly the same direction as the applied field, but in this work it is adequate to treat χ as a scalar. The random polarisation $\boldsymbol{P}_{spont}$ gives rise to spontaneous polarisation currents through a net polarisation current given from

$$\boldsymbol{J} = \partial \boldsymbol{P} / \partial t \quad (A5.2)$$

Now the helpfulness of this model comes in forming a heuristic link between the classical and quantum formalisms. In quantum modelling, one considers the electronic dipole moments p as being set up when an electron changes its wave function from an energy state $\mathscr{E}_1$ to $\mathscr{E}_2$ where the associated pair of wave functions do not have the same centre of symmetry. This quantum dipole oscillates at an angular frequency $\omega_0 = \omega_{12} = (\mathscr{E}_1 - \mathscr{E}_2)/\hbar$. The quantum interaction is that of a photon exchanging energy $\hbar \omega_{12} = \mathscr{E}_1 - \mathscr{E}_2$ with the two states. The corresponding classical picture is that of an electric field with angular frequency $\omega_0 = \omega_{12}$ interacting with the electronic dipole current at the same angular frequency ω_0. In both classical and quantum formalisms, the strongest interaction occurs only when the frequency associated with the oscillating dipole and the frequency associated with the field are equal. The susceptibility χ describes stimulated interactions as a function of frequency, while any increase in energy caused by the random spontaneous-emission energy is proportional to the square of |(random polarisation currents)| and so is always positive. As usual in work here, all material is nonmagnetic: $\mu_r = 1$.

The local instantaneous rate of energy-density exchange from the electric field into the polarisation currents takes place (see eqn. A5.13) through the local scalar product $(\boldsymbol{E}.\boldsymbol{J})$:

$$dU/dt = -(\boldsymbol{E}.\boldsymbol{J}) \quad (A5.3)$$

Using ideas from classical electrical-circuit analysis, if one uses a complex phasor field, the complex rate of energy density exchange gives

$$(E.J^*) = (j\omega\chi_r + \omega\chi_i)\varepsilon_0(E.E^*) \qquad (A5.4)$$

The Poynting vector $\mathcal{P} = (E \times H)$ gives the instantaneous power per unit area and, to give the power from the laser, must be integrated over the xy plane of the appropriate laser facet of effective area $\mathcal{A}$:

$$\mathcal{P}_z \mathcal{A} = \int\int (E \times H)_z \, dx \, dy \qquad (A5.5)$$

To see how these results can be reconciled with the particle model of Section 4.1, one must recognise that the electrical polarisation has three distinct components:

(i) A real susceptibility χ_r with $P = \chi_r \varepsilon_0 E$

In this first component, the dipole current $\{\partial P(t)/\partial t\}$ is in quadrature with the electric field and from eqn. A5.4 there is no mean real power exchange. There is reactive power where the power flows back and forth between the electric and magnetic fields, and so the effects of χ_r are reflected in the stored electric energy within the material which at any instant is given from

$$U_e = \tfrac{1}{2}(\varepsilon_0\varepsilon_{rr}E.E) = \{\tfrac{1}{2}(\varepsilon_0\chi_r E.E) + \tfrac{1}{2}(\varepsilon_0 E.E)\} \qquad (A5.6)$$

using the fact that $\varepsilon_{rr} = 1 + \chi_r$.

There is an important theorem in electromagnetism which states that, at resonance or in the normal mode of a guide, the mean electric-energy density is identical to the mean magnetic-energy density [1, 2]. The laser is both resonant and uses a guide for the optical fields so that, at the lasing frequencies, there is an equivalence of the mean values of magnetic and electric stored energies where, using the phasor notation:

$$U_e = \tfrac{1}{4}(\varepsilon_0\varepsilon_{rr} E.E^*) = U_m = \tfrac{1}{4}(\mu_0 H.H^*) \qquad (A5.7)$$

Here the fields refer to the peak values and the additional half has appeared because this is the *mean* of the instantaneous energy which peaks every half cycle.

(ii) An imaginary susceptibility χ_i with $P = j\chi_i \varepsilon_0 E$

This second component gives $[\partial P(t)/\partial t]$ directly in phase or in antiphase with the fields and results in a direct exchange of power. Here sign conventions are important. If one assumes, as here, a time variation $\exp(j\omega_o t)$ at an optical (angular) frequency ω_0, then the imaginary part of χ ($=j\chi_i$) gives a polarisation current $\boldsymbol{J} = -\omega_0 \chi_i \varepsilon_0 \boldsymbol{E}$, giving a mean rate of energy input *from the electrons into the field* determined from

$$\tfrac{1}{4}(j\omega_0 \boldsymbol{E}.\boldsymbol{P}^* + \text{c.c.}) = -\tfrac{1}{2}\chi_i \varepsilon_0 \omega_0 \boldsymbol{E}.\boldsymbol{E}^*$$

Now $\boldsymbol{E}.\boldsymbol{E}^*$ is proportional to the total mean electromagnetic energy in the guide which, in turn, is proportional to the photon density in the material. Consequently, χ_i determines the *stimulated emission* ($\chi_i > 0$) or *absorption* ($\chi_i < 0$). In a laser, χ_i is strongly dependent on the electron density. At the transparency density N_{tr} of the electrons, there is no net absorption of emission so that $\chi_i(N_{tr}) = 0$. Note that, for these imaginary components,

$$\chi_i = \varepsilon_{ri} \quad (A5.8)$$

The connection between optical gain and χ_i will be made shortly.

(iii) $[\partial \boldsymbol{P}(t)/\partial t]_{spont} = \boldsymbol{J}_{spont}$

The third component of energy exchange between the electrons and the electromagnetic field is the spontaneous emission. This quantum phenomenon has to be represented in a quasiclassical analysis by a random dipole current $[\partial \boldsymbol{P}(t)/\partial t]_{spont} = \boldsymbol{J}_{spont}$ which is calculated in such a way as to give the correct positive random emission of photons. The details of estimating this value are given in Appendix 9.

These three components of the polarisation current are put into eqns. A1.1 and A1.2, but now treating the fields and currents as functions of time with frequencies close to ω_0 rather than phasors,

$$\text{curl } \boldsymbol{E} = -\mu_0 \partial \boldsymbol{H}/\partial t \quad (A5.9)$$

$$\text{curl } \boldsymbol{H} = -\omega_0 \varepsilon_n \varepsilon_0 \boldsymbol{E} + \varepsilon_{rr} \varepsilon_0 \partial \boldsymbol{E}/\partial t + \boldsymbol{J}_{spont} \quad (A5.10)$$

The ground work has now been laid for finding rate equations from Maxwell's equations as follows next.

A5.2 Electromagnetic-energy exchange and rate equations reconciled

This section will develop rate equations corresponding to the particle interactions given in Chapter 4, but arriving at these through the classical electromagnetic-energy exchange. For the moment, it must be taken on trust that it is possible to design the material structures, such as the different slab layers, shown schematically in Figure A5.2, that are

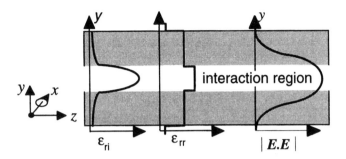

Figure A5.2 Optical fields and real and imaginary permittivities
Schematic profiles in a vertical direction

required in order to guide the optical fields to interact with the recombining carriers in the active region (see Chapter 4 and Appendix 2). Given such a guiding structure, even without knowing its details, it is possible to consider the energy exchange between the optical fields and the polarisation currents. The stored electromagnetic-energy density at any instant of time is given from $U = \frac{1}{2}(\varepsilon_0 \varepsilon_{rr} \mathbf{E}.\mathbf{E} + \mu_0 \mathbf{H}.\mathbf{H})$ with the local rate of change of energy density given from

$$\partial U/\partial t = \{\varepsilon_0 \varepsilon_{rr} \mathbf{E}.\partial \mathbf{E}/\partial t + \mu_0 \mathbf{H}.\partial \mathbf{H}/\partial t\} \qquad (A5.11)$$

Using the vector identity for the divergence of the Poynting vector, which gives the electromagnetic-power flux,

$$\text{div}(\mathbf{E} \times \mathbf{H}) = \mathbf{H}.\text{curl }\mathbf{E} - \mathbf{E}.\text{curl }\mathbf{H} \qquad (A5.12)$$

On substituting for $\partial \mathbf{H}/\partial t$ and $\partial \mathbf{E}/\partial t$ from eqns. A5.9 and A5.10, the instantaneous rate of energy-density exchange is given from

$$\partial U/\partial t = \mathbf{E}.(\text{curl }\mathbf{H} + \omega_0 \varepsilon_{ri} \varepsilon_0 \mathbf{E} - \mathbf{J}_{spont}) - \mathbf{H}.\text{curl }\mathbf{E}$$

$$= -\text{div}(\mathbf{E} \times \mathbf{H}) + \omega_0 \varepsilon_{ri} \varepsilon_0 \mathbf{E}.\mathbf{E} - \mathbf{E}.\mathbf{J}_{spont} \qquad (A5.13)$$

Next integrate this rate of change of energy over the cross-section of the laser so that, in the divergence, the terms $\partial/\partial x$ and $\partial/\partial y$ integrate over the cross-section of the laser to leave $(\mathbf{E} \times \mathbf{H})_x$ and $(\mathbf{E} \times \mathbf{H})_y$ which are zero at the sides of the laser guide, leading to the rate of longitudinal energy exchange:

$$\frac{\partial}{\partial t}\left(\int\int U\,dx\,dy\right)+\frac{\partial}{\partial z}\left(\int\int (E\times H)_z\,dx\,dy\right)$$

$$-\left(\int\int \omega_0\varepsilon_r\varepsilon_0 E.E\,dx\,dy\right)+\left(\int\int E.J_{spont}\,dx\,dy\right)=0 \quad (A5.14)$$

This gives the instantaneous energy exchange at any time.

It is possible to associate eqn. A5.14 with either the forward fields or the reverse fields. It is also more useful to consider the mean value of eqn. A5.14 over approximately one period of oscillation, when one may use the equivalence of mean electric and magnetic energy over the guide cross-section to write, for the forward field,

$$\int\int U_f\,dx\,dy = \int\int \varepsilon_r\varepsilon_0\overline{E_f.E_f}\,dx\,dy = \hbar\omega_0 S_f \mathcal{A} \quad (A5.15)$$

where $\mathcal{A}$ is the effective area containing a density S_f of photons with energy $\hbar\omega_0$ per photon travelling in the forward direction. The z component of the Poynting vector $\mathcal{P}_f$ gives the temporal mean forward-power flow $\mathcal{P}_f\mathcal{A}$ from

$$\mathcal{P}_f\mathcal{A} = \int\int \overline{(E_f\times H_f)_z}\,dx\,dy = \hbar\omega_0 S_f \mathcal{A} v_g \quad (A5.16)$$

where v_g is the group velocity. Here the important new point has been made which differs from the rate equations of Section 4.1, namely that power flow and energy stored can both be associated with the forward (and also the reverse) directions along the guide. Eqn. A5.14 may be recast to give the mean forward-power density and associated energy-density exchange as in eqn. 4.23 from

$$(\hbar\omega_0\mathcal{A})\left\{\frac{1}{v_g}\frac{\partial S_f}{\partial t}+\frac{\partial S_f}{\partial z}-2g(N)S_f\right\}=(\hbar\omega_0\mathcal{A})s_{spf} \quad (A5.17)$$

The factor of 2 for $2g(N)$ is again here because $g(N)$ is used as the field gain per unit distance as function of electron density N and $2g(N)$ refers to the power gain per unit distance. Here this net optical power gain is given from the temporal mean value of

$$(2\hbar\omega_0 \mathcal{A} v_g) g(N) S_f = \int \int \omega_0 \varepsilon_n \varepsilon_0 \overline{\mathbf{E}_f \cdot \mathbf{E}_f} \, dx \, dy \qquad (A5.18)$$

The spontaneous emission combined with its coupling factor β_{sp} into the forward-propagating field is determined from the mean temporal value of

$$(\hbar\omega_0 \mathcal{A} v_g) s_{sp\,f} = -\left(\int \int \overline{\mathbf{E}_f \cdot \mathbf{J}_{spont}} \, dx \, dy \right) \qquad (A5.19)$$

It may be thought, because the phase of $\mathbf{E}_f$ and $\mathbf{J}_{spont}$ must be random, that this spontaneous emission could take any sign. However, it turns out that any single spontaneous emission excitation $\mathbf{J}_{spont} \, dx \, dy$ is incoherent with all fields except those fields which are excited by this particular element itself. On averaging over time and space, the net energy gains can only be terms proportional to $|\mathbf{J}_{spont}|^2$ and the sign of this is such that spontaneous-emission power is always given from the material into the fields.

For energy propagating in the $-Oz$ direction, the above arguments are mirrored and there is, similarly, the equation (eqn. 4.25):

$$(1/v_g) \frac{\partial S_r}{\partial t} + \frac{\partial S_r}{\partial z} - 2g(N) s_r = s_{sp\,r} \qquad (A5.20)$$

Equations A5.17 and A5.20 combine to give $S = (S_f + S_r)$. The averaging over one cycle, as in eqn. A5.15, is kept in future equations but not shown explicitly. Integrating over the laser's length with $\mathcal{V} = \mathcal{A}L$ gives:

$$\int \int \int U \, dx \, dy \, dz =$$

$$\int \int \int \varepsilon_n \varepsilon_0 \mathbf{E} \cdot \mathbf{E} \, dx \, dy \, dz = \hbar\omega_0 S\mathcal{V} \qquad (A5.21)$$

The power flowing out at the 'right' and 'left' facet ends gives the rate of escaping photons from

$$\hbar\omega_0 S\mathcal{V}/\tau_p = \left\{\int\int (\boldsymbol{E}\times\boldsymbol{H})_z \, dx \, dy\right\}_R$$

$$- \left\{\int\int (\boldsymbol{E}\times\boldsymbol{H})_z \, dx \, dy\right\}_L \qquad (A5.22)$$

The mean rate of *net* stimulated emission over the whole volume is given from

$$\hbar\omega_0 S v_g 2g(N)\mathcal{V} = \int\int\int \omega_0 \varepsilon_n \varepsilon_0 \boldsymbol{E}.\boldsymbol{E} \, dx \, dy \, dz \qquad (A5.23)$$

Note that, because the calculation is concerned with the fields, it has by default included the confinement factor and any loss which was put in separately in Section 4.1. The net output of the radiative spontaneous emission (N/τ_{sp}) to the lasing mode taking into account *both* directions of propagation of energy is given from

$$\beta_{sp}(N/\tau_{sp})(\hbar\omega_0\mathcal{V}) = \left(-\int\int\int \boldsymbol{E}.\boldsymbol{J}_{spont} \, dx \, dy \, dz\right) \qquad (A5.24)$$

Combining both forward- and reverse-power-flow equations permits one to see the principles of assigning quantities from the electromagnetic-energy balance to give the equivalent 'photon'-particle rate equation as in eqn. 3.1 but taking the net gain and value of β_{sp} to include the appropriate confinement factors which do not now appear explicitly:

$$\frac{dS}{dt} = G(N)S - \frac{S}{\tau_p} + \beta_{sp}(N/\tau_{sp}) \qquad (A5.25)$$

A5.3 Electromagnetic-energy exchange and guided waves: field equations

Here the discussion concentrates on a TE index-guided field to demonstrate key points about energy exchange in waveguides and to develop the travelling-wave equations for forward and reverse waves in a guide to demonstrate why the group velocity appears.

Consider a TE field with E_x and

$$H_y = \frac{j}{\omega \mu_0} \frac{\partial E_x}{\partial z}$$

as the predominant components. The magnetic-field components may be eliminated from Maxwell's equations to obtain the wave equation for E_x, retaining a frequency-dependent real permittivity for the material but ignoring the frequency dependence of the much smaller imaginary permittivity:

$$\frac{\partial^2 E_x}{\partial x^2} + \frac{\partial^2 E_x}{\partial y^2} + \frac{\partial^2 E_x}{\partial z^2} - \frac{\partial}{\partial t}\left[\mu_0\{\varepsilon_{rr}(x, y, \omega) + j\varepsilon_{ri}(x, y)\}\varepsilon_0 \frac{\partial E_x}{\partial t}\right] = \mu_0 \frac{\partial J_{spont\, x}}{\partial t}$$

(A5.26)

A single-mode solution to this equation is solved for steady-state amplitudes with negligible loss and gain at one particular frequency ω and a corresponding value of β. These values of ω and β are the same over the whole cross-section of the guide, so that

$$E_x = E_f(z, t)\, u(x, y) \text{ where } E_f(z, t) = E_o \exp\{j(\omega t - \beta z)\}$$

For this steady-state solution E_o cancels with

$$\frac{\partial^2 u}{\partial x^2} + \frac{\partial^2 u}{\partial y^2} - \beta^2 u + \omega^2 \{\mu_0 \varepsilon_{rr}(x, y, \omega)\varepsilon_0\} u = 0 \qquad (A5.27)$$

After solving this equation for the real 'vertical' variation $u(x, y)$ with an index-guided wave (as, for example, in Chapter 3) relate ω and β giving an effective permittivity $\varepsilon_{rr\,eff}$:

$$\beta^2 = \omega^2 \{\mu_0 \varepsilon_{rr\,eff}(\omega)\varepsilon_0\} \qquad (A5.28)$$

Now suppose that the mode no longer has a steady-state amplitude but that

$$E_f(z, t) = F(z, t) \exp\{j(\omega_b t - \beta_b z)\} \qquad (A5.29)$$

where $F(z, t)$ varies slowly in space and time with respect to $\exp\{j(\omega_b t - \beta_b z)\}$. In using phasor notation, one can 'equate' ω with $(-j\partial/\partial t)$ but now one needs to separate the low frequencies, varying with rates noted by $(\partial/\partial t)$, from the optical frequency ω_b. This is done by writing $\omega \to \omega_b - j(\partial/\partial t)$ and $\omega^2 \to \omega_b^2 - 2\omega_b j(\partial/\partial t)$ to a first order in the slow changes denoted by $(\partial/\partial t)$. It follows from Taylor's theorem that the frequency-dependent permittivity becomes a temporally changing permittivity

350 *Distributed feedback semiconductor lasers*

$$\varepsilon_r(\omega) \rightarrow \varepsilon_r(\omega_b) + \varepsilon_r'(\omega_b) \times (-j\partial/\partial t) \qquad (A5.30)$$

where $\varepsilon_r'(\omega_b) = \partial \varepsilon_r(\omega)/\partial \omega |_{\omega_b}$.

Similarly, $\beta \rightarrow \beta_b + j(\partial/\partial z)$ where now $(\partial/\partial z)$ refers to the slow variations with space with $\beta^2 \rightarrow \beta_b^2 + 2\beta_b j(\partial/\partial z)$ to a first order in $(\partial/\partial z)$. Then eqn. A5.28 is generalised to a first order in $(\partial/\partial t)$ to become

$$\left(\frac{\partial^2 u}{\partial x^2} + \frac{\partial^2 u}{\partial y^2}\right) F - \beta_b^2 F u - j2\beta_b u \frac{\partial F}{\partial z}$$

$$+ \mu_0 \varepsilon_0 \varepsilon_r(x, y, \omega_b) u \left(\omega_b^2 F - j2\omega_b \frac{\partial F}{\partial t}\right) - j\omega_b^2 \mu_0 \varepsilon_0 \varepsilon_r'(x, y, \omega_b) u \frac{\partial F}{\partial t} = j\omega_b \mu_0 J_{spont}$$

(A5.31)

where $\varepsilon_r = \varepsilon_{rr} + j\varepsilon_{ri}$ with ε_{ri} ($\ll \varepsilon_{rr}$) giving the gain which determines the growth. The growth is usually sufficiently small that one may, to a first order, neglect ε_{ri}'. Note that one has multiplied the spontaneous emission by a factor $\exp\{j(\omega_b t - \beta_b z)\}$ but this does not alter its magnitude nor alter its random properties, and apart from a phase change there is no significant effect so it is not necessary to make this phase factor explicit.

The steady-state terms must cancel giving:

$$\beta_b^2 = \omega_b^2 \{\mu_0 \varepsilon_{rr\,eff}(\omega_b) \varepsilon_0\} \qquad (A5.32)$$

The group velocity is found from eqn. A5.32 using the expression $v_g = \partial \omega / \partial \beta$:

$$\frac{1}{v_g} = \frac{\partial \beta}{\partial \omega} = \frac{\omega_b}{\beta_b} \{\mu_0 \varepsilon_{rr\,eff}(\omega_b)\varepsilon_0\} + \frac{\omega_b^2}{\beta_b}\{\tfrac{1}{2}\mu_0 \varepsilon_{rr\,eff}'(\omega_b)\varepsilon_0\} \qquad (A5.33)$$

Removing the steady-state terms in eqn. A5.31 and with a little rearranging, such as cancelling the j operator, gives

$$u\frac{\partial F}{\partial z} + \mu_0 \varepsilon_0 \left(\frac{\omega_b}{\beta_b}\right) \{\varepsilon_{rr}(x, y, \omega_b) + \tfrac{1}{2}\varepsilon_{rr}'(x, y, \omega_b)\,\omega_b\} u\left(\frac{\partial F}{\partial t}\right)$$

$$- \tfrac{1}{2}\beta_b \left\{\frac{\varepsilon_{ri}(x, y, \omega_b)}{\varepsilon_{rr\,eff}}\right\} uF = -\left(\frac{\omega_b}{2\beta_b}\right)\mu_0 J_{spont} \qquad (A5.34)$$

The result is more useful if one uses the effective modal area:

$$\mathcal{A} = \int\int u(x, y)^2 \, dx \, dy \qquad (A5.35)$$

where u is real and one also writes

$$\int\int \frac{\varepsilon_n(x, y)}{2\varepsilon_{rr\,eff}} u(x, y)^2 \, dx \, dy = g\mathcal{A}/\beta_b \qquad (A5.36)$$

$$\int\int -\left(\frac{\omega_b}{2\beta_b}\right)\mu_0 J_{spont}(x, y) u(x, y) \, dx \, dy = i_{sp\,f} \mathcal{A} \qquad (A5.37)$$

where $i_{sp\,f}$ includes the spontaneous-coupling coefficient, i.e. only those spontaneous interactions feeding into the guide at the right frequency. Then, with eqn. A5.33, the equation for the forward field with gain becomes (eqn. 4.22 with $\delta=0$)

$$\frac{\partial F}{\partial z} + \frac{1}{v_g}\left(\frac{\partial F}{\partial t}\right) = gF + i_{sp\,f} \qquad (A5.38)$$

Similarly, for the reverse field (eqn. 4.24),

$$-\frac{\partial R}{\partial z} + \frac{1}{v_g}\left(\frac{\partial R}{\partial t}\right) = gR + i_{sp\,r} \qquad (A5.39)$$

A5.4 References

1 COLLIN, R.E.: 'Field theory of guided waves' (McGraw Hill, 1960)
2 SLATER, J.C.: 'Microwave electronics' (Van Nostrand, 1950)

Appendix 6
Pauli equations

This appendix gives the detailed calculations for finding the steady-state fields in a *uniform DFB laser* with uniform gain, and thereby finding the threshold conditions. The results are essential if one wishes to make comparisons with the numerical algorithms to estimate the accuracy of these algorithms. The appendix also provides a tutorial on the use of Pauli matrices for coupled differential equations which enables one to extend the concept of $\exp(\gamma z)$ to $\exp(Mz)$ where M is a matrix: a useful concept for linear differential equations.

Take the coupled-wave equations

$$\frac{1}{v_g}\frac{\partial F}{\partial t}+\frac{\partial F}{\partial z}=j\kappa R+(g-j\delta)F \tag{A6.1}$$

$$\frac{1}{v_g}\frac{\partial R}{\partial t}-\frac{\partial R}{\partial z}=j\kappa F+(g-j\delta)R \tag{A6.2}$$

Pauli matrices are widely used in the quantum theory of spin and polarisation of light and the way in which the fields are coupled, but they can also be used to provide neat solutions for other physical problems (see, for example, [1]) and in particular solutions for coupled linear equations. Define then the vector F and the Pauli matrices

$$F=\begin{bmatrix}F\\R\end{bmatrix}\quad \sigma_3=\begin{bmatrix}1&0\\0&-1\end{bmatrix}\quad \sigma_1=\begin{bmatrix}0&1\\1&0\end{bmatrix}\quad \sigma_2=\begin{bmatrix}0&-j\\j&0\end{bmatrix} \tag{A6.3}$$

to write eqns. A6.1 and A6.2 as

$$\frac{1}{v_g}\frac{\partial F}{\partial t}+\sigma_3\frac{\partial F}{\partial z}-jκ\sigma_1 F-(g-jδ)F=0 \tag{A6.4}$$

If, for example, there were no gain, phase or space variations, then

$$\frac{1}{v_g}\frac{\partial F}{\partial t}-jκ\sigma_1 F=0 \tag{A6.5}$$

The reader who has not met the concept of matrix exponentials will rapidly appreciate that one may, using a Taylor series, formally write

$$\exp(j\sigma_1 y)=1+j\sigma_1 y-(1/2!)(\sigma_1 y)^2-j(1/3!)(\sigma_1 y)^3+\cdots \tag{A6.6}$$

Using $\sigma_1^2=1$ (more strictly this is the identity matrix), one recovers the identity

$$\exp(j\sigma_1 y)=\cos(y)+j\sigma_1\sin(y) \tag{A6.7}$$

A solution then for eqn. A6.5 using Pauli matrices yields

$$F(t)=\cos(κv_g t)+j\sigma_1\sin(κv_g t) \tag{A6.8}$$

The object now is to integrate in space with a constant-frequency input with the offset frequency determined by $δ$ so that we rearrange eqn. A6.4 by multiplying through by σ_3 to obtain

$$\frac{\partial F}{\partial z}+\sigma_3\frac{1}{v_g}\frac{\partial F}{\partial t}-\sigma_3(g-jδ)F+κ\sigma_2 F=0 \tag{A6.9}$$

Neglecting the time variation for the moment, the formal solution of this equation is

$$F(z)=\exp\{-j(\sigma_3 D-\sigma_2 jκ)z\}F(0) \tag{A6.10}$$

where, for a shorthand in this appendix only, $D=δ+jg$.

However, because $\sigma_3\sigma_2=-\sigma_2\sigma_3$, it may be shown by using Taylor series as in eqn. A6.6 that $\exp\{-j(\sigma_3 D-\sigma_2 jκ)z\}$ cannot be the same as $\exp\{(-j\sigma_3 D)z\}\exp\{(-\sigma_2 κ)z\}$. The problem, stated briefly, is that the matrices σ_3 and σ_2 do not commute and so the matrices formed from $\exp\{(-j\sigma_3 D)z\}$ and $\exp\{(-\sigma_2 κ)z\}$ cannot commute. One has to consider

$$(\sigma_3 D-\sigma_2 jκ)(\sigma_3 D-\sigma_2 jκ)=(D^2-κ^2)-(\sigma_3\sigma_2+\sigma_2\sigma_3)jDκ=(D^2-κ^2) \tag{A6.11}$$

showing that

$$(\sigma_3 D-\sigma_2 jκ)^{2n}=(D^2-κ^2)^n \tag{A6.12}$$

$$(\sigma_3 D - \sigma_2 j\kappa)^{2n+1} = (D^2 - \kappa^2)^n (\sigma_3 D - \sigma_2 j\kappa) \tag{A6.13}$$

and hence, using the power-series expansions for cos and sin,

$$\exp\{-j(\sigma_3 D - \sigma_2 j\kappa)z\} = \cos(\beta_e z) - j\{(\sigma_3 D - \sigma_2 j\kappa)/\beta_e\}\sin(\beta_e z) \tag{A6.14}$$

where

$$\beta_e^2 = (D^2 - \kappa^2) = \{(\delta + jg)^2 - \kappa^2\}$$

Apply eqn. A6.14 to find the solutions for eqn. A6.9 for a length z:

$$F(z) = \cos(\beta_e z) F(0) + \frac{\sin(\beta_e z)}{\beta_e} \{(g - j\delta)\sigma_3 - \kappa\sigma_2\} F(0) \tag{A6.15}$$

where $\beta_e^2 L^2 = \{(\delta + jg)^2 - \kappa^2\} L^2$ and δ gives the 'frequency offset'.
Expanding out in full,

$$\begin{bmatrix} F(z) \\ R(z) \end{bmatrix} = \begin{bmatrix} a+d & jb \\ -jb & a-d \end{bmatrix} \begin{bmatrix} F(0) \\ R(0) \end{bmatrix} \tag{A6.16}$$

where $a = \cos(\beta_e z)$, $d = \{(g - j\delta)/\beta_e\}\sin(\beta_e z)$ and $b = (\kappa/\beta_e)\sin(\beta_e z)$.
Observe that $a^2 - d^2 - b^2 = 1$ so that

$$\det \begin{bmatrix} a+d & jb \\ -jb & a-d \end{bmatrix} = 1$$

This then permits eqn. A6.16 to be arranged more usefully so that the two outputs are in terms of the two inputs:

$$\begin{bmatrix} F(L) \\ R(0) \end{bmatrix} = \frac{1}{a-d} \begin{bmatrix} 1 & jb \\ jb & 1 \end{bmatrix} \begin{bmatrix} F(0) \\ R(L) \end{bmatrix} \tag{A6.17}$$

It is helpful to consider a unit input, $F(0) = 1$ with $R(L) = 0$, and then the reflection is given from $R(0) = jb/(a-d)$ while the transmission is $F(L) = 1/(a-d)$. With no gain or loss, d is always imaginary while b and a are always real so that $|R(0)|^2 + |F(L)|^2 = |F(0)|^2$.

Figure A6.1 then plots solutions using this approach. It is important to check the basic algebra of the calculation. One way is to look at the

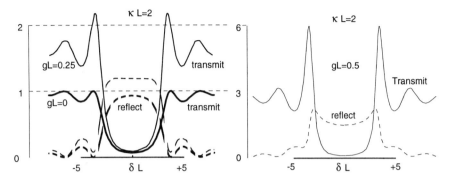

Figure A6.1 Power reflection and transmission for uniform DFB

Note that, when the structure is lossless and has no gain, $|T|^2+|R|^2=1$. Note also that the DFB acts as a stopband filter on transmission but the overall gain has peaks at the edges of this stopband. Eventually, as the gain increases, so the peak reflection and peak transmission tend to almost the same value. The classical oscillation condition is given when the reflection and transmission power gain become infinite so that zero power can excite a finite output

power reflection and the power transmission through a uniform section of DFB where $(\kappa L) = 2$. This is shown for three values of gain as the frequency deviation from the central structure, as measured by δL, varies. Typically the range of the stop-band has δL in excess of κL. The oscillation condition would give infinite reflection for a finite input.

With the uniform distributed-feedback laser of length L, there is no temporal growth rate for the inputs or outputs in the steady state. In the situation where the spontaneous emission is negligible, one has to have an output for zero input which means that $(a-d)=0$:

$$\cos(\beta_e L) = \{(g-j\delta)/\beta_e\} \sin(\beta_e L) \qquad (A6.18)$$

One can explore this threshold equation, for example with program dfbthr in directory **grating** from:

$$\text{error} = |(1/\beta_e L)\tan(\beta_e L) - 1/\{(g-j\delta)L\}| \qquad (A6.19)$$

At threshold with zero power output, the total intensity along the laser can be shown as being proportional to $|a(z)-d(z)|^2+|b(z)|^2$; however, this is not of much interest at operating power levels because spatial-hole burning, with variations in gain and photon density, play key roles. The field profiles given in Chapter 6 or from computation using the techniques of Chapter 7 are then more relevant.

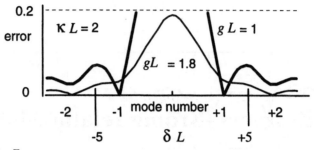

Figure A6.2 Error

Error is zero where oscillation modes appear. This is shown for $\kappa L=2$ and one can see here four modes. Two modes with gain $gL \sim 1$ with ± 1 modes while, if the gain is increased to $gL \sim 1.8$, these modes closest to the stopband grow rapidly with time and the ± 2 modes further away from the stopband just reach threshold

A6.1 Reference

1 GOLDSTEIN, H.: 'Classical mechanics' (Addison-Wesley, Reading, 1980), 2nd ed., p. 156 and exercises on pp. 185–186

Appendix 7
Kramers–Krönig relationships

A7.1 Causality

The Kramers–Krönig relationships provide fundamental rules for the relationships between the real and imaginary parts of the spectrum of any real physically realisable quantity when expressed as a complex function of frequency. One cannot, for example, design the optical gain spectrum to be any desired function of frequency without discovering that the phase spectrum is then closely prescribed. Frequently, such connections appear to be abstract and mathematically based. This appendix looks at three different ways of discovering these relationships which should help the reader to understand the fundamental nature and physics of the Kramers–Krönig relationships. The appendix includes at the end a collection of approximations for the real refractive indices of relevant laser materials.

The first way relies on a real 'causal' system where no real output $o(t)$ can occur from any system until *after* a real input $i(t)$ has occurred. An elegant way to proceed splits this real output $o(t)$ from such a system into even and odd parts, as in Figure A7.1, noting that

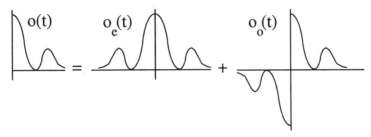

Figure A7.1 Splitting output into even and odd parts

An impulse response over positive time = even + odd functions over all time

$$o_o(|t|) = o_e(t) \tag{A7.1}$$

Now assume that the Fourier transform of $o(t)$ is $O_r(\omega) + jO_i(\omega)$ in the spectral domain so that the Fourier relationships give:

$$o_e(t) = \int_{-\infty}^{\infty} \{O_r(\omega') \cos(\omega't)\} \frac{d\omega'}{2\pi} \quad o_o(t) = \int_{-\infty}^{\infty} \{-O_i(\omega') \sin(\omega't)\} \frac{d\omega'}{2\pi} \tag{A7.2}$$

$$O_r(\omega) = \int_{-\infty}^{\infty} \{o_e(t) \cos(\omega t)\} \, dt \quad O_i(\omega) = \int_{-\infty}^{\infty} \{-o_o(t) \sin(\omega t)\} \, dt \tag{A7.3}$$

Using eqn. A7.1 links the real and imaginary parts of the spectrum [1] from

$$O_r(\omega) = \int_{-\infty}^{\infty} \left[\int_{-\infty}^{\infty} \{-O_i(\omega') \sin(\omega'|t|)\} \frac{d\omega'}{2\pi} \cos(\omega t) \right] dt \tag{A7.4}$$

$$O_i(\omega) = \int_{-\infty}^{\infty} \left[-\int_{-\infty}^{\infty} \{O_r(\omega') \cos(\omega't)\} \frac{d\omega'}{2\pi} \sin(\omega|t|) \right] dt \tag{A7.5}$$

The fact that cosine and sine are Hilbert-transform pairs permits one to show that $O_r = -\mathcal{H}(O_i)$ and $O_i = \mathcal{H}(O_r)$ where $\mathcal{H}$ corresponds to taking the Hilbert transform in frequency [2].

To apply this to optics, consider a short length h of optical medium with a real gain/loss $g(\omega)h$ and imaginary phase shift $\delta\phi(\omega)h$ as functions of frequency, and an input $I(\omega)$ (plane waves with *frequencies* ω) with a corresponding output $O(\omega)$ related to $I(\omega)$ by

$$O(\omega) \simeq I(\omega) \exp h\{g(\omega) + j\delta\phi(\omega)\} \quad \text{or}$$
$$\simeq \{1 + g(\omega)h + j\delta\phi(\omega)h\} I(\omega) \tag{A7.6}$$

Splitting into real and imaginary parts:

$$O_r(\omega) = \{1 + g(\omega)h\} I(\omega) \quad \text{and} \quad O_i(\omega) = \delta\phi(\omega) h I(\omega) \tag{A7.7}$$

If one allows a delta function or real impulse of field as an input in the time domain with $I(\omega) = 1$, there cannot be any real output $o(t)$ for

$t<0$. This therefore means that the results of eqns. A7.4 and A7.5 have to hold again and force an intimate connection between the gain and phase as a function of frequency. These relationships hold for *any* physically realisable outputs, of which an optical material's complex gain/phase is but one such quantity.

A7.2 Cauchy contours and stability

A second and equivalent form of these relationships is derived from the requirement that any realisable physical quantity has to be stable with time so that even optical gain which leads to lasing has to be stable with time. To explain this further, oscillations and instability occur only when the physical system with gain is placed into an environment with adequate feedback. The oscillation or lasing does not just happen on finding a semiconductor material with gain. Consider then the complex polarisation of a medium given as $\chi(\omega) = \{\chi_r(\omega) + j\chi_i(\omega)\}$ where changes in $\chi_r(\omega)$ and $\chi_i(\omega)$ provide the changes in refractive index and gain, respectively. Now $\chi(\omega)$ is a complex function of frequency and can be described through its poles and zeros in the complex frequency plane. It can only have 'stable' poles, i.e. poles in the complex frequency plane, where $\omega \to \omega' + j\alpha$ with $\alpha > 0$. Such poles, in the frequency domain, lead to a transient response in the time domain of the form $\exp(j\omega' t)\exp(-\alpha t)$ with $\alpha > 0$ for $t > 0$. Contrast this with a material where $\chi(\omega)$ has poles with $\alpha < 0$ which are associated with transients of the form $\exp(j\omega' t)\exp(+|\alpha| t)$. Then almost any disturbance, no matter how small, will grow without bound and *without any physical feedback*! Such material cannot therefore be found.

To discover the required restrictions, apply Cauchy's integral theorem to the complex susceptibility $\chi(\omega)$, taking an appropriate contour, as shown in Figure A7.2, where there are no poles inside the contour. The contour of radius $\delta_r \to 0$ gives the contribution $-j\pi\chi(\omega)$ while the contribution from the large contour goes to zero as $R \to$ infinity. Hence, in the limit on the real-frequency axis,

$$\int_{-\infty}^{\infty} \frac{\chi(\omega')}{(\omega' - \omega)} d\omega' = -j\pi\chi(\omega) \qquad (A7.8)$$

The next step is to take the real and imaginary parts so, with the contribution around the semicircles tending to zero, one has

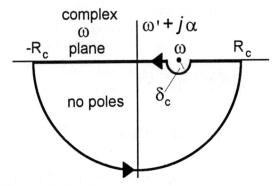

Figure A7.2   Contour for Cauchy
$R_c \to \infty$ and $\delta_c \to 0$

$$\oint \frac{\chi(\omega')}{(\omega' - \omega)} d\omega' = 0$$

$$\chi_r(\omega) = -\frac{1}{\pi} \int_{-\infty}^{\infty} \frac{\chi_i(\omega')}{(\omega' - \omega)} d\omega' \qquad (A7.9)$$

$$\chi_i(\omega) = \frac{1}{\pi} \int_{-\infty}^{\infty} \frac{\chi_r(\omega')}{(\omega' - \omega)} d\omega' \qquad (A7.10)$$

or $\chi_r = -\mathcal{H}(\chi_i)$ and $\chi_i = \mathcal{H}(\chi_r)$ where $\mathcal{H}$, as defined from the relationships above, corresponds to taking the Hilbert transform in frequency. Although this appears to be a different form from eqns. A7.4 and A7.5, it is the same result using different expressions for ways of finding Hilbert transforms.

A7.3 A proper physical basis builds in causality

Provided that one uses a physical model for a system, then one finds that the above difficult mathematics can be circumvented and one always satisfies the Kramers–Krönig relationships. For example, if each electron were 'bound' to a 'centre' with some damping and restoring force, then a *physical* equation of damped harmonic motion of an electronic dipole (see Appendix 5) is developed:

$$\frac{d^2x}{dt^2} + |\gamma| \frac{dx}{dt} + \Omega^2 x = q E(t) / m^* \qquad (A7.11)$$

With $E(t)$ varying as $E_\omega \exp(j\omega t)$, one finds a susceptibility (see Appendix 5 again) proportional to $x(\omega)/E_\omega$ which, for a density of electrons N, is of the form

$$\chi_{loss}(\omega) = \Omega_{plasma}^2 / (\Omega^2 + j\omega\gamma - \omega^2) \qquad \text{where } \Omega_{plasma}^2 = (Nq^2/\varepsilon_0 m^*)$$
$$= \Omega_{plasma}^2 \{(\Omega^2 - \omega^2) - j\omega\gamma\}/\{(\Omega^2 - \omega^2)^2 + (\gamma\omega)^2\} \qquad (A7.12)$$

The even and odd relationships between the real and imaginary parts of the susceptibility appear naturally, and can be shown [3] to satisfy the Hilbert relationships that are given in eqns. A7.9 and A7.10.

There can be several different components for the susceptibility with different resonances, losses etc. For example, if $\Omega \to 0$ and $\gamma \to 0$, appropriate for a free electron plasma with no damping forces, one finds a negative component of real susceptibility at frequencies $\omega \gg \Omega$ giving a negative refractive index. This is the component responsible for the *plasma* effect referred to in relation to the linewidth-broadening factor:

$$\chi_{free\ electron}(\omega) = -\Omega_{plasma}^2 / \omega^2 \qquad (A7.13)$$

This is a useful guide to the contribution to the susceptibility given by the free electrons in the conduction band, where one notes that increasing the electron or hole density reduces the permittivity or real refractive index of the material. Some damping or absorption might be added to simulate the free electron absorption which occurs in practice, but this has been neglected here.

Suppose the material changes from loss to gain; then the (Kramers–Krönig) stability of the material requires that one still has $(\gamma/\Omega) > 0$ so that the material gain on inversion would, in this dipole model, have to come from a 'negative' effective mass (the equivalent of inversion) giving

$$\chi_{gain} = -\Omega_{gain}^2 \{(\Omega'^2 - \omega^2) - j\omega\gamma'\}/\{(\Omega'^2 - \omega^2)^2 + (\gamma'\omega)^2\} \qquad (A7.14)$$

Hence, in general, there are at least three main forms of the variation with frequency around any local frequency: plasma, loss and gain. These components have to be added appropriately in terms of their susceptibilities using the Clausius–Masotti (also known as the Lorenz–Lorentz) relation [4,10], but the broad overall effect locally around any one frequency is to be able to split the susceptibility into general forms such as:

362 Distributed feedback semiconductor lasers

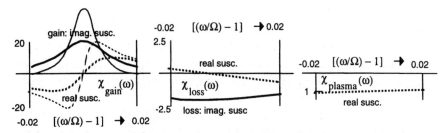

Figure A7.3 *Shape of contributions for real susceptibility and imaginary susceptibility*

The same normalising frequency Ω is taken for all components for the centre of gain or centre of loss and the plasma effect. Parameters $\gamma_{gain}/\Omega = 0.05$ for broader gain and $\gamma_{gain}/\Omega = 0.025$ for narrower gain line; $\gamma_{loss}/\Omega = 0.45$. Note that χ is not shown to any physical scale; the sketch is simply to show the shape of the contributions with changing frequency

$$\chi = \chi_{gain}(\omega) + \chi_{loss}(\omega) + \chi_{free\ electron}(\omega) \tag{A7.15}$$

$$= -\Omega_{gain}^2\{(\Omega'^2 - \omega^2) - j\omega\gamma'\}/\{(\Omega'^2 - \omega^2)^2 + (\gamma'\omega)^2\}$$

$$+ \Omega_{plasma}^2\{(\Omega^2 - \omega^2) - j\omega\gamma\}/\{(\Omega^2 - \omega^2)^2 + (\gamma\omega)^2\} - \Omega_{plasma}^2/\omega^2 \tag{A7.16}$$

It is important not to expect eqn. A7.16 to give valid results over a wide range of frequencies, because it is not possible by these simple methods to produce the asymmetric gain curves that necessarily arise from the existence of the bandgap (e.g. no gain below the bandgap frequency), nor does eqn. A7.16 give the correct asymptotic behaviour [5]. The results give a few key dependences that are physically permitted and sketched in Figure A7.3. (If one knows the spectrum over the whole frequency range, one can apply the full Kramers–Krönig relationships.)

Figure A7.3 uses eqn. A7.15 to produce a schematic change of complex susceptibilities for a narrow range of frequencies around the central gain peak. Note how, as the gain narrows, so the associated excursion in the real permittivity increases [6]. This is the essential basis to reductions in α_H by use of high-gain, narrow-bandwidth quantum-well materials.

A7.4 Refractive index of transparent quaternary alloys

More elaborate models for the permittivity of materials are found in the literature, basing the translation from the imaginary part of the permittivity into the real part through the use of Kramers–Krönig relationships. One example is given here of the single-effective-oscillator model which can be used to provide an approximate expression for the refractive index of the quaternary alloys used to fabricate the waveguide structure for semiconductor lasers. One notes that, if one takes the permittivity as in eqn. A7.12 then, as the term γ becomes smaller and smaller but the resonance remains, so $\varepsilon_{imaginary} \to \frac{1}{2}\varepsilon_{loss}(0)\ \Omega\pi\delta(\omega-\Omega)$. The more sophisticated models then suggest taking the single-oscillator model and superimposing a range of such single oscillators over a finite range of frequencies to give more realistic matches to the physically observed refractive index. One such idea is discussed in [7] where putting $\mathscr{E}=\hbar\omega$ gives the photon energy, the loss part of the imaginary component of permittivity is taken to be $\varepsilon_{imaginary}=\eta\mathscr{E}^4$ ($\mathscr{E}_g<\mathscr{E}<\mathscr{E}_f$) with $\varepsilon_{imaginary}=0$ elsewhere and the parameter η is a 'constant' of proportionality. The Kramers–Krönig relationships are then applied to this term using standard integrals to give a general equation describing the dependence of the refractive index n (using $n^2=\varepsilon_{real}$) on the bandgap energy $\mathscr{E}_g$ and the energy of the incident radiation $\mathscr{E}$ from:

$$n^2 = 1 + \frac{\mathscr{E}_d}{\mathscr{E}_o} + \frac{\mathscr{E}^2\mathscr{E}_d}{\mathscr{E}_o^3} + \frac{\eta\mathscr{E}^4}{\pi}\ln\frac{(2\cdot\mathscr{E}_o^2-\mathscr{E}_g^2-\mathscr{E}^2)}{(\mathscr{E}_g^2-\mathscr{E}^2)} \qquad (A7.17)$$

where

$$\eta = \pi\mathscr{E}_d/\{2\mathscr{E}_o^3(\mathscr{E}_o^2-\mathscr{E}_g^2)\} \qquad (A7.18)$$

and

$$\mathscr{E} = 1.2398/\lambda \qquad (A7.19)$$

with $\mathscr{E}$ in electron volts and λ in micrometres. $\mathscr{E}_o$ and $\mathscr{E}_d$ are parameters dependent on the composition of the alloys. This expression is valid provided that the energy of the incident radiation is less than the bandgap energy; in other words, the material is transparent. Hence, it cannot be used for the active region material in a laser which, in any case, will also be influenced by the injected-carrier concentration. We can now detail the specific parameter values applicable to the following alloy material systems:

(i) $In_{1-x}Ga_xAs_yP_{1-y}$ lattice-matched to InP [7,8]

$$\mathscr{E}_g = 1.35 - 0.72y + 0.12y^2 \qquad (A7.20)$$

$$x = 0.1894y/(0.4184 - 0.013y) \qquad (A7.21)$$

where

$$\mathscr{E}_o = 0.595x^2(1-y) + 1.626xy - 1.891y + 0.524x + 3.391 \qquad (A7.22)$$

$$\mathscr{E}_d = (12.36x - 12.71)y + 7.54x + 28.91 \tag{A7.23}$$

and x can only have values between 0 (InP) and 0.4672.

(ii) $In_{1-x-z}Ga_xAl_zAs$ lattice-matched to InP ($x=0.4672$) [9]

$$\mathscr{E}_g = 0.75 + 1.06z + 0.87z^2 \tag{A7.24}$$

where

$$\mathscr{E}_o = 2.48 + 1.427z + 2.366z^2 \tag{A7.25}$$

$$\mathscr{E}_d = 25.26 + 11.631z + 7.844z^2 \tag{A7.26}$$

Note that z can only take on values between 0 (InGaAs) and 0.48 for alloys lattice-matched to InP.

(iii) $In_{1-x}Ga_xAs_yP_{1-y}$ lattice-matched to GaAs [7]

$$\mathscr{E}_g = 1.918 - 0.593y + 0.171y^2 - 0.072y^3 \tag{A7.27}$$

$$x = 0.51 + 0.49y \tag{A7.28}$$

where:

$$\mathscr{E}_o = 3.813 - 0.663y + 0.643y^2 - 0.143y^3 \tag{A7.29}$$

$$\mathscr{E}_d = 32.755 - 2.715y + 6.056y^2 \tag{A7.30}$$

and x can take on values between 0.51 (GaInP) and 1.

Please note that in [7], (p. L138) the equivalent equation to eqn. A7.27 has a negative sign for the $0.758x^2$ term, and not positive as it should be.* The above expressions are useful for determining the refractive index of all the layers in a laser structure for waveguiding calculations, with the exception of the active region.

A7.5 References

1 LEE, Y.W.: 'Statistical theory of communication' (John Wiley, 1960)
2 DEUTSCH, R.: 'System analysis techniques' (Prentice Hall, Englewood Cliffs, 1969), p. 108
3 ERDELYI, A. (Ed.): 'Tables of integral transforms, Vol. 2' (McGraw-Hill, 1954) Use partial fractions and entries 6 and 7, Chap. XV, p. 244
4 ASHCROFT, N.W., and MERMIN, N.D.: 'Solid state physics' (Saunders College, Philadelphia, 1976), pp. 539–544
5 GREENHOW, M. 'High-frequency and low-frequency asymptotic consequences of the Kramers–Kronig relations' *J. Eng. Math.*, 1986, **20**, pp. 293–306

---

* GREENE, P.D.: Private communication.

6 YAMANAKA, T., YOSHIKUNI, Y., YOKOYAMA, K., and LUI, W.: 'Theoretical study on enhanced differential gain and extremely reduced linewidth enhancement factor in quantum well lasers', *IEEE J. Quantum Electron.*, 1993, **29**, pp. 1609–1616
7 UTAKA, K., SUEMATSU, Y., KOBAYASHI, K., and KAWANISHI, H.: 'GaInAsP/InP integrated twin-guide lasers with first order distributed Bragg reflectors at micron wavelength', *Jpn. J. Appl. Phys.*, 1980, **19**, L137–140
8 BROBERG, B., and LINDGREN, S.: 'Refractive index of $In(1-x)Ga(x)As(y)P(1-y)$ layers and InP in the transparent wavelength region', *J. Appl. Phys.*, 1984, **55**, pp. 3376–3381
9 GREENE, P.D., WHITEAWAY, J.E.A., HENSHALL, G.D., GLEW, R.W., LOWNEY, C.M., BHUMBRA, B., and MOULE, D.J.: 'Optimisation and comparison of InP-based quantum well lasers incorporating InGaAlAs and InGaAsP alloys'. International symposium on *GaAs and related compounds*, Jersey, 1990, *Inst. Phys. Conf. Ser.*, 112, chap. 8, p. 555
10 GUENTHER, R. D.: 'Modern optics' (Wiley, 1990) pp. 272–279

Appendix 8
Relative-intensity noise (RIN)

The relative-intensity noise (RIN) is an important quantity in determining whether lasers are acceptable for use in optical-communication systems. Its analytic study can require extensive algebra [1], but in this appendix the emphasis is on the physical significance of RIN and simulations using time-domain modelling to estimate its value for DFB lasers.

The discovery of the optical-fibre amplifier, with its relatively low-noise optical amplification, has maintained the interest in amplitude modulation for optical communication. Although both analogue and digital modulation are of interest, the discussion here focuses on digital intensity modulation. The net photon stream is turned from a digital-signal pattern (Figure A8.1a) imposed on a random photon stream (Figure A8.1b), the latter being created by (i) the random spontaneous emission and (ii) shot noise in the electron current driving the laser. So far this shot-noise term has not been considered, but it is easily added into the numerical model, as discussed shortly. The signal and noise combine (Figure A8.1c) and, taking the idealised case, with 100% efficiency for the photon detector, the photon stream is translated directly into electronic current in the load of the photodetector:

$$I(t) = (qP_{carrier}/\mathrm{h}f) + (qP_{noise}/\mathrm{h}f) \tag{A8.1}$$

where $P_{carrier}$ and P_{noise} give the photon power for the optical carrier (which is being digitally modulated) and optical noise, respectively. The current is matched into a resistor so that the ratio of equivalent electrical carrier power to noise power is

$$\mathrm{CNR}_{electrical\ equivalent} = \overline{(P_{carrier})^2} / \overline{(P_{noise})^2} \tag{A8.2}$$

where $\overline{(P)^2}$ indicates a temporal average. The electrical relative-intensity noise is the inverse, so that one writes

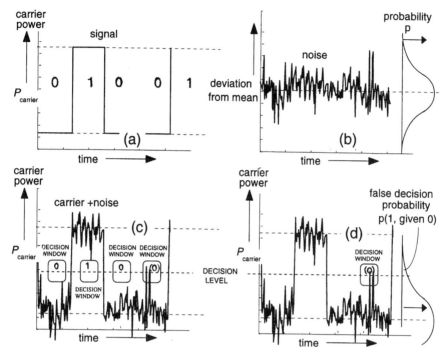

Figure A8.1 Importance of relative-intensity noise

(a) Shows an ideal digital signal
(b) Photon detection within a decision window varies with a Gaussian distribution
(c) Shows carrier power plus noise power. Decisions are made as to whether one or zero has been transmitted according to a detection of photons above or below the decision level within the decision window of time
(d) Noise peaks give a probability of a false decision being made so that a 1 instead of 0 is detected. Scale of both noise and probability are enhanced manyfold to illustrate the problem

$$\mathrm{RIN} = \overline{(P_{noise})^2} / \overline{(P_{carrier}^2)} \qquad (A8.3)$$

The importance of the RIN can be illustrated in Figure A8.1c where a decision has to be made as to whether a one or a zero has been detected. The probability distribution of the signal, within the time window required to make a decision, is given by $p(P_{carrier})$ (Figure A8.1b sketched as approximately Gaussian). The decision level is taken here to be at the 'midpoint' between the 'one' and 'zero' levels and the decision is made within the time window, controlled by the bit-rate clock (Figure A8.1c). Occasionally, on account of the contamination of the signal or carrier by noise, a signal which is correctly a 'zero' will be detected as a 'one' because the signal is above the threshold level for long enough within the decision window to return a 'one' state. One can increase the decision level until there is negligible probability of this happening; however, one then has to consider the probability of the detector returning a 'zero' when a 'one' has been transmitted. The decision level has to be chosen to minimise both probabilities and need not be exactly at the 'midpoint' as indicated here.

The probability of a false decision giving a 'one' is indicated by the small shaded area under the probability curve in Figure A8.1d indicating the integral over the relevant section of the probability distribution of the signal. A useful model is to take this probability distribution to be Gaussian (Figure A8.1b) with a variance which is determined by the mean-square deviation of the photon count within the decision window given by S_{count}:

$$\overline{(S_{count})^2} = \sigma^2 \qquad (A8.4)$$

The Gaussian probability function $p(S_{count})$ for finding a photon level S_{count} given a mean level $S_{count\,0}$ is then

$$p(S_{count})\,dS_{count} = (1/\sqrt{\pi})\,\exp\{-(S_{count} - S_{count\,0})^2/2\sigma^2\}\,(dS_{count}/\sigma\sqrt{2}) \qquad (A8.5)$$

Taking for simplicity $S_{count\,0} = 0$ for a zero, and take the decision level to be $S_{decision}$, then the probability of the signal rising above the decision level when a zero has been transmitted is, say, $p_{one\,given\,zero}$

$$p_{one\,given\,zero} = \int_{S_{decision}}^{\infty} (1/\sqrt{\pi})\,\exp(-S_{count}^2/2\sigma^2)\,d\{S_{count}/\sigma\sqrt{2}\} \qquad (A8.6)$$

$$p_{one\,given\,zero} = \int_{Q}^{\infty} (1/\sqrt{\pi})\,\exp(-x^2)\,dx \qquad \text{where } Q = S_{decision}/(\sigma\sqrt{2})$$

$$(A8.7)$$

A similar result holds for probability $p_{zero\,given\,one}$ of the signal falling below the decision level when a one has been transmitted. Assuming transmission probabilities p_{one} and p_{zero} for ones and zeros, then the net probability of a false signal at any detection decision is

$$p_{net} \approx p_{one} \cdot p_{zero\,given\,one} + p_{zero} \cdot p_{one\,given\,zero} \qquad (A8.8)$$

In general, the decision level is adjusted to give the minimum value for p_{net}. The analysis is simplified here by assuming that the noise at both the 'one' and 'zero' levels is the same and the decision level is adjusted to the midlevel so that $Q = S_{one}/(\sigma\, 2\sqrt{2})$ where S_{one} is the maximum output-photon count. For most digital systems, a maximum value of this false probability is usually taken as 10^{-9} or even smaller, so that the probability of a false decision is then

$$p_{net} \sim p_{one\ given\ zero} < 10^{-9} \rightarrow Q > 6 \text{ (as a useful guide)} \qquad (A8.9)$$

Hence at the peak power of the laser one requires a signal-to-noise ratio which is given from

$$20\log_{10}[(S_{count\ max})/\sigma\sqrt{2}] > 20\log_{10}(12) \sim 21.5 \text{ dB} \qquad (A8.10)$$

Analogue-television transmission using amplitude modulation is particularly demanding on the carrier-to-noise ratio for good-quality pictures, but these demands can be alleviated by using frequency modulation. Bearing in mind that photodetectors are not 100% efficient, one can show that, even with frequency modulation, one is likely to require the laser to have a net RIN of around -40 dB or lower [2]. Analogue applications may require RIN requirements below -50 dB for the net RIN integrated over the signal bandwidth. Because the signal bandwidth varies with different systems, it is considered to be more useful to plot the RIN of the laser against frequency in dB/(Hz) terms. In a laser, it is found that the RIN peaks at the photon–electron resonance f_r discussed in Section 3.2.3, which is typically in the gigahertz range with $f_r \propto (I_{drive} - I_{threshold})^{0.5}$.

To model RIN correctly, the laser model given in Chapter 7 should be modified to provide shot noise into the electron current. Shot noise should be modelled as a Poisson stream with a probability of one electron arriving in $\tau_{arrival}$ seconds where $\tau_{arrival}$ then determines the mean current. The mean arrival rate in any time interval T is then $N_{mean} = T/\tau_{arrival}$ with a variance that is also N_{mean}. However, with practical currents of a few milliamperes or more, the rate of arrival of electrons is so high that the Poisson stream is well approximated by having a Gaussian distribution with a mean and variance equal to the equivalent Poisson stream, even for subpicosecond time intervals. Hence if the mean number of electrons within any interval is N_e then the probability of $N_{mean} + N_e$ electrons is approximately $\{1/\sqrt{(2\pi N_{mean})}\}\exp\{-\tfrac{1}{2}(N_e^2/N_{mean})\}$. The time-step interval is s so that, if the mean current is I_0 and there are M laser sections, then there is a mean electron number per section per step given as $N_{mean} = sI_0/(qM)$. The variance in this number is also N_{mean} so that the current at each interval in each section is therefore

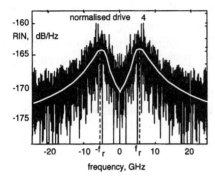 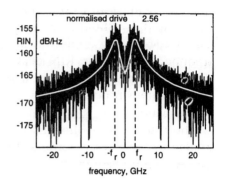

Figure A8.2 RIN simulations

$$I = (q/s)(N_{mean} + \sqrt{N_{mean}}\text{randn}) \quad (A8.11)$$

where randn is a normally distributed random number with variance of unity. A caveat has to be written, because it is possible for I to be negative if randn is less than $-\sqrt{N_{mean}}$. On such occasions, it would be more appropriate to return $I=0$ than a negative current. The frequency with which this error occurs gives a rough measure of the appropriate use of the Gaussian approximation to the Poisson distribution for large numbers. In certain physical cases, one may be able to provide some smoothing for the drive current [3–6] so that it has a variance less than a Poisson stream.

The estimation of RIN using time-domain modelling demonstrates the link between measuring low-frequency effects and the time of calculation. The model DFB programs which are provided only resolve frequencies down to about 3 GHz—this limitation can be removed by running a modified program for longer. The model RIN program (Figure A8.2) resolves down to frequencies around 25 MHz and takes about 90 min per run on a 166 MHz Pentium PC, calculating nearly a quarter of a million time steps and averaging appropriately. The time-domain model, by its very nature, gives a stochastic output so that one has to take the smoothed values to estimate the RIN but one can see how noise fluctuates as well as causing the signal to fluctuate with time. No two runs will be identical. The model described below uses the program dfbrin within directory dfb and looks at RIN for a $\lambda/4$-phase-shifted DFB laser using the standard parameters within the model.

The standard numerical laser model for a $\lambda/4$-phase-shifted DFB laser (that is provided with this book) is adapted to take two runs changing the drive levels. The differences in the resonance frequencies f_r are found to be proportional to $(I_{drive} - I_{threshold})^{0.5}$ with the normalised threshold of 2.1 and the normalised drives shown here of 4 and 2.56, giving approximately a twofold change in the resonance frequency where the RIN peaks. The mean intensity output with a normalised drive level of 4 is approximately 6 dB up on the drive level at 2.56. Intensity noise is always symmetrical around the zero frequency at baseband.

The resonance in the RIN occurs because of the photon–electron resonance, and is the same resonance as that discussed in Chapter 4 and Appendix 4.

A8.1 References

1 COLDREN, L.A., and CORZINE, S.W.: 'Diode lasers and photonic integrated circuits' (Wiley, 1995), pp. 230–241
2 GREEN, P.E.: 'Fibre optic networks' (Prentice Hall, Englewood Cliffs, 1993), chap. 9 gives an overview about the relative importance of different forms of noise on different systems
3 GOLUBEV, Y.M., and SOLOKEV, I.V.: 'Photon anti-bunching in a coherent light source and suppression of the photorecording noise', *Sov. Phys.–JETP*, 1984, **60**, pp. 234–238
4 MACHIDA, S., YAMAMOTO, Y., and ITAYA, Y.: 'Observation of amplitude squeezing in a constant current driven semiconductor laser', *Phys. Rev. Lett.*, 1987, **58**, pp. 1000–1003
5 RICHARDSON, W.H., MACHIDA, S., and YAMAMOTO, Y.: 'Squeezed photon number noise and sub-Poissonian electrical partition noise in a semiconductor laser', *Phys. Rev. Lett.*, 1991, **66**, pp. 2867–2870
6 LI, Y-Q., EDWARDS, P.J., LYNAM, P., and CHEUNG, W.N.: 'Quantum-correlated light from transverse junction stripe laser diodes', *Int. J. Optoelectron.*, 1996, **10**, pp. 417–421

Appendix 9
Thermal, quantum and numerical noise

A9.1 Introduction

There are at least three 'different' principles that can be used to calculate the spontaneous emission in a laser:
(i) the method given in Section 2.4.1;
(ii) a method which considers the spontaneous emission which has to occur in an ideal waveguide amplifier to obtain the correct 'noise' output [1, 2]. This amplifier can also be turned into an attenuator to see that the results are still consistent; and
(iii) the spontaneous emission which can also be found by the Einstein treatment of counting photon states [3].

All these methods are tied together to show consistency with one another. The appendix is concluded with essential material for interpreting the spontaneous emission in a numerical formalism.

The appendix first considers thermal and quantum noise and the ideal signal-to-noise power ratio that can be measured using a 100%-efficient photodetector. An ideal optical amplifier followed by an ideal detector is then considered. At the output of this ideal amplifier, the signal-to-noise-power ratio depends on the spontaneous emission and it is argued that this ratio has to be the same as the signal-to-noise-power ratio for the ideal direct detection of optical fields at the input. The ideal amplification cannot improve (or worsen) the signal-to-noise ratio. This argument gives one measure of the spontaneous emission. The same device considered as an attenuator in thermal equilibrium with its surrounds gives yet another measure of spontaneous emission, and finally spontaneous emission can be estimated through counting photon states.

A9.2 Thermal and quantum noise

Thermal noise at microwave frequencies provides an available power of $kT\Delta f$ flowing between two matched power absorbers, connected by a single-mode transmission line, all at a temperature T (where k is Boltzmann's constant, T is the absolute temperature and Δf is the effective frequency bandwidth of the measurement or the system). In thermal equilibrium, the noise power into a resistor balances the noise power flowing out. If the two absorbers are at different temperatures, net power flow is from the hotter to the cooler. The 'Nyquist' relationship $kT\Delta f$ forms the basis of calculations of signal-to-noise ratios at microwave and lower frequencies [4]. However, at room temperature, $kT\Delta f \sim 40$ pW of thermal power for every gigahertz of bandwidth so that, with the unlimited bandwidth of free space, astronauts would be incinerated were it not for the Bose–Einstein correction [5] giving the available thermal noise power as

$$P_{available} = hf\Delta f / \{\exp(hf/kT) - 1\} \quad (A9.1)$$

Below, say, 1000 GHz at room temperature, the Nyquist formula is recovered because $\{\exp(hf/kT) - 1\} \sim (hf/kT)$, but now eqn. A9.1 suggests that the noise at optical frequencies is negligible! Then quantum noise [6] has to be introduced and an additional term is conventionally included of the form

$$P_{noise} = \tfrac{1}{2}hf\Delta f + hf\Delta f / \{\exp(hf/kT) - 1\} \quad (A9.2)$$

This again suggests a catastrophe as the bandwidth Δf increases, but the additional term of $\tfrac{1}{2}hf\Delta f$, unlike the Nyquist term, is not available power but a measure of uncertainty in any measurement. Uncertainty may not be recovered as usable power and is not amplified by an ideal amplifier, but is simply present in any measurement.

The magnitude of this uncertain power depends in detail on how one measures the field [6]. With direct detection of the optical-field intensity by an ideal photodiode, $P_{uncertainty} = hf\Delta f$, so one may write

$$P_{uncertainty} / P_{signal} = hf\Delta f / P_{signal} \quad (A9.3)$$

The factor of $1/2$ in eqn. A9.2 requires coherent rather than direct detection. One way of understanding the result in eqn. A9.3 is to consider a source of light where there are, on average, p photons per second emitted over a signal bandwidth δf (the connection between δf and Δf emerges shortly). Then the detector receives an optical power

$$P_{signal\ input} = \mathsf{p}hf \quad (A9.4)$$

where f is the mean optical frequency. The direct detection process is random, and the electrons (one per photon if the detector is 100% efficient) emerge from the detector in a Poisson stream. The mean 'signal' current is $I_0 = q\mathsf{p}$ where q is the electronic charge and the random arrival of the electrons gives 'shot noise' [4], with a mean-square noise current:

$$i_n^2 = 2qI_0\delta f \quad (A9.5)$$

The electrical power into an ideal load of R ohms for the photo-detector gives

$$(P_{noise\ electrical}/P_{signal\ electrical})_{output} = (2qI_0\delta f)/(q\mathsf{p})^2 = 2\delta f hf/P_{signal\ input} \quad (A9.6)$$

Now recall that detection takes the signal down to baseband where the noise bandwidth runs from $-\delta f$ to $+\delta f$, and so the net bandwidth $\Delta f = 2\ \delta f$, so that eqn. A9.6 confirms the result of eqn. A9.3.

With coherent optical signals, it is possible to recover a factor of 2 by using homodyne detection which requires that the phase of the output is known. This additional information about the phase reduces the minimum uncertainty to $\frac{1}{2}hf\Delta f$ rather than the value of $hf\Delta f$ obtained by power (photon) detection. The minimum uncertainty in measurement of the optical-field power then directly relates to the best possible signal-to-noise-power ratio, given that one uses ideal measurement methods.

A9.3 Ideal amplification

Now turn to assessing the spontaneous emission which arises in the ideal amplifier of a forward optical field where the travelling-wave equation is given from (see eqn. 4.20)

$$\frac{1}{v_g}\frac{\partial E}{\partial t} + \frac{\partial E}{\partial z} - gE = i_{spf}(t, z) \quad (A9.7)$$

Here i_{spf} is proportional to the random dipole 'current' as a function of time that excites the spontaneous emission for the *forward* field and g is the field gain, at present taken to be a constant. To simplify the analysis, redefine the field variables:

$E = F \exp\{j(\omega t - \beta z)\}$ and $\bar{\imath}_{spf}(t, z) = i_{spf}(t, z) \exp\{-j(\omega t - \beta z)\}$ (A9.8)

so that $\bar{\imath}_{spf}(t, z)$ varies relatively slowly with time and space like F around some central frequency ω with $\beta = \omega/v_g$, so that eqn. A9.7 becomes adequately approximated by

$$\frac{\partial \{F \exp(-gz)\}}{\partial z} = \bar{\imath}_{spf}(t, z) \exp(-gz) \quad (A9.9)$$

Because $\bar{\imath}_{spf}(t, z)$ and $i_{spf}(t, z)$ differ only by a phase factor, the autocorrelations of each of the two excitations are equal (denoting ensemble/time averages by ⎯⎯):

$$\overline{\bar{\imath}_{spf}(t'', z'') * \bar{\imath}_{spf}(t', z')} = \overline{i_{spf}(t'', z'') * i_{spf}(t', z')} \quad (A9.10)$$

The random properties of the spontaneous emission are not changed fundamentally by the change of phase factor. The forward field and then the 'power flow' are next found from integration and using the fact that the spontaneous emission is random so that one is, in effect, just adding random spontaneously generated components to the field as the energy propagates along the guide:

$$F = F_{in} \exp(gz) + \exp(gz) \int_0^z \bar{\imath}_{spf}(t, z') \exp(-gz') \, dz' \quad (A9.11)$$

Squaring up on either side and ignoring terms like $F^* \bar{\imath}_{sf}$ which average to zero

$$F^* F = F_{in}^* F_{in} \exp(2gz) +$$

$$\exp(2gz) \int_0^z dz'' \int_0^z \bar{\imath}_{spf}(t, z')^* \exp(-gz') \bar{\imath}_{spf}(t, z'') \exp(-gz'') \, dz'$$

(A9.12)

Spontaneous excitations are assumed to be spatially uncorrelated, such that

$$\overline{\bar{\imath}_{spf}(t, z'') * \bar{\imath}_{spf}(t, z')} = r_{spf}(t)^2 \delta(z' - z'') \quad (A9.13)$$

The fields can be normalised so that F^*F gives the forward power flow:

$$F^* F = P_f = h f S_f \mathcal{A} v_g \quad (A9.14)$$

along the guide of effective area $\mathcal{A}$ where S_f is the photon density associated with the forward power flow. Then $\bar{r}_{spf}(t)^2$ averaged over time gives the mean-square noise power per unit length of amplifier. Integration over the length L gives the output at the end of the amplifier region from

$$F^*F = F_{in}^* F_{in} \exp(2gL) + \exp(2gL) \int_0^L \bar{r}_{spf}(t)^2 \exp(-2gz') \, dz' \quad (A9.15)$$

For future comparison, note that the rate of change of the mean power flow given by eqn. A9.15 can also be written at a position z as a sum of stimulated growth plus addition of spontaneous emission:

$$dP_f/dz = 2gP_f + \bar{r}_{spf}(t)^2 \quad (A9.16)$$

Writing $G_L = \exp(2gL)$ as the power gain along the guide gives

$$P_{fout} = G_L \left\{ P_{fin} + \frac{(G_L - 1)}{G_L} \frac{\bar{r}_{spf}(t)^2}{2g} \right\} \quad (A9.17)$$

For large enough gains with no losses of photons from input to output,

$$P_{fout} = G_L \left[P_{fin} + \frac{\bar{r}_{spf}(t)^2}{2g} \right] \quad (A9.18)$$

Now consider the interesting situation. At the input there was a power P_{in} but with a quantum uncertainty of the power (using direct detection) given from $hf\delta f$ over a bandwidth of δf. However, this uncertainty is not power that can get amplified but is merely an uncertainty in the measurement of the fields. At the end of an ideal quantum-amplification process with a sufficiently large gain G_L, the quantum uncertainty can be negligible compared with output power levels, but now it appears that there is an added available noise power of $\bar{r}_{spf}(t)^2/2g$ produced by amplifying the available thermal noise. The measurements at the input and the output are ideal measurement processes and, if the amplification is ideal quantum amplification with no absorptive loss of photons but only quantum spontaneous emission along with ideal stimulated gain, it is argued that the signal-to-noise ratio at the output must equal the signal-to-noise ratio $P_{fin}/(hf\delta f)$ at the input. Quantifying this result gives

$$2gP_{fin}/\bar{r}_{spf}(t)^2 = P_{fin}/hf\delta f \quad \text{or} \quad \bar{r}_{spf}(t)^2 = 2ghf\delta f \quad (A9.19)$$

Here then is a statistical measure of the mean-square spontaneous excitation at each point in the guide linking the spontaneous emission with the gain, as discussed in Section 2.3. Note that, in this work, $2g$ is the net power gain per unit length making allowances for any confinement factor Γ; some authors use g as the power gain. If there is a spectral filter, there may be only a narrow band δf of laser power that is relevant in determining the spontaneous noise but, taking the spontaneous emission over its whole range $\Delta_{sp}f$, then g has to be averaged appropriately.

A9.4 The attenuator

The calculation is now repeated with the field gain per unit length g changed into a loss per unit length $(-a)$ where one then finds that the output power is now given from:

$$P_{fout} = AP_{fin} + (1-A)\frac{\bar{r}_{spf}(t)^2}{2a} \tag{A9.20}$$

where $A = \exp(-2aL)$.

Consider what happens if the input power is just the thermal power given from

$$hf\Delta_{sp}f/\{\exp(hf/kT_{na}) - 1\} \tag{A9.21}$$

where T_n is the input noise temperature and $\Delta_{sp}f$ is the whole spontaneous-noise bandwidth. If the attenuator is in thermal equilibrium with its surrounds, T_n must be the same 'noise temperature' for the whole amplifier input ensuring that

$$\frac{\bar{r}_{spf}(t)^2}{2a} = hf\Delta_{sp}f/\{\exp(hf/kT_{na}) - 1\} \tag{A9.22}$$

The transition from attenuation to gain can be made via the concept of a negative reciprocal temperature where $(1/T_n)$ is large near absolute zero but becomes small by pumping sufficient electrons and holes into the interaction region. This pumping initially effectively increases their equivalent temperature with the attenuation a tending to zero until transparency is reached when $a = 0$. Then with still harder pumping, a changes sign to become gain $g(=-a)$ with $(1/T_n) \to -|(1/T_n)|$ and with yet harder pumping still to give large enough 'population inversion' $(1/T_n) \to -\infty$ with $\exp(hf/kT_n) \to 0$ and one recovers eqn. A9.17. The conclusion that one arrives at is that eqn. A9.17 should be modified to allow for an arbitrary degree of pumping of the laser by writing

$$\bar{r}_{sp\,f}(t)^2 = (2g)\, n_{inv}\, hf \Delta_{sp} f \qquad (A9.23)$$

where $n_{inv} = 1/\{1 - \exp(-hf/k|T_n|)\}$ gives the measure of the pumping of the laser or the 'population inversion' of the charge carriers in the laser. The strongest inversion makes n_{inv} unity, but more typically $n_{inv} \sim 2$ or so. This discussion owes much to Henry [1] and Marcenac [2]. Ghafouri–Shiraz [7] has a more detailed treatment of the laser amplifier including losses. In the present work, the important question that is addressed is the magnitude of the ideal spontaneous-emission sources.

In eqn. 5.9 no allowance was made for any difference in the spontaneous noise in the forward *and* the reverse directions because, from symmetry, the mean values for both the forward and reverse emission are equal in magnitude: $\bar{r}_{sp\,f}(t)^2 = \bar{r}_{sp\,r}(t)^2 = r_{sp}^2$. However, the equality of mean magnitude must not be mistaken for correlation.

A9.5 Einstein treatment: mode counting

Next consider the Einstein treatment of stimulated and spontaneous interactions with charge carriers in a uniform waveguide of effective cross-sectional area $\mathcal{A}$ using the two-energy interaction model of Section 2.3.1 for the interaction between photons and the charge carriers in the medium. Recalling eqn. 2.10,

$$dS/dt = K_{stim} S(NN_2 + PN_1 - N_1 N_2) + \beta_{sp} K_{sp} PN \qquad (A9.24)$$

Make the translation that $d(v_g t) \to dz$ and use eqn. A9.14 where $P_f = hf S_f \mathcal{A}\, v_g$ and follow Section 2.3.1 using the Fermi–Dirac statistics but with the inversion factor n_{inv} then eqn. A9.24 becomes

$$dP_f/dz = \{K_{stim}(PN/n_{inv})/v_g\} P_f + hf \beta_{sp} K_{sp} PN \mathcal{A} \qquad (A9.25)$$

The spontaneous term $\beta_{sp} K_{sp} PN$ in eqn. A9.24 gives only that spontaneous emission per unit length which is actually coupling to the waveguide and being subsequently amplified. Comparing eqn. A9.25 with eqn. A9.16 shows the links

$$2g = (K_{stim}/v_g n_{inv}) PN \qquad r_{sp\,f}^2 = hf \beta_{sp} K_{sp} PN \mathcal{A} \qquad (A9.26)$$

From eqns. A9.19 and A9.26,

$$r_{sp\,f}^2 = (2g)\, n_{inv}\, hf\, \delta f = (K_{stim}/v_g) PN hf\, \delta f = hf \beta_{sp} K_{sp} PN \mathcal{A} \qquad (A9.27)$$

From eqn. 2.12, $K_{sp} = B$ where B is the bipolar recombination constant for the material. The connection between K_{sp} and K_{stim} was found from eqn. 2.19 considering the equilibrium conditions and black-body-radiation density to give eqn. 2.20 written as

$$(K_{sp}/K_{stim}) = 8\pi\Delta_{sp} f(f^2/c_m^2 v_g) \tag{A9.28}$$

On rearranging eqn. A9.27 with the result of eqn. A9.28, explicit values for the spontaneous excitation and spontaneous-coupling coefficients are obtained as

$$r_{sp}^2 = hf\beta_{sp}BPN\mathcal{A} \tag{A9.29}$$

$$\beta_{sp} = (1/2)(\lambda_m^2/4\pi\mathcal{A})(\delta f/\Delta_{sp} f) \tag{A9.30}$$

A9.6 Aperture theory

Now from basic aerial theory [8], any aerial with an aperture $\mathcal{A}$ embedded wholly in a material where the central wavelength is λ_m has an aerial 'gain' G_{aerial} given from

$$G_{aerial} = (\mathcal{A}4\pi/\lambda_m^2) \tag{A9.31}$$

Here it is assumed that the whole aerial is immersed in the medium so that λ_m is the effective wavelength appropriate to that material. The 'gain of an aerial' can also be found by finding the average solid angle of emission and comparing this with the total 4π steradians so that the effective solid angle of emission Ω from the aerial is given from $G_{aerial} = 4\pi/\Omega$ and hence $(\lambda_m^2/4\pi\mathcal{A}) = \Omega/4\pi$.

At first sight, the effective aperture of all the radiating dipoles calls for a full analysis of the radiation of a dipole within the complex waveguide. However, one already knows that the net field pattern of all this spontaneous emission that actually couples to the waveguide mode, by definition of the mode, has to have an effective cross-section of $\mathcal{A}$ (the effective cross-section of the waveguide), so $\mathcal{A}$ is also the effective aperture of each dipole so that $\Omega = \lambda_m^2/\mathcal{A} = \lambda^2/\mathcal{A}\varepsilon_{rr\,eff}$ where $\varepsilon_{rr\,eff}$ is the real effective relative permittivity of the guide. This is exactly the same answer as finding the solid angle of emission outside the waveguide into air and then recognising that Snell's law reduces this solid angle by $(1/n_{rr\,eff})^2$ as derived in Section 2.4. Hence, for a single forward propagating mode,

$$r_{sp}^2 = \tfrac{1}{2}(\delta f/\Delta_{sp} f)(\Omega/4\pi)(B\mathcal{A}PNhf) \tag{A9.32}$$

As pointed out in Section 2.4, the factor of $\tfrac{1}{2}$ recognises that only one polarisation couples to the waveguide; the factor $(\delta f/\Delta_{sp} f)$ recognises

380 Distributed feedback semiconductor lasers

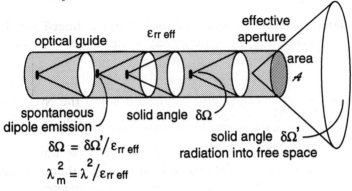

Figure A9.1 Schematic of spontaneous dipole emission

Solid angle Ω determined by area of guide because this becomes the effective aperture for the whole system of spontaneous dipoles. Comparison is made with the solid angle into free space to show that internal solid angle is correctly reduced by $\varepsilon_{\text{effective}}$

that only a fraction of the spontaneous emission over its whole spectrum couples into the lasing-mode frequencies; the factor $(\Omega/4\pi)$ recognises that spontaneous emission covers the whole 4π steradians but only couples into the appropriate solid angle for the waveguide (here Ω is taken as appropriate for the solid angle of, say, the forward emission alone); while the term $(B\mathcal{A}PNhf)$ gives the spontaneous emission over all possible polarisations. With a spontaneous-confinement factor Γ_{sp}, this emission rate is modified to $(\Gamma_{sp} B\mathcal{A}PN\,hf)$ with $\Gamma_{sp}=\Gamma$ in this work.

A9.7 Numerical modelling of spontaneous noise

It is now essential to discuss the modifications required for the finite-difference time-domain analysis and how to identify and model this spontaneous-emission noise numerically, given a space step of s. A straightforward method modifies eqn. A9.12; one finds with $g=0$:

$$F_{out}^{*}F_{out} = F_{in}^{*}F_{in} + \sum s\bar{i}_{spf}(Ts/v_g, Z's) \sum s\bar{i}_{spf}(Ts/v_g, "Zs)$$

(A9.33)

where spontaneous excitation $\bar{i}_{spf}(Ts/v_g, Zs)$ is inserted at each of the space points [T, Z] which are separated by s so that Z and T are integers defining the step numbers. The lack of correlation between each space and time point requires for distinct Z and Z' and distinct T and T':

$$\overline{s\tilde{i}_{spf}(Ts/v_g, Z's)s\tilde{i}_{spf}(Ts/v_g, Zs)} = 0 \quad (A9.34)$$

$$\overline{s\tilde{i}_{spf}(T's/v_g, Zs)s\tilde{i}_{spf}(Ts/v_g, Zs)} = 0 \quad (A9.35)$$

If one wishes to restrict the noise bandwidth below the Nyquist limit, one should introduce temporal correlation, as will be discussed shortly.

Now similar results can be formed for the reverse wave: there is nothing special about the forward direction other than the spontaneous excitation for the forward wave is uncorrelated with the excitation for the reverse wave. Hence one has

$$\overline{s\tilde{i}_{spr}(Ts/v_g, Z's)s\tilde{i}_{spr}(Ts/v_g, Zs)} = 0 \quad (A9.36)$$

$$\overline{s\tilde{i}_{spr}(T's/v_g, Zs)s\tilde{i}_{spr}(Ts/v_g, Zs)} = 0 \quad (A9.37)$$

with

$$\overline{s\tilde{i}_{spr}(T's/v_g, Z's)s\tilde{i}_{spf}(Ts/v_g, Zs)} = 0 \quad (A9.38)$$

but now for any T, T', Z and Z', even though these may be equal. This ensures that there is no coupling between the forward and reverse emission when there is no scattering from the forward field into the reverse field.

When space and time are equal in both excitation terms

$$\overline{s\tilde{i}_{spf}(T/v_g, Zs)s\tilde{i}_{spf}(Ts/v_g, Zs)} = sr_{spf}(t)^2 \quad (A9.39)$$

$$\overline{s\tilde{i}_{spr}(T/v_g, Zs)s\tilde{i}_{spr}(Ts/v_g, Zs)} = sr_{spr}(t)^2 \quad (A9.40)$$

with

$$\overline{r_{spf}(t)^2} = \overline{r_{spr}(t)^2} = r_{sp}^2 \quad (A9.41)$$

As the length of the laser waveguide $L=N\,s$ increases (retaining $g=0$ for this calculation), the mean spontaneous forward power increases as

$$F_{out}^*F_{out} = F_{in}^*F_{in} + Nsr_{sp}^2 \quad (A9.42)$$

These requirements are met for the forward field by inserting, at each space step and at every time step, a random excitation given by

$$s\tilde{i}_{spf}(Ts/v_g, Zs) = \{\tfrac{1}{2}(\Omega/4\pi)\,(\delta f/\Delta f)\,(\Gamma_{sp}\mathcal{A}sBPNhf)\}^{1/2}\mathcal{R} \quad (A9.43)$$

where $\mathcal{R} = \mathcal{X} + j\mathcal{Y}$ with $\mathcal{X}$ and $\mathcal{Y}$ normally distributed random real numbers such that the mean-square value $\overline{\mathcal{R}^2} = 1$. The randomness of the complex $\mathcal{R}$ gives the correct zero-mean correlation in amplitude and in phase from point to point and time to time, as required for

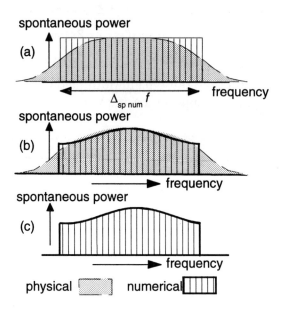

Figure A9.2   Spontaneous spectrum

    a Approximated by rectangular distribution with uncorrelated impulses at each time point
    b Slightly curved distribution achieved by partial correlation of impulses at adjacent time points
    c Offset central frequency by adding a phase factor

spontaneous emission. A similar but *independent* random value of $\mathcal{R}$ is used to excite the reverse field. This independence of random excitation of the forward and reverse fields is essential. The value given for the bandwidth δf in the computation needs consideration because, in modelling, the bandwidth is determined by the step length s.

With a step length for the finite-difference scheme of s, a single 'impulse' at one point in an array of points has a 'white' spectrum covering frequencies $\pm\frac{1}{2}$ (v_g/s) about the central frequency. The numerical bandwidth for the spontaneous emission is then $\Delta_{sp\,num}f = (v_g/s)$ determined from the spatial step length, and this is also the computing-system bandwidth. Retaining the approximation that the spontaneous spectrum is white over this range (v_g/s), one should replace $\delta f/\Delta_{sp}f$ with $(v_g/s\,\Delta_{sp}f)$:

$$s\bar{\iota}_{sp\,f}(\mathsf{T}s/v_g,\ \mathsf{Z}s) = \{\tfrac{1}{2}(\Omega/4\pi)\,(v_g/s\Delta_{sp}f)\,(\Gamma_{sp}\mathcal{A}sBPNhf)\}^{1/2}\mathcal{R} \quad (A9.44)$$

Figure A9.2*a* indicates schematically the approximation that such a 'white' spontaneous spectrum gives.

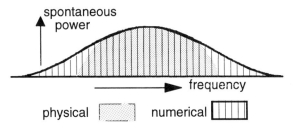

Figure A9.3 Numerical-noise spectrum

With short enough space step, the numerical-noise spectrum may extend beyond the physical spectrum but can be tailored to approximate to the physical spectrum in the central region

Band limiting of the spontaneous spectrum can be achieved by providing correlations between the spontaneous emissions and time. This can be done with a numerical filter as in Chapter 7 or, more simply, if approximately, by correlating the noise at just two adjacent time steps by having, for example:

$$s\bar{i}_{spf}(Ts/v_g, Zs)_{new} = \bar{i}_{spf}(Ts/v_g, Zs)\cos\phi + \bar{i}_{spf}\{(T-1)s/v_g, Zs\}\sin\phi \quad (A9.45)$$

With $\sin\phi \sim 0.1$, for example, then the spontaneous spectrum alters as in Figure A9.2b. An offset central frequency is achieved by introducing a phase shift; for example, writing $\sin\phi \to \exp(j\,\delta)\sin\phi$ where in Figure A9.2c $\sin\phi = 0.1$ and the phase factor is $\delta = 0.25$. The factors δ and ϕ then constitute parameters which can be fitted to the requirements. The filtering in time and space can be done so as to fit with eqns. A9.36 and A9.37. Readers are encouraged, as an exercise, to set up a MATLAB program to test for themselves this band limiting and spectral shift of the spontaneous emission.

If one normalises $F^*F \to (S_f \mathcal{A}\, v_g)$ so as to give the photon numbers per second rather than the optical power level, one omits hf from eqn. A9.44 to give

$$s\bar{i}_{spf}(Ts/v_g, Zs) = \{\tfrac{1}{2}(\Omega/4\pi)(v_g/s\Delta_{spf})(\Gamma_{sp}\mathcal{A}sBPN)\}^{1/2}\mathcal{R} \quad (A9.46)$$

$$= \{\Gamma_{sp}V_{sp}BPN\}^{1/2}\mathcal{R} \quad (A9.47)$$

where $V_{sp} = (K_{tr}\lambda^2 v_g/8\pi\,\varepsilon_{rr\,eff}\,\Delta_{spf})$ is the effective volume for the spontaneous emission per step length. The transverse Petermann factor K_{tr} (see Section 2.4.2) has been included for generality to cover gain guides not discussed above, but for index guides as discussed here it is set at 1. The shorter the step length, the wider is the bandwidth of spontaneous noise, and this increase in noise is taken into account automatically by having more steps per given length. This theory assumes only a single transverse mode, so that the larger the cross-sectional area or effective aperture of the guide, the higher is the aerial gain for a small dipole radiating into that guide and the smaller is the coupling angle of spontaneous emission into that mode. This is why the effective area $\mathcal{A}$ cancels in its effect on the noise. With gain guides, the Petermann factor can well be a function of $\mathcal{A}$.

The values for the reverse spontaneous excitation are identical in formulation to those for the forward excitation given above, but generated with completely independent random generator $\mathcal{R}$ from the generator in eqn. A9.46.

If the step length s is made so short that $v_g/\text{s} > \Delta_{sp}f$, one has to reconsider eqn. A9.45 and is forced to introduce a spectral filter, for example by correlating the spontaneous noise at two adjacent step points. If, for example, one had

$$s\bar{i}_{spf}(\text{Ts}/v_g, \text{Zs})_{new} = (1/\sqrt{2})\bar{i}_{spf}(\text{Ts}/v_g, \text{Zs}) + (1/\sqrt{2})\bar{i}_{spf}\{(\text{T}-1)\text{s}/v_g, \text{Zs}\}$$

(A9.48)

and s is sufficiently short, the numerical-noise spectrum, may extend beyond the physical-noise spectrum, as in Figure A9.3, but could approximate the significant central region adequately.

A9.8 Higher-order noise statistics

It has so far been assumed that the only terms which matter in the modelling of the noise are the mean and second moment, so that there is zero mean correlation between adjacent points of spontaneous emission and also the correct mean-square spontaneous emission power. The current evidence is that this is often sufficient for much laser modelling, but one can see where the numerical model departs from the physics. For example, if the time step is so long that, during each time step, 100 electrons and holes on average recombine, the amplitude fluctuations will be different in detail from the situation where the time step is so short that on average only 1/100 electrons and holes recombine (modelling admits of fractional electrons!). This problem of balancing the time step and spontaneous emission to obtain the correct granularity of emission has not been explored in this modelling primer, and has to be left for further research.

A9.9 References

1 HENRY, C.H.: 'Theory of spontaneous emission noise in open resonators and its application to lasers and optical amplifiers', *J. Lightwave Technol.*, 1986, **4**, pp. 288–297
2 MARCENAC, D.D., and CARROLL, J.E.: 'Quantum mechanical model for realistic Fabry–Perot lasers', *IEE Proc.–J.*, 1993, **140**, pp. 157–171
3 THOMPSON, G.H.B.: 'Physics of semiconductor laser devices' (Wiley, 1980)
4 FREEMAN, J.J.: 'Principles of noise' (Wiley, 1958)
5 KITTEL, C., and KROEMER, H.: 'Thermal physics' (W.H. Freeman and Co., 1980), 2nd ed.
6 MARCUSE, D.: 'Principles of quantum electronics' (Academic Press, 1980)
7 GHAFOURI-SHIRAZ, H.: 'Fundamentals of laser diode amplifiers' (Wiley, Chichester, 1995)
8 RAMO, S., WHINNERY, J.R., and VAN DUZER, T.: 'Fields and waves in communication electronics' (Wiley, 1994), 3rd ed.

Appendix 10
Laser packaging

A10.1 Introduction

The extra requirements of packaging optoelectronic rather than just electronic devices are challenging [1], and real packages involve numerous compromises between different aspects of performance, as well as manufacturing costs. To make the best of a well designed device, DFB-laser chips must be packaged so as

(i)   to protect them from the environment;
(ii) to enable monitoring of the optical output;
(iii) to provide thermal heat sinking;
(iv) to provide a rigid and stable optical coupling platform to a fibre (or other output system); and then
(v)   to allow electrical interfaces between the appropriate 'DC' and modulated RF current supplies and the laser.

A10.2 Electrical interfaces and circuits

The need to obtain stability and precision of the mechanical alignment in laser packages often results in a metal package with glass-to-metal seals for electrical connecting pins, and glass windows or carefully designed fibre interfaces. Rapid progress is being made with ceramic packages incorporating stripline interfaces, and some laser packages are now partly injection moulded. The variability of package and laser parameters is large, and the parameters relevant to any particular laser and its package need to be measured appropriately and then used in any modelling. The values quoted here are for a typical metal package with glass-to-metal seals.

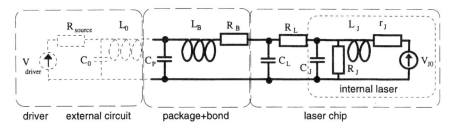

Figure A10.1 Schematic of laser package

Showing electrical parasitic capacitances and inductances and interfaces

Figure A10.1 indicates such typical parasitic circuit components which need to be considered in evaluating the overall frequency response of a laser to direct modulation. The glass-to-metal seal of a package may present a lead-through capacitance $C_P \sim 0.5$ pF, and the fine gold or aluminium bond wire to the laser chip usually contributes an inductance $L_B \sim 1$ nH, giving a characteristic filter impedance $\sim 50\ \Omega$ to match the driver's source impedance but with a cutoff frequency as low as 7 GHz. The laser capacitance C_L may or may not help compensate to a cutoff around 10 GHz, but very careful microwave design of special packages is needed for still higher frequencies. 'SPICE' models are likely to be available for components within the electronic driver, and sophisticated 3D modelling and measurement [2] may be necessary to establish the characteristic of the laser package.

The RF equivalent circuit of lasers [3] suggests that the internal laser dynamics adds an inductance to a series dynamic resistance r_J which, with C_J, helps to give a first-order account of the photon–electron resonance within the laser along with the electrical-to-optical power conversion determined by r_J. The shunt resistance R_J along with C_J determines the damping of the modulation and relaxation rate to equilibrium. While this gives an approximation to the laser's dynamic characteristics, the steady current through the laser determines a junction voltage V_{J0} which, in turn, determines a steady recombination current through R_J. The internal recombination rate should be different dynamically from the static value, requiring more elements than are shown here. More complete modelling of laser and package strays becomes complex [4], and distributed equivalent circuits may have to be used instead of lumped circuits at high enough microwave frequencies where the laser's length becomes comparable with the modulation wavelength [5]. The package elements are indicated in Figure A10.1 as a 'lumped' equivalent-circuit model and are often simplified further for modelling (as in Appendix 11). Laser series resistances are typically 3–5 Ω for buried heterostructure devices, and 1–2 Ω for ridge-type devices. The parallel laser capacitance C_L varies strongly with the laser's structure: a typical buried-heterostructure laser may have $C_L \sim 200$ pF, whereas comparable 'ridge'-type lasers could have 20 pF, and specially designed lasers for high-speed operation may have $C_L < 5$ pF.

A10.3 Thermal considerations

Lasers are always sketched in this book as having their junctions close to the upper surface. In practice, they are usually soldered junction-side down directly onto a good heatsink (such as type-IIA diamond) which can then give thermal impedances of around 30 degC/W, reducing to, perhaps, 10 degC/W for 'long' lasers such as those used as pumps for erbium-fibre amplifiers. Chips mounted with the junction side uppermost will typically have thermal impedances 2–4 times greater, although 10 times greater is possible in some modules. Excessive heating makes almost all laser parameters worse; this is a particular problem with multilaser modules, where thermal crosstalk also exists. At some frequencies, thermally induced chirp is a serious effect, and can cause changes in wavelength >0.1 nm on timescales of 0.1–10 μs [6]. The dynamic thermal properties of the laser are controlled by several time constants, characteristic of parts of the laser and its heatsinking arrangement; however, to a first order, GaAs-based lasers may often be characterised by a single time constant of about 1 μs, while long-wavelength InP-based lasers suffer more from direct heating of their active regions, and often show significant heating effects in about 0.1 μs. These times appear long compared with 1 Gbit/s data, but many standards for data transfer allow long strings of '1's or '0's, and so thermally induced chirp and other effects may cause major problems.

A10.4 Laser monitoring

Many laser packages contain a photodiode designed to monitor the laser output, often merely detecting the slow variations in the average laser power. First, the monitor diode usually detects light from the back facet (which commonly has a coating to increase its reflection to ensure that more power emerges from the front facet). Power for the fibre is extracted from the front facet [often antireflection (AR) coated]. In Fabry–Perot lasers, the front and back powers track each other very well, but in DFB lasers spatial-hole burning and longitudinal shifts of the lasing power can result in considerable divergences (see Figure 5.18). In addition, efficient coupling of optical power from laser to fibre requires submicron tolerances, and is likely to vary with temperature, whereas power coupled to the monitor photodiode is stable due to the relatively large sensitive area. Hence monitoring of power in the system fibre is likely only to be accurate to ± 1–2 dB, and long-term 'creep' of the package may increase this figure. Caution is therefore needed in analogue or other high-linearity applications in using the monitor diode to control the laser's precise performance.

A10.5 Package-related backreflections and fibre coupling

As seen in Chapter 1, the proximity of the laser round-trip gain to unity makes the laser very sensitive to backreflections within its coherence length, which certainly applies to most packages. For example, Petermann [7] calculates that, at 5 mW output power, feedback levels below 10^{-4} are required to achieve linewidths of 70 MHz starting from an intrinsic linewidth of 12.5 MHz, assuming $\alpha_H = 6$. The critical effects noted for facet reflections in Chapter 6 can also occur with reflections from elements within the package. Monitor photodiodes are therefore usually angled with respect to the laser axis, and are often AR coated in order to have a negligible effect on the laser. 'Selfoc' lenses are occasionally employed to couple light from the laser-output spot to the cleaved end of a single-mode fibre and in these cases, if all the glass/air interfaces are flat, the typical 4% power reflection at each interface can create serious problems of interference, and interaction with the laser. Use of a fibre end which is rounded to form an integral lens (Figure A10.2) not only reduces the number of glass-to-air interfaces to one but also defocuses any reflected power, greatly reducing any coupling back into the laser of reflected light. Typical packages employing either discrete lenses or lens-ended fibres achieve coupling efficiencies of 20–40% for laser power into a single-mode fibre, although values above 70% are possible with more expensive assemblies and mode transformers [8]. The relatively large circular emission areas of VCSELs combined with appropriate lenses should achieve yet higher coupling efficiencies. These figures should be compared with only 5–10% coupling directly from a conventional edge-emitting laser

390 Distributed feedback semiconductor lasers

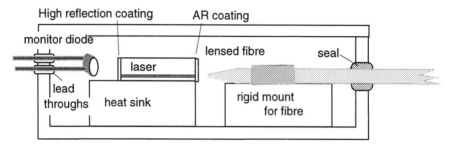

Figure A10.2   Schematic laser package

> Lensed fibre must be rigidly mounted, but the monitor diode is not quite so critical provided that it does not reflect light back into the laser. The laser is mounted with its junction closest to the heat sink; the package is normally hermetically sealed. Laser connections are not shown, but need to have low inductances with low lead-through capacitances. Rigid mount for fibre can be formed from a V-groove etched in silicon [10–12]

into a flat, cleaved single-mode fibre end, which also results in significant back reflections. Lasers without fibre pigtails but incorporating demountable optical connectors into their packages have an acute problem for high-performance use because the laser output should be coupled optimally to the fibre end face and therefore, by reciprocity, any reflection from this flat end will couple optimally back into the laser. While angling of the laser axis with respect to the axis of the output fibre alleviates this, the better solution for critical applications is to use a laser package incorporating a permanent fibre pigtail, and then splice this pigtail to the system fibre.

To reduce back reflections, 'optical isolators' can be employed in systems. These make use of the Faraday effect [9,10] of polarisation rotation under magnetic fields in suitable materials, and can attenuate any reflections coming back to the laser by ~30 dB for a single-stage unit, or 50–60 dB for a double-stage one with transmission losses of 0.5–1 dB; but some of them generate significant reflections from their own internal interfaces. Critical analogue applications such as multi-channel cable television will certainly require isolators, and the rest of the package will also need very careful design, probably involving antireflective coating to all optical interfaces.

In WDM systems where lasers with several different wavelengths are required, the coupling of many devices to many fibres in one package is even more challenging than coupling a single fibre to a single device. An ingenious solution has been found using anisotropic wet etching of silicon along crystal planes to produce well defined V-grooves. The lasers can be flip-chipped onto the silicon motherboard containing such V-grooves which can align the fibres to the laser with great precision [11–13], and as such are also useful in volume production for single lasers.

A10.6 References

1 MATTHEWS, M.R., MACDONALD, B.M., and PRESTON, K.R.: 'Optical components—the new challenge in packaging', *IEEE Trans. Compon., Hybrids Manuf. Technol.*, 1990, **13,** pp. 798–806
2 XENG, J-X.: 'A 3D electromagnetic simulator for high frequency applications', *IEEE Trans. Compon., Packag. Manuf. Technol. B, Adv. Packag.*, 1995, **18,** pp. 578–595
3 TUCKER, R.S., and POPE, D.J.: 'Microwave circuit models of semiconductor injection lasers', *IEEE Trans.*, 1983, **MTT—31,** pp. 289–294
4 DELPIANO, F., PAOLETTI, R., AUDAGNOTTO, P., and PULEO, M.: 'High-frequency modeling and characterization of high-performance DFB laser modules', *IEEE Trans. Compon. Packag. Manuf. Technol. B, Adv. Packag.*, 1994, **17,** pp. 412–417
5 TIWARI, S.: 'Transmission line delay limitations of laser bandwidths', *IEE Proc. Optoelectron.*, 1994, **141,** pp. 163–166
6 NEIFELD, M.A., and CHOU, W-C.: 'Electrical Packaging Impact on Source Components in Optical Interconnects', *IEEE Trans. Compon., Packag. Manuf. Technol., B, Adv. Packag.*, 1995, **18,** pp. 578–595
7 HELMS, J., KURTZKE, C., and PETERMANN, K.: 'External feedback requirements for coherent optical communication systems', *J. Lightwave Technol.*, 1992, **10,** pp. 1137–1141
8 ROBERTSON, M.J., LEALMAN, I.F., and COLLINS, J.V.: 'The expanded mode laser – a route to low cost optoelectronics', *IEICE Trans. Electron.*, 1997, **E80C,** pp. 17–23
9 GREEN, P.E.: 'Fibre optic networks', (Prentice Hall, Englewood Cliffs, 1993), p. 87
10 GOWAR, J.: 'Optical communication systems' (Prentice Hall, 1993), 2nd. edn., p. 438
11 HALL, S.A., LANE, R., WANG, H.C., and GARERI, A.: 'Assembly of laser-fibre arrays', *J. Lightwave Technol.*, 1994, **12,** pp. 1820–1826
12 PEALL, R.G., SHAW, B.J., AYLIFFE, P.J., PRIDDLE, H.F.M., BRICHENO, T., and GURTON, P.: '1×8, 8 Gbit/s transmitter module for optical space switch applications', *Electron. Lett.*, 1997, **33,** pp. 1250–1252
13 CANN, R., HARRISON, P., and SPEAR, D.: 'Use of silicon vee groove technology in the design and volume manufacture of optical devices', *Proc. SPIE*, 1997, **3004,** pp. 170–173

Appendix 11
Tables of device parameters and simulated performance for DFB laser structures

Table A11.1 Summary of material and device parameters used for large-signal dynamic modelling of uniform-grating DFB laser

Vertical and horizontal waveguide parameters

$W_{eff,\,vert}$	=	0.564 (0.548)* μm	effective width (vertical) for optical mode
d_{qw}	=	0.05 μm	total width of quantum wells
θ_{vert}	=	37.6° (40.4°)*	full-width half-maximum, FWHM for emission from facet
$W_{eff,\,hor}$	=	1.519 (1.587)* μm	effective width (horizontal) for optical mode
W	=	2.0 μm	ridge width
θ_{hor}	=	24.8°	FWHM for emission from facet
Γ_{MQW}	=	0.085 (0.088)*	overall confinement factor
$\mathcal{A}$	=	1.519 × 0.564 = 0.857 (0.869)* μm² effective optical area	
Ω	=	2π (37.6°/180) × (24.8°/180) = 0.181 (0.191)* emission solid angle (steradians)	

* The static modelling used similar parameters but where they differ from those used for the dynamic modelling they are shown in brackets

Gain, loss and linewidth parameters

dg/dN	=	3.0×10^{-16} cm²	differential field-gain
N_{tr}	=	1.5×10^{18} cm⁻³	transparency density
α_H	=	3.0	linewidth-enhancement factor
ϵ	=	1.0×10^{-17} cm³	nonlinear-gain parameter
a_{wg}	=	30 cm⁻¹	waveguide power loss

Carrier recombination and IVBA parameters

A	=	0	linear recombination (also written as $1/\tau_r$)
B	=	1×10^{-10} cm^3/s	'bimolecular' recombination—assumed radiative
C	=	1.3×10^{-28} cm^6/s	assumed nonradiative (Auger recombination)
b	=	2.0×10^{-17} cm^2	IVBA coefficient

Effective refractive indices

n_{eff}	=	3.28	giving phase velocity $v_p = c/n_{eff}$
$n_{eff\,g}$	=	3.70	giving group velocity $v_g = c/n_{eff\,g}$

Uniform grating parameters

$\kappa = 100$ cm^{-1}, $\kappa L = 3$, and $L = 300$ μm

Direct-current drive with parasitic resistances/capacitors in general ignored

Table A11.2 Material and device parameters used for modelling unstrained (InGa)As-(InGa)(AsP) 6QW, $2 \times \lambda_m/8$ DFB laser structure

Vertical and horizontal waveguide parameters

$w_{eff,\,vert}$	=	0.56 μm	effective width (vertical) for optical mode
d_{qw}	=	0.039 μm	6 QWs of 0.0065 μm thickness
θ_{vert}	=	47.3°	full-width half-maximum, FWHM for emission from facet
$w_{eff,\,hor}$	=	1.83 μm	effective width (horizontal) for optical mode
W	=	2.5 μm	ridge width
θ_{hor}	=	22.7°	FWHM for emission from facet
Γ_{MQW}	=	0.0687	overall confinement factor
$\mathcal{A}$	=	$1.83 \times 0.56 = 1.02$ μm^2	effective optical area
Ω'	=	$2\pi(47.3°/180) \times (22.7°/180) = 0.21$	solid angle of emission in steradians outside laser

Gain, loss and linewidth parameters

dg/dN	=	3.0×10^{-16} cm^2	differential field gain
N_{tr}	=	1.5×10^{18} cm^{-3}	transparency density
α_H	=	3.0	linewidth enhancement factor
ϵ	=	1.0×10^{-17} cm^3	nonlinear-gain parameter
a_{wg}	=	30 cm^{-1}	waveguide power loss

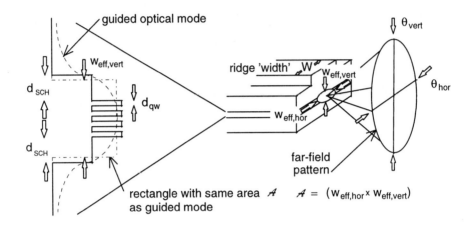

Figure A11.1 Waveguide parameters

Separate confinement parameters on left are relevant only to multiquantum-well devices
Total thickness of active regions $d_{mqw} = N_{qw} d_{qw}$ (N_{qw} = number of quantum wells)
Total confinement factor of mode to quantum wells $\Gamma_{mqw} = N_{qw} \Gamma_{qw}$ (Γ_{qw} = confinement factor per well)

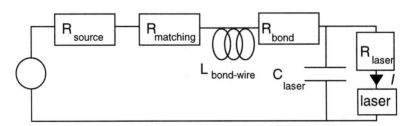

Figure A11.2 Laser and submount parasitics: simplified model

I is drive current into the laser. The dynamic-laser-model circuit parameters indicating the photon–electron resonance, which are shown schematically in Figure A10.1, are not shown here because these are included through the dynamic-modelling process

Carrier recombination and IVBA parameters

A	=	0	linear recombination (also written as $1/\tau_r$)
B	=	1×10^{-10} cm³/s	'bimolecular' recombination—assumed radiative
C	=	1.3×10^{-28} cm⁶/s	assumed nonradiative (Auger recombination)
b	=	2.0×10^{-17} cm²	IVBA coefficient

Effective refractive indices

n_{eff}	=	3.28	giving phase velocity $v_p = c/n_{eff}$
$n_{eff\,g}$	=	3.70	giving group velocity $v_g = c/n_{eff\,g}$

Grating parameters

$2 \times \lambda_m/8$ grating, $\kappa = 60$ cm⁻¹, $\kappa L = 2.1$, $L = 350$ μm
$\lambda_m/8$ phase shifts positioned symmetrically in cavity and spaced 245 μm apart, i.e. $\Delta L = 122.5$ μm

Carrier-transport parameters (see Appendix 4)

τ_{cap}	=	70 ps	diffusion and capture time constant
τ_{esc}	=	140 ps	escape and diffusion time constant
d_{SCH}	=	0.21 μm	separate confinement heterostructure width with two sections of equal length to give 0.42 μm total
Γ_{SCH}	=	0.56	confinement factor for SCH region
$dn/dN\vert_{SCH}$	=	-1.5×10^{-20} cm³	rate of change of refractive index with electron density in SCH region

Laser and submount parasitics

R_{laser}	=	5 Ω	C_{laser}	=	8.25 pF
$L_{bondwire}$	=	1.5 nH	$R_{bondwire}$	=	0.2 Ω
$R_{matching}$	=	43 Ω	R_{source}	=	50 Ω

Appendix 12
About MATLAB programs

A12.1 Instructions for access

The programs accompanying this book were written in MATLAB 4, but they have been designed to be compatible with the student edition [1] so that a recommended 'minimum' specification for a system is:

MATLAB 4 Student Edition, 100 MHz PC/Mac with 32 MB RAM.

Users should note that these programs are supplied 'as is', on a free basis and may be copied and modified by individual users but may not be redistributed or used commercially without permission of the copyright holder. While care has been exercised by the authors in producing and testing these programs, the publishers and the authors make no warranty for them, nor do they offer any maintenance, support or advice for using these programs. The programs are for tutorial purposes and not for commercial designs.

Before obtaining the programs from the World Wide Web, users are recommended to create, within their main MATLAB directory, a subdirectory (or folder), say **laser**, with further sub-directories **dfb**, **diff**, **fabpero**, **fftexamp**, **filter**, **grating**, **slab**, **spontan**. Within MATLAB, the effective initiation file is the M-file called **matlabrc.m** and this should be backed up for future reference into a second file, say **matlabrb.m**. On opening matlabrc.m, one looks for the lines of code with the entries:

    'C:\MATLAB\toolbox\matlab\demos;',...
    ]);

and inserts, as in Table A12.1, the entries for interrogating the new subdirectories. The punctuation marks at the start and ends of the lines need to be followed in exact detail. MATLAB will then automatically interrogate these directories/subdirectories on start up

Table A12.1 Modifying matlabrc.m file

```
'C:\MATLAB\toolbox\matlab\demos;',...
'C:\MATLAB\laser\dfb;',...
'C:\MATLAB\laser\diff;',...
'C:\MATLAB\laser\fabpero;',...
    etc.
'C:\MATLAB\laser\spontan;',...
]);
```

and the M-files entered into these subdirectories should be usable directly. The instructions for automatically interrogating directories may change with MATLAB versions and readers in general are advised to consult the MATLAB user guide or help menu before starting. Users who already make extensive use of MATLAB may have to ensure that the names of directories/M-files downloaded from the WWW do not clash with directories/M-files which are already installed.

To access the programs, users may click, in their Web browser, onto the main IEE WWW books information page by entering appropriately:

http://www.iee.org.uk/publish/books/circsyst.html

Further links will lead users to the appropriate information about this book and the directories containing the programs and detailed information on how to load them. Different Web-browsers offer different facilities and no information is given about these.

The information contained on the Web site combined with Table A12.2 should aid the selection of the programs. Although the authors have accessed these MATLAB programs from the Web site with both Macintosh and PC hardware using MATLAB 4 and MATLAB 5, the instructions cannot cover the complete variety of hardware and software that readers may use. When accessing the MATLAB code from the web, operators could make use of **Microsoft Word, Notepad** or other text editor as the initial text editor and this can influence the optimum method of access from the Web—e.g. cut-and-paste or direct-save.

Within directories **dfb** or **slab**, it is unwise to attempt to run M-files until *all* the interlinked M-files within each of these directories have been collected. Within other directories most M-files are stand alone and can be run without the aid of related M-files. M-files **colortog** and **delfig** appear in a number of directories for convenience as discussed below.

Table A12.2 Directories (shown in bold) and M-files (*.m)

dfb	**diff**	**fabpero**	**ffexamp**	**filter**	**grating**	**slab**	**spontan**
colortog.m	advec.m	fpdyna.m	delfig.m	colortog.m	delfig.m	group.m	amp1.m
delfig.m	delfig.m	fpstat.m	spect1.m	delfig.m	dfbthr.m	p1new.m	amp1f.m
dfb14.m	stepj.m	delfig.m	spect2.m	filt1.m	dispbrag.m	p2thick.m	amp2.m
dfb218.m	stepr.m		spect3.m	filt1n.m	refl.m	p3refind.m	amp3.m
dfb238.m			spect4.m	filt2.m		p4gain.m	spont.m
dfbamp.m			spect5.m	gain.m		p5layer.m	delfig.m
dfbgain.m			spect6.m			p6bragg.m	
dfbrin.m			spect7.m			p7layer.m	
dfbuni.m						p8calc.m	
p1las1.m						p9search.m	
p1las2.m						p10refl.m	
p2comp.m						p11serch.m	
p3nein.m						p12layer.m	
p4run.m						p13field.m	
p4run2.m						p14far.m	
p5spont.m						p15kappa.m	
p6plot.m						slabexec.m	
p7plot.m						colortog.m	
p8plot.m							
p9plot.m							
sponto.m							

A12.2 Introduction to the programs

To offer any program is to leave oneself open to criticism about programming style, choice of language etc. The authors do not claim to have any special flair in writing programs, and these are not polished commercial code but endeavour to give the readers tutorial text to help to develop their own work. There is an attempt to keep the programs straightforward and compartmentalised and to have substantial annotation within each program. There is a small number of programs within a suite of completely independent directories. In each directory there is a **readme** file (stored for ease of identification as **1readme.txt** within each directory) which will provide information about the programs within the directory.

The programs are within the scope of the inexpensive student version of MATLAB 4 and usually use only the basic MATLAB features even though the student version may allow more sophistication. The authors have worked mainly with IBM-compatible PCs with clock rates above 100 MHz, and it is worth remembering that enthusiasm can die when results take too long to appear.

Random-access memory (RAM) is important. The MATLAB figures produced by the code appear on our machines to use very roughly 1–2 Mbyte of memory, so that it is easy to find 16 Mbyte of memory being used simply in storing a relatively small number of figures, and it seems that 32 Mbyte of RAM is highly desirable for easy computation. For any serious future user of this code, 64 Mbyte and 166 MHz clock rate is recommended, remembering that even at the time of writing this is already out of date! The M-file delfig, which is in several subdirectories, permits the rapid deletion of figures to avoid clogging memory.

Each machine, in our experience, can have slight differences in its preferences about the directories in which the MATLAB codes are stored. It appears to be safest to store the working programs actually within a subdirectory within the MATLAB main directory. However, above all, store the core programs in a back-up version on your own machine to avoid continual reliance on the Net. Modify only copied versions of this back-up and then back-up the modified versions!

Remember that programs are stored with the extension .m (e.g. slabexec.m) but to run the program slabexec.m, for example, one only types slabexec at the command-window prompt. The MATLAB command window must, of course, be open and address the right directory or it will simply return:

⩾ ??? Undefined function or variable slabexec

Ensuring that one has the right window is best accomplished by clicking on

File , Run M-file, Browse

and then switching into the required directory and program as per menu.

This appendix lists, under the different directories, the programs which have been made available along with a very brief description of them, and where sample outputs can be found within the book. Opening the directory will give access to a README file about programs in the directory. Opening the program as if to modify the M-file will give access to a short account of the program, along with 'running' comments with the line header given by the symbol % . Within the programs, there are default settings so that one usually presses <return> or types return when asked to make an entry from the keyboard. The default setting is then automatically returned and the program should run. In this way it is hoped that the reader will at least get started, but the real exercise for the reader is to both run *and* modify the programs.

The contents of the various directories are briefly recounted below, in approximately the order the programs appear in the book.

slab

The main program in the directory **SLAB** is **SLABEXEC** which permits one to analyse slab waveguides with TE- (or TM-) field distributions for an arbitrary number (~30) of layers of different complex permittivities including gain guides. Figure A12.1 shows a sequence of outputs. The default mode of operation of **slabexec** plots out:

(*a*) five layers with slightly different refractive indices; then in the next graph
(*b*) the log of the *'reflection coefficient'* as defined in Figure 3.6, where the troughs in the reflection indicate the region of the effective refractive index for a more detailed search for the solution. If this plot does not show clear sharp minima, it is likely that the guide will not guide and a solution will not be found. Given clear minima, the program looks for the detailed solution and plots
(*c*) the *TE (TM) field intensity*; and then finally plots
(*d*) the *far-field pattern* along with the values of the confinement factors for the different modes.

Figures 3.7, 3.8 and 3.11 are typical of the output available from this program.

Within **slabexec** there is the option to ask for a Bragg grating. This follows the routines of Chapter 4 and Appendix 2, using a format of input suggested by Figure A2.6. It will simply print appropriate values of κ on the waveguide patterns as estimated from the routines discussed in Chapter 4.

Within directory **slab** there are two other minor programs:

(i) a program **colortog** which returns

Paper type has been set to A4

Figure set to black and white

This sets figures to black on white rather than white on black, and can be helpful when producing figures for papers. It is also in other directories.

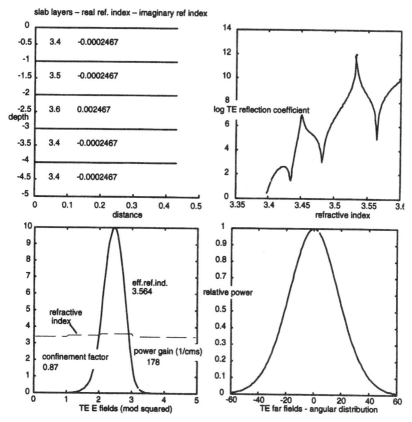

Figure A12.1 Sequence of plots obtained from the default mode for running slabexec

(ii) a program called **group** which provides data as in Appendix 3, Figures. A3.1 and A3.2.

The other (linked) M-files within directory **slab** are used within **slabexec** and are annotated either in the 1readme file or within their own code.

grating

There are three programs within a directory **grating**:

(i) Program **dispbrag** can help the reader to investigate the real and imaginary parts of the propagation coefficient normalised to the grating-coupling coefficient $\beta_e/|\kappa|$ as a function of the detuning from the Bragg condition normalised to the grating-coupling coefficient $\delta/|\kappa|$. This will print out figures similar to those in Figures 5.2 to 5.4.

(ii) Program **refl** allows the reader to explore $\rho = R/F$ at the input for

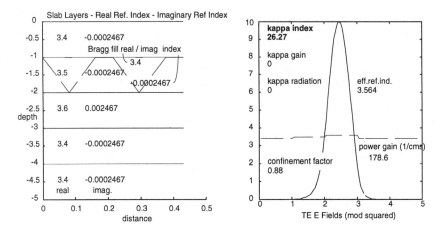

Figure A12.2 Modified output from slabexec *when inserting a Bragg grating etched into the waveguide with an infill of different material*

Note that guide is too wide and the overlap with the grating is inadequate, leading to a low κ

real κ similar to Figure 5.5 which shows this reflectivity as a function of $\delta/|\kappa|$ for a real κ, so that one can see how the reflectivity falls off as $\delta/|\kappa|$ diverges from zero.

(iii) Program **dfbthr** returns two numbers: the approximate normalised gain gL and offset-frequency parameter δL for a given $\kappa L > 0.5$ required to get to threshold in a uniform DFB laser. It can be helpful in ensuring that, when running **dfbamp** in directory **dfb**, one is operating with a gain that is perhaps half that required for lasing.

diff

(i) Programs **stepr** and **stepj** in directory **diff** integrate $dF/dz = zF$ and $dF/dz = jzF$, respectively, showing how step length alters the accuracy. Figure 7.2 gives a sample of the output expected.

(ii) Program **advec** provides a basic test program for the advection equations where the forward field is stored at all space points but for only two time steps. Figure 7.3 provides a sample of the output expected.

spontan

(i) Program **spont** in directory **spontan** gives the reader an indication of the modelling process for spontaneous emission. It provides a dynamic output of the static display, shown in Figure 7.5.

(ii) Programs amp1, amp2 and amp3 provide three short demonstrations of generation and amplification based on an 'ideal' uniform laser amplifier, demonstrating that spontaneous emission and gain are linked. Program amp1f is similar to amp1 but has a filtered noise input rather than relying on white spontaneous noise.

fftexamp

Programs spect1 through to spect7 in directory fftexamp are unremarkable, but may help to refresh the reader about using the fast Fourier transform. The associated 1readme file gives more details.

filter

(i) Program filt1 models a numerical-frequency filter, and the use of a white-noise input to demonstrate filtering is shown in the program filt1n. The display output is similar to that in Figure 7.9. An alternative filtering is demonstrated in program filt2.
(ii) Program gain implements the application of the filter theory to a laser amplifier with gain over a controllable narrow band. The display output is similar to that in Figure 7.10.

dfb

This suite of programs gives a dynamic output from different types of DFB lasers showing the output building up with time as the laser is switched on; and then one can take the spectrum and/or ask for the electron density and optical-field patterns with or without spontaneous noise.
(i) Program dfbamp in the directory dfb models a DFB amplifier with uniform gain with zero input on the right and a unit-amplitude field entering on the left, so that the device acts as a coherent forward-wave 'amplifier'. The output is a dynamic version of the static display shown in Figure 7.4 with a step field propagating into the laser from the left-hand end and then becoming reflected as it travels along through the laser so that one can see the process of the distributed feedback dynamically while the laser is settling down to its steady state.
(ii) Program dfbuni brings the elements of the DFB-laser modelling together for a uniform device with real κ. Although this is the main program about distributed-feedback lasers it is probably easier if one simply wants to run a program to use dfb14, as discussed below. As

outlined in the text, uniform-DFB lasers have a number of problems and these physical problems of being unsure what mode will lase, as discussed in Chapter 6, are reflected in the numerical programs. This uncertainty in whether the higher- or lower-frequency mode of the uniform-DFB laser will appear can be disconcerting to someone starting to run a program and finding that the output differs from run to run. The output form is similar to that shown in Figure 7.11.

The program drives a number of subprograms which are common to the other forms of DFB laser, and these subprograms are best dealt with in the 1readme file. For example, there are material-parameter subprograms, and a lookup table for spontaneous noise is used in program spont2 in the directory dfb and called up as required.

(iii) Program dfb14 uses much of dfbuni but now has a $\lambda/4$ phase shift in its centre. Figure 7.11 shows a typical output against time, a spectrum and a field pattern (with the spontaneous noise on and spontaneous noise off).

(iv) Program dfb218 in directory dfb again uses much of dfbuni but now has two phase shifts of $\lambda/8$, one on either side and offset from the laser's centre.

(v) Program dfb238 also uses much of dfbuni but now has two phase shifts, as for dfb218 but now each of $3\lambda/8$.

(vi) Program dfbgain yet again uses much of Dfbuni but now has a complex grating with a weak gain coupling (that is carrier independent for simplicity) but enables the reader to explore the mode change as the sign and magnitude of the gain component of the grating change.

(vii) Program dfbrin looks at RIN for a $\lambda/4$-phase-shifted DFB laser using the standard parameters within the model and gives an output similar to that in Figure A8.2.

fabpero

There are two MATLAB programs fpstat and fpdyna in directory fabpero which are able to give the static and dynamic characteristics of a Fabry–Perot-type laser using normalised rate equations discussed at the end of Appendix 4. The outputs are similar to those in Figures 4.2 and 4.3.

A12.3 Reference

1 'MATLAB 1995, Student Edition' (Prentice Hall and The MathWorks Inc, 1995)

Index

absorption 46, 51–2
  free carrier 60–61
  intervalence band 70
acceptors 48
advection equations 112–13, 214–17
amplifiers
  erbium doped 15–16
  ideal 374–7
  integration, *see* integration
  noise theory, *see* noise
aperture theory, *see* spontaneous emission
AR (antireflection) coating, *see* facet reflectivity
arrays of laser diodes 30, 287
astigmatic pattern of emission from laser 392–4
attenuators 377–8
Auger, *see* recombination
autocorrelation, *see* spontaneous emission,
    correlation in space-time

band-gap, *see* semiconductors
band tails 52–3
bands, *see* energy bands
beam formation 76
BER (bit-error-rate) 17
Bernard and Duraffourg lasing condition 41, 48
black body radiation 49, *see* thermal radiation
blaze of grating 18, 22
Bragg gratings 18–24, 114, 116–25
  complex-index variations 131–2, 183–9
    fabrication 183–4
  coupling of waves, *see* coupled-wave equations
  diffraction order 18–22
  dispersion relationship 133–5

  fabrication 22–4
  Fourier components 20–22, 118
  introduction 18–22
  order 20
  period 130
  periodic gain 116–17
  periodic permittivity 116, 121
  phase matching 117–120
  phase shifts 137–45
  radiation loss 20–22, 121
  reflection per period 22, 25–6, 129
  reflectivity 137–9
  second order 20–21, 24
    Fourier components 120
    radiation loss 121–4
  shape of teeth 124–5
  stopbands 133–5
Bragg wavelength 20
Brewster angle 311
buried heterostructures 10

carrier induced index changes, *see* refractive index
carrier transport, *see* quantum wells
causality, *see* Kramers-Krönig relationships
charge carriers
  clamping of density 105–6
  variation with Fermi-level 48, 50–51, 153–4
  diffusion damping 107
  in crystals
    electrons 37–8
    holes 38
  lifetime of, *see* recombination time
  rate equation 100, 243–4
  thermalisation of distribution 59–60
  transport 102
chirp 6, 61–3, 71, 108–9
  clamping, *see* charge carriers

406 *Distributed feedback semiconductor lasers*

coherent detection 15
complex
  dielectric constant 61, 78
  refractive index 61, 78
conduction band, *see* energy bands
confinement
  carrier 6–7
  factor 11–12, 79–81, 98–100
  horizontal 9–11, 77, 90–91
  photon 7–8
  vertical 77
connectorised lasers 390
contact resistance, *see* resistance
contour integral, *see* Kramers–Krönig relationships
coupled cavity lasers, *see* laser diode (types)
coupled-mode equations, *see* coupled-wave equations
coupled-wave equations 119–20, 129–32, 222–8, 243
  analytic/numeric solutions 224–5
  eigen-modes 132–3
  finite difference scheme 219, 222–3
  gain coupling 220–21
  MATLAB code 223–4
  matrix formulation 135–7, 219–21, 352–6
coupling fibres to laser diodes 389–91
  V-groove technology 390–91
critical angle 8
current spreading 9, 31
cut-off in waveguide 95

DBR, *see* laser diodes (types)
delay in turn on of light 108
density of states, *see* energy bands
detuning, *see* frequency deviation
device parameters for laser diodes 392–5
  carrier transport parameters 395
  confinement factors 392–4
  effective mode widths 392–4
  FWHM for emission 392–4
  gain saturation 392–3
  gain/loss parameters 392–3
  linewidth enhancement factor 392–3
  recombination parameters 393, 395
  separate confinement layer thickness 394–5
DFB (distributed feedback) laser 17, 26–8
  see also Bragg gratings
  asymmetric grating 191
  dynamic modelling 194–202

gain grating
  complex index 116–17, 183–7, 190–91
  dynamic effects 187–9
  facet reflectivity effects 189
  static effects 184–7
  high power 189–94
  high–low reflectivity 190
  integrated amplifier 191–4
  effects of carrier transport 197–202
  light-current characteristic 155–8
  phase shifts 139–45, 221
    single $\lambda/4$ 137–9, 140–41, 142–3
    two $\lambda/8$ 138–9, 143–4, 149–52, 197–202
  simulated emission spectrum 159–61
  simulated static performance 155–61
  uniform structure 140, 142, 147–8, 194–7
    matrix analysis 135–7
    oscillation modes 355–6
    Pauli matrix formulation 352–3
    reflectivity 355
    threshold conditions for lasing 355–6
    transmission 355
dielectric waveguide, *see* waveguides, slab
diffraction 18–22, 79, 90
diffusion, effect on direct modulation 107
digital modulation, effects of facet reflections 181–3
direct bandgap, *see* semiconductors
direct modulation 1, 70, 176, 258–9
discretisation 230–32
dispersion in fibres 16–17, 195–6, 258–9
dispersion relationship for grating 133–5
distributions
  black body 49
  Boltzmann 41
  Fermi–Dirac 40–41

edge/surface emission comparison 287–8
effective area 79–81, 98, 351
effective index method for confinement 90–91
effective mass 38–9, 50–51
  for electrons 38–9
  for holes 39
effective permittivity 78
effective refractive index 78, 84–6
efficiency of a laser 13, 31, 244
eigen-modes 132–3

Einstein relations, *see* spontaneous emission
electrical parasitics, *see* package for laser diode
electroabsorption modulator 259–60
electromagnetic energy exchange 341–51
electron beam lithography 24, 285
electron concentration 48, 50–51, 153–4
emission, spontaneous, *see* spontaneous emission
emission, stimulated, *see* stimulated emission
energy, *see also* energy exchange
  electromagnetic 81, 341–51
  exchange between energy levels 47
  optical 97–126
  rate of change 345
energy bands
  conduction 37–8
  density of states 39
  heavy-hole 45
  valence 38
energy exchange 341–51
  from rate equation 344–8
  reconciliation with field equation 348–51
energy states 37–40
energy–momentum diagrams 38, 40, 42, 44, 45
erbium doped fibre amplifier (EDFA) 15
error control (numerical) 211, 216–17, 221–2
error program for threshold 355–6
evanescence 8, 86–7
extinction ratio 261

fabrication of DFB lasers 22–4
Fabry–Perot resonator 2–4, 209
facet reflectivity 2, 3–4, 311–12
  antireflection coatings 15, 28
  high–low reflection 12–13, 28
  spectral effects 179–82
far fields 92–3, 319–22
Faraday effect 390
feedback 4–5, 13, 14, 20–22, 25, 140, 276, 278, 280, 284, 285,
  see also DFB
Fermi–Dirac statistics 46, 48
FFT (fast Fourier transform) 210, 232–3
  Parseval's theorem 232
fibre attenuation 15
fibre dispersion, *see* dispersion in fibres
filamentation 9–10

filter theory
  application to gain filtering 237–41
  electron rate equation 245–6
  Lorentzian 233–7
  numerical implementation 235–7, 240–41
  travelling waves 238–40
FM (frequency modulation) response 336–8
forward waves 2–3, 79, 111–15, 117–19, 130, 136, 141–2, 214–16, 219–23, 305
free carrier absorption 60–61
frequency chirp, *see* chirp
frequency deviation 114, 130
frequency shift with carrier-density, *see* chirp

gain 2–5, 37–55
  field 3–4, 101
  guiding, *see* waveguiding mechanisms
  per unit distance 100–101
  per unit time 99–101
  power 101
  round-trip 4–5
  wave propagation with 3
gain-grating, *see* DFB
gain saturation 58–60, 102, 107–8
  parameter values 59
  quantum wells 68–9
Green's function 122
group refractive index 324–8
group velocity 327, 350

Henry's alpha factor, *see* linewidth enhancement factor
heterojunction, *see* laser diode (types)
Hilbert transforms 61, 357–62
hole concentration expression 48, 51, 153–4
holes, light and heavy 40, 45
homojunctions, *see* laser diode (types)

ideal optical transmitter 254–5
imaginary dielectric constant, *see* complex dielectric constant
indirect energy gap, *see* semiconductors
index, *see* refractive index
InGaAs, strained (980nm) 69
instability 145, 158, 191, 195, *see also* pulsations
integration
  laser and amplifier 191–3
  laser and modulator 203, 259–60
inversion factor 378

408 Distributed feedback semiconductor lasers

k-selection 43–4, 50–53
Kane theory, *see* transition rates
Kramers–Krönig relationships 61, 357–62
    Cauchy contour theory 359–60
    causality theory 357–9
    physical basis 360–62
laser diodes
    chirp 6, 28, 29, 61–4, 71, 108–9
    desirable features 1
    early developments 1–6
    electron-density-time characteristics 108–9
    impedance/resistance 102
    light-current characteristics 5–6, 105–6, 108
    longitudinal modes 2, 5
    SMSR (side mode suppression ratio) 17, 167–8
    spectra 5
    temperature effects 64–6
    threshold-temperature coefficient 66
    transverse modes 2
    tuning mechanisms 276
    wavelength-temperature-coefficient 64–5
laser diode (types)
    cleaved-coupled-cavity 13–14
    DBR (distributed Bragg reflector) 25
    distributed feedback, *see* DFB
    double heterojunction 2, 6–7
    external cavity 13–14
    external grating 14–15
    Fabry–Perot 2–6
    Fabry–Perot variations 12–17
    heterojunctions
        double 2, 6–8
        single 2
    homojunctions 1–2
    injection locked 14
    surface emitting, *see* surface emitting lasers
    tunable 29, 274–85
    VCSEL, *see* surface emitting lasers
lattice matching 69
Lax averaging 213, 215, 218
leakage current 7, 31
lifetime
    carrier, *see* recombination time
    intraband 59
    operational 1, 22
    photon 99, 104–5
    spontaneous 54, 58

linewidth
    enhancement 171–2
        dynamics 176
    homogeneous broadening 53–4
    influence of reflections 179
    inhomogeneous broadening 53–4
    Lorentzian 54, 167–8
    modulation increases 166–7
    power output effect 170
    rebroadening 176–7
    Schawlow–Townes formula 170
    static calculations 168–70
    stored energy effect 169–70
    systems effects 166–8
    tunable lasers 284
linewidth enhancement factor 61–4, 108, 171
    changes with frequency 63–4
    definition 62
    diagrams of variations 62, 63
    effect of complex grating 174
    effective value 172–6
        estimates 175
        measurements 173–4
    numerical effects 242
    quantum well effects 64
loss, optical 3

Mach–Zehnder interferometer 263–5
MATLAB 396–404
    accessing programs 396–8
    computing system requirements 396
    initiating file 397
    program listings 398
    version 4, student limitations 233
MATLAB directory titles
    dfb 191, 224, 241, 370, 403–4
    diff 213, 217, 402
    fabpero 340, 404
    fftexamp 232, 403
    filter 235–6, 403
    grating 135, 137, 401–2
    slab 87, 95, 119, 125, 183, 322, 400–1
    spontan 226, 229, 402–3
MATLAB directories (subject matter)
    advection equations, *see* **diff**
    Bragg gratings, *see* **grating**
    DFB numerical code, *see* **dfb**
    differential equations, *see* **diff**
    FFT tutorials, *see* **fftexamp**
    filter theory, *see* **filter**
    gain guiding, *see* **slab**
    gain–frequency filtering, *see* **filter**

Index   409

light–current characteristics for FP laser, *see* fabpero
rate equations for FP laser, *see* fabpero
slab waveguide, *see* slab
spontaneous emission, *see* spontan
MATLAB programs 396–404
  advec 217
  amp1–3 229
  amp1f 230
  dfbamp 224
  dfbgain 191
  dfbrin 370
  dispbrag 135
  filt1 235
  filt2 236
  fpdyna 340
  fpstat 340
  listings 398
  refl 137
  slabexec 22, 87, 95, 119, 124, 125, 183, 322
  spect1–7 232
  stepj 213
  stepr 213
matrix element for band-to-band transitions 43
Maxwell's equations 83, 111, 304
MBE (molecular beam epitaxy) 22
MOCVD (metal organic chemical vapour deposition) 22
mode locking 14, 253, 266
mode spacing 5, 279
modelling methods, *see* optical components
modulation of lasers, *see* direct modulation
modulators, external
  electroabsorption 259–60
  prechirp effects 259
momentum
  electron 38
  phonon 41–2
  photon 41

near-field pattern 94
noise 372–84
  attenuation 377–8
  CNR (carrier to noise ratio) 17
  ideal amplifier 374–7
  Nyquist formula for thermal noise 373
    ultra-violet catastrophe 373
  RIN (relative intensity noise) 17, 366–71
    decision level 367–9

  drive level 370
  photon-electron resonance effects 370–71
  shot noise 366
  uncertainty as noise 373
normalised rate equations 106, 108, 244, 339–40
numerical distortion 216–7
numerical stability 216–7, 223
  effects of discretisation 230–32
  Fourier checks 221–2
Nyquist formula for noise, *see* noise
Nyquist spectral-band limits 230–32
  for FFT 232–3

obliquity factor 92–3, 320–22
optical absorption 99
optical component modelling methods 253–5
  beam propagation 253
  finite difference 209–48
  finite elements 253
  Fourier techniques 255
  power matrix methods 161
  TLLM (transmission line laser modelling) 210, 253
  transfer matrix methods 141–5
optical gain, *see* gain
optical isolators 390
optical systems 15–17, 252
  BER (bit-error-rate) 17
  modelling 255–6
  power budgets 17
  sources for WDM systems 285–6
  WDM (wavelength division multiplexing) 17, 285–6
    with DFB lasers 189–90
ordinary differential equations 211–14
  accuracy of solution 213-4
  central difference methods 212–13
  Euler methods 211–212
  oscillation condition 2–4
  package for laser diode 386–9
    equivalent circuit 387, 394
    values for 395
    monitor diode 388–9
    thermal considerations 388
parabolic bands 39
particle balance 110
Pauli matrices 352–3
Petermann's factors 58, 175, 383
phase fluctuations 172
phase jumps, *see* DFB laser, phase shifts
phase matching 4, 27

phase velocity 325–6
photon density (black body) 49
photon lifetime 99, 103,104–5, 329–30
photon–electron resonance 107–8
plane waves 304–5
plasma effect 63
polarisation 341–2
power amplification 193, 256–8
propagation constant 3, 78, 91, 111–15, 324
pulsations 148, 177, 188, 266
push–pull electronics 265–6
push–pull laser 265–74
  asymmetry 269–72
  speed of response 272–4
  stopband–distance diagrams 267
  symmetry 266–9
  time resolved chirp 269–72
  transient modes 272–4

quantum wells
  carrier capture 70–71
  carrier transport effects 334–8
    AM and FM response 337–8
  carrier transport, parameter values 395
  density of states 66–7
  effect on waveguiding 93–5
  effects of strain 69–70
  enhancement of tuning-range 274–5
  gain saturation 68–9
  intervalence band absorption 70
  introduction 66–8
  multiple 68–9
  SCH (separate-confinement heterostructure) 70–71
quasi-Fermi level 40–41, 48

rate equations 98–110
  carrier transport delay 334–8
  damping 332–3
  dynamic analysis 106–10
  electronic 100–103, 330
  large signal analysis 339–40
  peak magnification of AM response 333
  photon-electron resonance 332–3
  photonic 98–100, 329
  small signal analysis 106–7, 329–38
    FM response 336–7
  steady state solutions 104–6, 330–31
rates of change
  electric fields 111–16, 130
  electron density, *see* rate equations
  energy, *see* energy

numerical formulation 214 *et seq.*
photon density, *see* rate equations
recombination
  Auger 44–6, 100–101
    process diagram 45
    temperature changes 46
  bandgap states 44
  bandtails 44
  nonradiative 48, 100–101
  radiative 41–3, 100–101
  total 100–101
recombination time (τ_r) 100–101, 107, 330, 332
reflections, external
  coherence collapse 178–9
  laser stability 177–9
  packaging 389–90
reflections, internal, *see* facet reflectivity
refractive index, *see also* effective refractive index
  carrier induced changes 171–2
  complex value 61
  model for quaternary materials 362–4
  plasma effect 361
  single oscillator model 362–3
  table of values 328
  temperature coefficient 65
resistance
  contact 48–9, 154
  series 152–4
reverse waves 2–3, 7, 115, 120,130, 136, 141–2, 214–15, 219–23, 305

saturation of gain, *see* gain saturation
scattering losses 60
SCH (separate confinement heterostructure) 70–71, 94–5, 394–5
selection rule for k, *see* k-selection
semiconductor optical amplifier (SOA) 253, 256–8, 260–64
  cross-phase wavelength conversion 263–4
  four wave mixing 261–2
  ideal optical transmitter 254–5
  instantaneous frequency 255, 257–8
semiconductors
  band-gap
    direct 41–3
    indirect 41–2
  GaAlAs 2, 42
  GaAs 1, 43
  GaPAs 2
shot noise, *see* noise

side mode suppression ratio 150–52,
    167–8, 180–83, 197–201
  effect of phase shifts 150–52
slab waveguide, *see* waveguides
Snell's law 8, 57, 305–7
SOA, *see* semiconductor optical amplifier
spatial hole burning 181
  changes with time 197–202
  emission frequency effects 146, 158
  κL effects 148
  lateral 145
  light-current characteristic effects 146, 158
  longitudinal 145–52, 158
    spectral effects 150–52
  phase-shift effects 148–50
spectra
  facet reflectivity effects 179–82
  gain 50–54, 56
  optical gain 52, 53, 56
  spontaneous emission 53–5
spectral hole burning 59–60
spectral-temporal interchange 245, 349–50
spontaneous emission 4, 5, 43, 47, 54–5, 99, 113–14, 130–31, 372–85
  aperture theory 379–80
  autocorrelation, *see* correlation in space–time
  correlation in space–time 227–8, 375, 381
  counting photon states 378–9
  Einstein treatment 46–50, 372, 378–9
  magnitude 229
  numerical modelling 226–9, 380–84
    in DFB 241–2
    space-time correlation 381
  numerical/physical bandwidth 382
  statistics 384
  travelling-waves 226–8
  tutorial programs 229–30
spontaneous emission coupling
  explanatory diagram 57
  factor 55–8,105–6, 108
spontaneous excitation, *see* spontaneous emission
stimulated emission 47–9, 51–2
stopband-distance diagram 147–8
stored electromagnetic energy 343
strained material 69–70
stripe geometry lasers 9
surface emitting lasers
  advantages 286–7
  comparison with edge-emitters 287–9

grating coupled DBR 287–8
VCSEL (vertical cavity surface-emitting laser) 30, 286–94
  construction 291–4
  polarisation 294
  speed of response 291
susceptibility 342–4
  imaginary 343–4
  real 343
switch on delay 108

TE (transverse electric) fields 308
  special cases of reflection 309
temperature, effects 64–6
thermal noise 373
thermal radiation 49, 373 *et seq*
threshold (for lasing) 55, 66, 70
  conditions 104
  current 46, 67
  spectra 55
TLLM (transmission line laser modelling) 210, 253
TM (transverse magnetic) fields 309–10
  special cases of reflection 310–11
total internal reflection (TIR) 8, 78, 305–7
transfer matrix methods 141–5, 179–80, 209
transition rates 43–4
  Kane theory 43
transparency electron-density 48, 68
transport, diffusion effects 70–71
tunable laser diodes
  carrier density tuning 276–7
  Franz-Keldysh effect 279
  linewidth 284
  modelling 284–5
  nonuniform gratings 279–81
  quantum confined Stark effect 279
  temperature tuning 276–7
  Vernier tuning 29, 280–82
tunable laser diodes (types) 274–85
  GAVCF (grating assisted vertical-coupler filter) 283
  interrupted gratings 279–80
  multicontact 277–9
  quantum well material 274–6
  superstructure gratings 280
  tunable twin guide 283
  Y-laser 282–3

uncertainty
  amplification and uncertainty 374–7
  Heisenberg 44

noise 373–4
uniform DFB, see DFB, uniform structure

V-parameter 90, 325
valence band, see energy bands
VCSEL, see surface emitting lasers
velocity
  group 112–13, 324–8, 350
  phase 112–13, 324–8
vernier tuning, see tunable laser diodes

wave equation 83, 304
  advection equation 214–18
  central difference methods 215–18
wave propagation 111–16
waveguide absorption values 392–3
waveguide modes
  different orders 87–9
  orthogonality 92
waveguides
  modelling 76–96
  multilayer slab guides 313–18
  reflection coefficient at surface 316
  ridge 10–11
  slab 8, 81–90
    multilayer 82–3
    reflection coefficient 86–7
    TE slab modes 313–16
    TM slab modes 317–18
  slab-waveguide program 322–3
waveguiding mechanisms
  gain guiding 9, 78–9, 89–90
  horizontal guiding 90–91
  index guiding 9, 77–9
  quantum well effects 93–5
waveguiding modes
  scaling 90
  TE (tranverse electric) 81–4
  TM (transverse magnetic) 81–4
WDM (wavelength division multiplexing), see optical systems

Y-laser 29, 282–3

zero dispersion 16